Relative Index Theory, Determinants and Torsion for Open Manifolds

Relative Index Theory, Determinants and Torsion for Open Manifolds

Jürgen Eichhorn
Universität Greifswald, Germany

Published by
World Scientific Publishing Co. Pte. Ltd.
5 Toh Tuck Link, Singapore 596224
USA office: 27 Warren Street, Suite 401-402, Hackensack, NJ 07601
UK office: 57 Shelton Street, Covent Garden, London WC2H 9HE

British Library Cataloguing-in-Publication Data
A catalogue record for this book is available from the British Library.

RELATIVE INDEX THEORY, DETERMINANTS AND TORSION FOR OPEN MANIFOLDS

Copyright © 2009 by World Scientific Publishing Co. Pte. Ltd.

All rights reserved. This book, or parts thereof, may not be reproduced in any form or by any means, electronic or mechanical, including photocopying, recording or any information storage and retrieval system now known or to be invented, without written permission from the Publisher.

For photocopying of material in this volume, please pay a copying fee through the Copyright Clearance Center, Inc., 222 Rosewood Drive, Danvers, MA 01923, USA. In this case permission to photocopy is not required from the publisher.

ISBN-13 978-981-277-144-5
ISBN-10 981-277-144-1

Printed in Singapore.

Contents

Introduction vii

I Absolute invariants for open manifolds and bundles 1
1 Absolute characteristic numbers 4
2 Index theorems for open manifolds 46

II Non-linear Sobolev structures 62
1 Clifford bundles, generalized Dirac operators and associated Sobolev spaces 63
2 Uniform structures of metric spaces 89
3 Completed manifolds of maps 121
4 Uniform structures of manifolds and Clifford bundles .. 124
5 The classification problem, new (co–)homologies and relative characteristic numbers 141

III The heat kernel of generalized Dirac operators 169
1 Invariance properties of the spectrum and the heat kernel 169
2 Duhamel's principle, scattering theory and trace class conditions 180

IV Trace class properties 192
1 Variation of the Clifford connection 192
2 Variation of the Clifford structure 203
3 Additional topological perturbations 223

V Relative index theory 239
1 Relative index theorems, the spectral shift function and the scattering index 239

VI Relative ζ-functions, η-functions, determinants and torsion ... 252
1 Pairs of asymptotic expansions ... 252
2 Relative ζ-functions ... 256
3 Relative determinants and QFT ... 267
4 Relative analytic torsion ... 269
5 Relative η-invariants ... 272
6 Examples and applications ... 277

VII Scattering theory for manifolds with injectivity radius zero ... 299
1 Uniform structures defined by decay functions ... 299
2 The injectivity radius and weighted Sobolev spaces ... 307
3 Mapping properties of $e^{-t\Delta}$ ... 317
4 Proof of the trace class property ... 322

References ... 331

List of notations ... 338

Index ... 340

Introduction

It is one of the main goals of modern mathematics to describe a mathematical subject, situation, result by a sequence of honest numbers. We remind e.g. in topology / global analysis of the rank of (co–)homology groups, homotopy groups, K–(co–)homology, characteristic numbers, topological and analytical index, Novikov–Shubin–invariants, analytical torsion, the eta invariant, all these numbers defined in the compact case. Including bordism and Wall groups, which are also of finite rank, one has an appropriate approach to the classification problem for compact manifolds.

For open manifolds, all these numbers above are not defined in general. The ranks of the group of algebraic topology can be infinite, the integrals to define characteristic numbers can diverge, elliptic operators must no longer be Fredholm, the spectrum of self–adjoint operators must not be purely discrete, etc.

We will prove at the beginning of chapter I that there are no non–trivial number valued invariants which are defined for all oriented (including open) manifolds and which behave additively under connected sum. Moreover, we prove, that for any $n \geq 2$ there are uncountably many homotopy types of open manifolds. Hence a classification, essentially relying on number valued invariants, probably should not exist.

The main idea of our approach – brought to a point – is as follows. We consider pairs (P, P'), where P e.g. stands for a triple (manifold, bundle, differential operator), and we define relative invariants $i(P, P')$, where P' runs through a so called generalized component $\operatorname{gen comp}(P)$ which consists of all P' with finite distance from P. The distance comes from a metrizable uniform structure. To define the corresponding metrizable uniform structure is the content of chapter II and is one of the columns of our approach. Then the classification of the Ps amounts to the classification of the generalized components and the classification of the Ps inside $\operatorname{gen comp}(P)$.

This treatise is organized as follows. In chapter I section 1 we present classes of open manifolds for which the classical characteristic numbers via Chern–Weil construction are defined, study their invariance and meaning. Here we include the important contributions of Cheeger/Gromov from [15], [16]. We call them absolute invariants since they are defined for single objects and not for pairs with one component fixed. Section two is devoted to some index theorems for certain classes of open manifolds and elliptic differential operators. It is visible that all these are very special classes and that the wish for a corresponding theory for all open manifolds requires a now, another approach. For us this is the relative index theory, applied to pairs. Then the main question is, what is an admissible pair of Riemannian manifolds or Clifford bundles with associated generalized Dirac operators?

In a local classical language this would mean, what are the admissible perturbations of the coefficients in questions – and of the domains?

We answer these questions in a very general and convenient language, the language of metrizable uniform structures. We define for a pair of objects under consideration a local distance, define by means of this local distance a neighbourhood basis of the diagonal and finally a metrizable uniform struture. In all our cases, the distance contains a certain Sobolev distance. For this reason, we give in chapter II 1 a brief outline of the needed facts and refer to [27] for more. II 2 is devoted to a very brief outline on uniform structures of proper metric spaces. Since diffeomorphisms enter into the definition of our local distances, we collect in II 3 some definitions and facts on completed diffeomorphism groups for open manifolds. In II 4, we introduce those uniform structures of manifolds and Clifford bundles and their generalized components, which are fundamental in the central chapters IV, V, VI. The final section II 5 contain the first steps of our approach to the classification problem for open manifolds. We introduce bordism for open manifolds, several bordism groups, reduce their calculation to that for generalized components, introduce relative characteristic numbers and establish generators for bordism groups of manifolds with non–expanding ends.

Chapter III is the immediate preparation for the chapters IV, V, VI. Section III 1 is devoted to the invariance of the essential spectrum under perturbation inside a generalized component, to heat kernel estimates, and we introduce in III 2 standard facts of scattering theory as wave operators, their completeness and the spectral shift function of Birman / Krein / Yafaev.

As it is clear from our approach and the criteria in section III 2, the absolute central question is the trace class property of $e^{-tD^2} - e^{-t\tilde{D}'^2}$, $E \in \mathrm{gen\,comp}(E)$. Moreover, the expression $\mathrm{tr}(e^{-tD^2} - e^{-t\tilde{D}'^2})$ enters into the integral for the relative index, relative zeta and eta functions. We establish this trace class property step by step in sections IV 1 – IV 3, admitting larger and larger perturbations. The proof of the trace class property is the heart of the treatise, really rather complicated and the technical basis for what follows.

In section V 1, we prove several relative index theorems and in V 2 properties of the scattering index. We remark that there are other well known relative index theorems e.g. in [8], [9] for exponentially decreasing perturbations. Both these theorems are very special cases of our more general result. In chapter VI, we apply our achievements until now to define relative zeta and eta functions, relative analytic torsion, relative eta invariants and relative determinants, which are in particular important in QFT. Section VI 6 presents numerous examples, special cases and applications of these notions. In particular, we present classes of open manifolds which satisfy the geometric and spectral assumptions which we assumed in the preceding sections. A particular simple case are manifolds with cylindrical ends for which we describe the scattering theory. Here we essentially rely on [3], [49]. An interpretation of our relative determinants in the case of cylindrical ends is given by theorem 6.17 in chapter VI.

Until now, we always assumed bounded geometry, i.e. injectivity radius > 0 and bounded curvature together with a certain number of derivatives. Clearly, this restricts the classes of metrics under consideration. In [54], W. Müller and G. Salomonsen established a scattering theory without the assumption injectivity radius > 0 but they admit other perturbations. The difference $g-h$ together with certain derivatives must be of so-called moderate decay. We reformulate their appraoch in our language of uniform structures and generalized components and extend their results to arbitrary vector bundles with bounded curvature.

Part of this book has been written at the Wuhan University (China). The author thanks the DFG and Chinese NSF for support. The other part has been written at the MPI in Bonn.

The author is particularly grateful to Gesina Wandt for her permanent engagement and patience in printing the manuscript.

Greifswald, December 2008

I Absolute invariants for open manifolds and bundles

For closed manifolds, there is a highly elaborated theory of number valued invariants. Examples are the characteristic numbers like Stiefel–Whitney, Chern and Pontrjagin numbers, the Euler number, the dimension of rational (co–)homology and homotopy groups, the signature. Moreover, we have invariants coming from surgery theory etc. Taking into account a Riemannian metric, we obtain global spectral invariants like analytic torsion, the eta invariant. On general open manifolds, more or less all of this fails. Characteristic numbers are not defined, (co–)homology groups can have infinite rank etc.

We have the following simple

Proposition 0.1 *Let $\mathfrak{M}^n$ be the set of all smooth oriented manifolds and V a vector space or an abelian group. There does not exist a nontrivial map $c : \mathfrak{M} \longrightarrow V$ such that*

1) $M^n \cong M'^n$ *by an orientation preserving diffeomorphism implies $c(M) = c(M')$ and*

2) $c(M \# M') = c(M) + c(M')$.

Proof. Assume at first $M^n \not\cong \Sigma^n$, fix two points at M^n, then $M_\infty = M_1 \# M_2 \# \ldots$, $M_i = (M, i) \cong M$ has a well defined meaning. We can write $M_\infty = M_1 \# M_{\infty,2}$ with $M_{\infty,2} = M_2 \# M_3 \# \ldots$ and get $c(M_\infty) = c(M) + c(M_{\infty,2}) = c(M) + c(M_\infty)$ hence $c(M) = 0$.
Assume $M^n = \Sigma^n$, and ord $\Sigma^n = k > 1$ which yields

$$c(\Sigma^n \# \ldots \# \Sigma^n) = k \cdot c(\Sigma^n) = c(S^n), \quad c(\Sigma^n) = \frac{1}{k} c(S^n),$$

$$c(\Sigma^n) = c(\Sigma^n \# S^n) = \left(1 + \frac{1}{k}\right) c(S^n),$$

$$c(S^n) = 2c(S^n), \quad c(S^n) = 0, \quad (\Sigma^n) = 0.$$

$\square$

The only real invariant, defined for all connected manifolds M^n known to the author is the dimension n. If one characterizes orientability / nonorientability by ± 1, then there are two such invariants. That is all. Looking at the classification theory, we see a deep distinction between the case of closed or open manifolds, respectively.

Denote by $\mathfrak{M}^n([cl])$ the set of all diffeomorphism classes of closed n–manifolds. Then we have

Proposition 0.2 $\#\mathfrak{M}^n([cl]) = \aleph_0$.

Proof. According to Cheeger, there are only finitely many diffeomorphism types for (M^n, g) with diam $(M^n, g) \leq D$, $r_{\text{inj}}(M^n, g) \leq i$, and bounded sectional curvature with bound K. Setting $D_\nu = K_\nu = i_\nu = \nu$ and considering $\nu \to \infty$, we count all diffeomorphism types of closed Riemannian n–manifolds, in particular all diffeomorphism types of closed manifolds. $\square$

On the other hand, for open manifolds there holds

Proposition 0.3 *The cardinality of $\mathfrak{M}([open])$ is at least that of the continuum for $n \geq 2$.*

Proof. Assume $n \geq 3$, n odd, let $2 = p_1 < p_2 < \ldots$ be the increasing sequence of prime numbers and let $L^n(p_\nu) = S^n/\mathbb{Z}_{p_\nu}$ be the corresponding lens space. Consider $M^n := d_1 \cdot L(p_1) \# d_2 \cdot L(p_2) \# \ldots$, $d_\nu = 0, 1$. Then any $0, 1$–sequence $(d_1, d_2, \ldots)$ defines a manifold and different sequences define non diffeomorphic manifolds. If $n \geq 4$ is even multiply by S^1. For $n = 2$ the assertion follows from the classification theorem in [59]. $\square$

There are simple methods to construct from only one closed manifold $M^n \neq \Sigma^n$ infinitely many nondiffeomorphic manifolds. This, proposition 0.3 and other considerations support the naive imagination, that "measure of $\mathfrak{M}([open])$: measure $\mathfrak{M}([cl]) =$

∞ : 0". We understand this as an additional hint how difficult any classification of open manifolds would be.

The deep distinction between the propositions 0.2 and 0.3 indicates that the chance to classify open manifolds (at least partially) by means of number valued invariants is very small. This is additionally supported by proposition 0.1.

Concerning number valued invariants, there are two ways out from this situation,

1. to consider only those Riemannian manifolds for which absolute characteristic numbers and other invariants are defined,

2. to give up the concept of absolute characteristic numbers and invariants and to establish a theory of relative invariants for pairs of manifolds and bundles.

In this chapter, we give an outline of absolute number valued invariants. In section 1 we introduce and discuss absolute characteristic numbers for open manifolds associated to a Riemannian metric. These numbers are invariants of the component of the Sobolev manifold of metric connections. In the compact case, there is only one such a component and one gets back the well known independence of the metric. To define the Sobolev component we use the language of uniform structures. A comprehensive treatise of Sobolev uniform structures will be given in chapter II. We conclude section 1 with a short discussion of the Novikov conjecture for open manifolds.

Many characteristic numbers appear as the topological index of certain differential operators. An outline of index theorems for open manifolds will be the content of section 2. To define relative number valued invariants for pairs of manifolds and bundles will be the topic of the chapters IV, V and VI.

1 Absolute characteristic numbers for open manifolds

Let (M^{4k}, g) be closed, oriented, g an arbitrary Riemannian metric, $p_i(M, g)$ the associated by Chern–Weil construction Pontrjagin classes, $e(M, g)$ the Euler class, L_k the Hirzebruch polynomial. Then there are the well known equations

$$\sigma(M^{4k}) = \int L_k(M, g) = \int L_k(p_1(M, g), \ldots, p_k(M, g)) = \sigma(M, g) \tag{1.1}$$

and for (M^n, g) oriented

$$\chi(M^n) = \int e(M, g) = \chi(M, g). \tag{1.2}$$

These equations express in particular that the r.h.s. are in fact independent of g and are homotopy invariants. We proved that the space of Riemannian metrics on a manifold splits w.r.t. a canonical uniform structure into "many" components and that on a compact manifold there is only one component (cf. e.g. [32]). The independence of g can be reformulated as the r.h.s. depend only on comp(g), since the space of Riemannian metrics on closed manifolds consists only of one component. We will extend the definitions of the l.h.s. and the r.h.s. to certain classes of open manifolds. In some cases there even holds equality. The main questions connected with such an extension are

1) the invariance properties,

2) b applications, the geometrical meaning.

It is clear that the definition of characteristic numbers via Chern–Weil construction can be extended to an open manifold if the Chern–Weil integrand is $\in L_1$, as a very special case if this integrand is bounded and (M^n, g) has finite volume.

We present in chapter II a comprehensive discussion of Sobolev uniform structures. For our purpose here we briefly introduce the notion of a basis $\mathfrak{B} \subset \mathfrak{P}(X \times X)$ for a uniform structure $\mathfrak{U}$ on a set X. $\mathfrak{B}$ is a basis if it satisfies the following conditions.

(B_1) If $V_1, V_2 \in \mathfrak{B}$ then $V_1 \cap V_2$ contains an element of $\mathfrak{B}$.
(U'_1) Each $V \in \mathfrak{B}$ contains the diagonal $\Delta \subset X \times X$.
(U'_2) For each $V \in \mathfrak{B}$ there exists $V' \in \mathfrak{B}$ s.t. $V' \subseteq V^{-1}$.
(U'_3) For each $V \in \mathfrak{B}$ there exists $W \in \mathfrak{B}$ s.t. $W \circ W \subset V$.

A uniform structure $\mathfrak{U}$ is metrizable if and only if $\mathfrak{U}$ has a countable basis. For a tensor field t on a Riemannian manifold (M^n, g) we denote

$$^{b,m}|t| := \sum_{\mu=0}^{m} \sup_x |\nabla^\mu t|_x,$$

where $||_x \equiv ||_{g,x}$ denotes the pointwise norm with respect to g and we set $^b|t| \equiv {}^{b,0}|t|$. By $|t|_{p,r}$ we denote the Sobolev norm

$$|t|_{p,r} \equiv |t|_{\nabla,p,r} = \left(\int_M \sum_{i=0}^{r} |\nabla^i t|_x^p \, \mathrm{dvol}_x(g) \right)^{\frac{1}{2}}$$

and set $||_p \equiv ||_{p,0}$. The same definitions hold for tensor fields t with values in a Riemannian vector bundle E.

Let (M^n, g) be an open complete manifold, G a compact Lie group with Lie algebra $\mathfrak{G}$, $\varrho : G \longrightarrow U_N$ or $\varrho : G \longrightarrow SO_N$ a faithful representation, $P = P(M, G)$ a principal fibre bundle and $E = P \times_G E_N$ the associated vector bundle which is endowed with a Hermitean or Riemannian metric. According to the faithfulness of ϱ, the connections on P and E are in a one–to–one relation, $\omega \longleftrightarrow \nabla^\omega = \nabla$. Denote by $\mathcal{C}(P, B_0, f, p) = \mathcal{C}(E, B_0, f, p)$ the set of all connections $\omega \longleftrightarrow \nabla^\omega = \nabla$ with bounded curvature, i.e. satisfying $(B_0) : |R| \leq C$, where R denotes the curvature and $||$ the pointwise norm, and having finite p–action

$$\int |R^{\nabla^\omega}|^p \, \mathrm{dvol}_x(g) < \infty.$$

We fix P and E and write therefore simply $\mathcal{C}(B_0, f, p)$. Let $\delta > 0$ and set

$$\begin{aligned}
V_\delta &= \{(\nabla, \nabla') \in \mathcal{C}(B_0, f, p)^2 \,|\, {}^{b,1}|\nabla - \nabla'|_{\nabla,p,1} \\
&= {}^b|\nabla - \nabla'| + {}^b|\nabla(\nabla - \nabla')| \\
&+ |\nabla - \nabla'|_p + |\nabla(\nabla - \nabla')|_p < \delta\}
\end{aligned}$$

Lemma 1.1 $\mathfrak{B} = \{V_\delta\}_{\delta>0}$ *is a basis for a metrizable uniform structure* $^{b,1}\mathfrak{U}^{p,1}(\mathcal{C}(B_0, f, p))$.

Proof. We start with (U_2'): For each $V \in \mathfrak{B}$ there exists $V' \in \mathfrak{B}$ such that $V' \subseteq V^{-1}$.

$$^{b,1}|\nabla' - \nabla|_{\nabla',p,1} =\,^b|\nabla' - \nabla| +\,^b|\nabla'(\nabla' - \nabla)| + |\nabla' - \nabla|_p + |\nabla'(\nabla' - \nabla)|_p.$$

Hence we have to estimate only

$$\begin{aligned}^b|\nabla'(\nabla' - \nabla)| &\leq\,^b|(\nabla' - \nabla)(\nabla' - \nabla)| +\,^b|\nabla(\nabla' - \nabla)| \\ &\leq C^{b,1}|\nabla' - \nabla|^2 +\,^{b,1}|\nabla' - \nabla|\end{aligned}$$

and

$$\begin{aligned}|\nabla'(\nabla' - \nabla)|_p &\leq |(\nabla' - \nabla)(\nabla' - \nabla)|_p + |\nabla(\nabla' - \nabla)|_p \\ &\leq C_2^{\,b}|\nabla' - \nabla||\nabla' - \nabla|_p + |\nabla(\nabla' - \nabla)|_p,\end{aligned}$$

i.e.

$$^{b,1}|\nabla' - \nabla|_{\nabla',p,1} \leq P_1(^{b,1}|\nabla' - \nabla|_{\nabla,p,1}),$$

where P_1 is a polynomial without constant term. (U_2') is done. For (U_3') : For each $V \in \mathfrak{B}$ there exists $W \in \mathfrak{B}$ such that $W \circ W \subseteq V$ we have to estimate in

$$^{b,1}|\nabla_1 - \nabla_2|_{\nabla_1,p,1} \leq\,^{b,1}|\nabla_1 - \nabla|_{\nabla_1,p,1} +\,^{b,1}|\nabla - \nabla_2|_{\nabla_1,p,1} \quad (1.3)$$

only the term $^{b,1}|\nabla - \nabla_2|_{\nabla_1,p,1}$. But

$$\begin{aligned}^{b,1}|\nabla - \nabla_2|_{\nabla_1,p,1} &=\,^b|\nabla - \nabla_2| +\,^b|\nabla_1(\nabla - \nabla_2)| \\ &\quad + |\nabla - \nabla_2|_p + |\nabla_1(\nabla - \nabla_2)|_p \\ &\leq\,^b|\nabla - \nabla_2| +\,^b|(\nabla_1 - \nabla)(\nabla - \nabla_2)| \\ &\quad +\,^b|\nabla(\nabla - \nabla_2)| + |\nabla - \nabla_2|_p + \\ &\quad + |(\nabla_1 - \nabla)(\nabla - \nabla_2)|_p + |\nabla(\nabla - \nabla_2)|_p \\ &\leq\,^{b,1}|\nabla - \nabla_2|_{\nabla,p,1} \\ &\quad + 2^{b,1}|\nabla_1 - \nabla|_{\nabla_1,p,1} \cdot\,^{b,1}|\nabla - \nabla_2|_{\nabla,p,1},\end{aligned}$$

together with (1.3)

$$^{b,1}|\nabla_1 - \nabla_2|_{\nabla_1,p,1} \leq P_2(^{b,1}|\nabla_1 - \nabla|_{\nabla_1,p,1}, |\nabla - \nabla_2|_{\nabla,p,1}),$$

where P_2 is a polynomial without constant term. (U_3') is done. □

Denote by $^{b,m}\Omega^q(\mathfrak{G}_E)$ or $\Omega^{q,p,r}(\mathfrak{G}_E)$ or $^{b,m}\Omega^{q,p,r}(\mathfrak{G}_E)$ the completion of

$$^b_m\Omega^q(\mathfrak{G}_E) = \{\eta \in \Omega^q(\mathfrak{G}_E) \mid {}^{b,m}|\eta| := \sum_{\mu=0}^m \sup_x |\nabla^\mu \eta|_x < \infty\}$$

or

$$\Omega^{q,p}_r(\mathfrak{G}_E) := \{\eta \in \Omega^q(\mathfrak{G}_E) \mid |\eta|_{p,r}$$

$$:= \left(\int \sum_{i=0}^r |\nabla^i \eta|_x^p \, \mathrm{dvol}_x(g)\right)^{\frac{1}{p}} < \infty\}$$

$$^b_m\Omega^{q,p}_r(\mathfrak{G}_E) = {}^b_m\Omega^q(\mathfrak{G}_E) \cap \Omega^{q,p}_r(\mathfrak{G}_E)$$

with respect to $^{b,m}|\,|$ or $|\,|_{p,r}$ or $^{b,m}|\,|_{p,r} = {}^{b,m}|\,| + |\,|_{p,r}$, respectively. We obtain $\Omega^{q,p,d}$ etc. by replacing $\nabla \longrightarrow d$ and $\Omega^{\cdots}(E,D)$ by replacing $\nabla \longrightarrow D$. Here $\Omega^*(\mathfrak{G}_E)$ are the forms with values in $\mathfrak{G}_E = P \times_G \mathfrak{G}$.

Denote by $^{b,1}\mathcal{C}^{p,1}(B_0, f, p)$ the completion w.r.t. $^{b,1}\mathfrak{U}^{p,1}$.

Theorem 1.2 a) $^{b,1}\mathcal{C}^{p,1}(B_0, f, p)$ *is locally arcwise connected.*
b) *In* $^{b,1}\mathcal{C}^{p,1}(B_0, f, p)$ *coincide components and arc components.*
c) $^{b,1}\mathcal{C}^{p,1}(B_0, f, p)$ *has a decomposition as a topological sum*

$$^{b,1}\mathcal{C}^{p,1}(B_0, f, p) = \sum_{i \in I}^{b,1} \mathrm{comp}^{p,1}(\nabla_i).$$

d)
$$^{b,1}\mathrm{comp}^{p,1}(\nabla) = \{\nabla' \in {}^{b,1}\mathcal{C}^{p,1}(B_0, f, p) \mid {}^{b,1}|\nabla - \nabla'|_{\nabla,p,1} < \infty\}$$
$$= \nabla + (\text{completion of } {}^b_1\Omega^1(\mathfrak{G}_E, \nabla) \cap \Omega^{1,p}_1(\mathfrak{G}_E, \nabla)$$
$$w.r.t. {}^{b,1}|\,|_{\nabla,p,1}) = \nabla + {}^{b,1}\Omega^{1,p,1}(\mathfrak{G}_E, \nabla).$$

Proof. The only fact to prove is a). b) and c) are consequences of a) and d) follows from $\nabla' = \nabla + (\nabla' - \nabla)$. Let $\varepsilon > 0$ be so small that in $U_\varepsilon(\nabla)$ $^{b,1}|\cdot - \cdot|_{\nabla,p,1}$ and the metric of $^{b,1}\mathcal{C}^{p,1}(B_0,f,p)$ are equivalent. Put for $\nabla' \in U_\varepsilon(\nabla)$, $^{b,1}|\nabla - \nabla'|_{\nabla,p,1} < \varepsilon$, $\nabla_t := (1-t)\nabla + t\nabla' = \nabla + t(\nabla' - \nabla)$. If $\nabla_\nu \in {}_1^b\Omega(\mathfrak{G}_E, \nabla) \cap \Omega_1^{1,p}(\mathfrak{G}_E, \nabla)$ and $^{b,1}|\nabla_\nu - \nabla|_{\nabla,p,1} \xrightarrow[\nu \to \infty]{} 0$ then $\nabla_{\nu,t} = \nabla + t(\nabla_\nu - \nabla) \longrightarrow \nabla + t(\nabla' - \nabla) = \nabla_t$, i. e. $\nabla_t \in {}^{b,1}\mathcal{C}^{p,1}(B_0,f,p)$. Moreover, $^{b,1}|\nabla_{t_1} - \nabla_{t_2}|_{\nabla,p,1} = |t_1 - t_2| \cdot {}^{b,1}|\nabla' - \nabla|_{\nabla,p,1} \xrightarrow[t_1 \to t_2]{} 0$. □

Lemma 1.3 *The elements ∇ of ${}^{b,1}\mathcal{C}^{p,1}(B_0,f,p)$ satisfy (B_0) and*

$$\int |R^\nabla|_x^p \, \mathrm{dvol}_x(g) < \infty.$$

Proof. By the definition of $^{b,1}\mathcal{C}^{p,1}$ its elements are C^1 (since they arise by uniform convergence of 0-th and 1rst derivatives) hence R^∇ is defined. If $\nabla_\nu \longrightarrow \nabla$, $\nabla_\nu \in \mathcal{C}(B_0,f,p)$, $\nabla = \nabla_\nu + (\nabla - \nabla_\nu)$, then, for fixed ν,

$$R^\nabla = R^{\nabla_\nu + (\nabla - \nabla_\nu)} = R^{\nabla_\nu} + d^\nabla(\nabla - \nabla_\nu) + \frac{1}{2}[\nabla - \nabla_\nu, \nabla - \nabla_\nu]. \quad (1.4)$$

Each term of the r. h. s. of (1.4) is bounded, hence R^∇. Moreover $|R^{\nabla_\nu}| \in L_p$, $d^\nabla(\nabla - \nabla_\nu)| \in L_p$ and $[\nabla - \nabla_\nu, \nabla - \nabla_\nu] \leq C \cdot {}^b|\nabla - \nabla_\nu| \cdot |\nabla - \nabla_\nu| \in L_p$. □

Now let $\omega \longleftrightarrow \nabla^\omega = \nabla$ be given. After choice of a bundle chart with local base $s_1, \ldots, s_N : U \longrightarrow E|_U$ the curvature Ω will be described as $\Omega s_i = \sum_j \Omega_{ij} \otimes s_j$, where (Ω_{ij}) is a matrix of 2–forms on U, $\Omega_{ij}(s_k,s_l) = \Omega_{ij,kl} = R_{ij,kl}$. An invariant polynomial $P : \mathrm{Mat}_N \longrightarrow \mathbb{C}$ defines in well known manner a closed graded differential form $P = P(\Omega) = P_0 + P_1 + \cdots$, where P_ν is a homogeneous polynomial, $P_r(\Omega) = 0$ for $2r > n$. The determinant is an example for P. If ω is not smooth then $P(\Omega)$ is closed in the distributional sense. Let $\sigma_r(\Omega)$ be the $2r$–homogeneous part (in the sense of forms) of $\det(1 + \Omega_{ij})$.

Lemma 1.4 *Each invariant polynomial is a polynomial in $\sigma_1, \ldots, \sigma_N$.*

Lemma 1.5 *If* $\omega \in {}^{b,1}\mathcal{C}^{p,1}(B_0, f, p)$ *and* $r \geq 1$ *then*

$$\int |\sigma_r(\Omega)|_x^p \, \mathrm{dvol}_x(g) < \infty. \tag{1.5}$$

Proof. For the pointwise norm $|\ |_x$ there holds $|\Omega|_x^2 = \frac{1}{2}\sum_{i,j}\sum_{k<l}|\Omega_{ij,kl}|_x^2$, where $\Omega_{ij,kl} = \Omega_{ij(e_k,e_l)}$ and $e_1, \ldots e_n$ is an orthogonal base of T_xM. According to our assumption we have $|\Omega|_p^p = \int |\Omega|_x^p \, \mathrm{dvol} < \infty$ and $|\Omega|_x \leq b$ for all $x \in M$. The proof is done if we could estimate $|\sigma_r(\Omega)|_x$ from above by $|\Omega|_x$. By definition

$$\sigma_r(\Omega) = \frac{1}{r!} \sum \varepsilon_{j_1\ldots j_r}^{i_1\ldots i_r} \Omega_{i_1 j_1} \wedge \cdots \wedge \Omega_{i_r j_r}, \tag{1.6}$$

where summation runs over all $1 \leq i_1 < \cdots < i_r \leq N$ and all permutations $(i_1 \ldots i_r) \longrightarrow (j_1 \ldots j_r)$. ε denotes the sign of this permutation. We perform induction. For $r = 1$ follows $\sigma_1(\Omega) = \sum \Omega_{ii}$. The inequality

$$|\Omega_{ij}|_x^2 \leq \sum_{s,t}|\Omega_{st}|_x^2 = 2|\Omega|_x^2 \tag{1.7}$$

implies in particular $|\sigma_1(\Omega)|_x^2 \leq 2N|\Omega|_x^2$. For arbitrary forms φ, ψ there holds

$$|\varphi \wedge \psi|_x \leq |\varphi|_x \cdot |\psi|_x. \tag{1.8}$$

For forms with values in a vector bundle we have to multiply the r.h.s. of (1.8) with a constant. (1.6), (1.7), (1.8) and the induction assumption thus give

$$|\sigma_r(\Omega)|_x^2 \leq a \cdot |\Omega|_x^{2r}, \tag{1.9}$$

together with $|\Omega|_x^2 \leq b^2$ finally $|\sigma_r(\Omega)|_x = c \cdot |\Omega|_x$. □

Corollary 1.6 *Let P be an invariant polynomial, $\omega \in {}^{b,1}\mathcal{C}^{p,1}(B_0, f, p)$, $r \geq 1$. Then each form $P_r(\Omega)$ is an element of ${}^{b,1}\Omega^{2r,p,1}$.*

Proof. This follows form 1.4, 1.5 and (1.8). □

Denote by $H^{*,p,\{d\}} = Z^{*,p,\{d\}}/B^{*,p,\{d\}}$ or $^bH^* = {^bZ^*}/{^bB^*}$ the L_p- or bounded cohomology, respectively, where $\{d\}$ refers to the closure of d as coboundary operator.

Corollary 1.7 *Under the assumptions of 1.6, P and ω define well defined classes* $[P_\varrho(\Omega^\omega)] \in H^{2\varrho,p,\{d\}}(M)$, $[P_\varrho(\Omega^\omega)] \in {^b H^{2\varrho}}(M)$.
□

Now arises the natural question, how does $[P_\varrho(\Omega^\omega)]$ depend on ω? We denote $I = [0,1]$, $i_t : M \longrightarrow I \times M$ the imbedding $i_t(x) = (t,x)$ and furnish $I \times M$ with the product metric $\begin{pmatrix} 1 & 0 \\ 0 & g \end{pmatrix}$. Here we write $H^{q,p,\{d\}} \equiv H^{q,p}$ etc..

Lemma 1.8 *For every $q \geq 0$ there exists a linear bounded mapping* $K : \Omega^{q+1,p,d}(I \times M) \longrightarrow \Omega^{q,p,d}(M)$ *resp.* $K :^b \Omega^{q+1,d}(I \times M) \longrightarrow {^b \Omega^{q,d}}(M)$ *such that* $dK + Kd = i_1^* - i^* - 0$.

Proof. Since $g_{I \times M} = \begin{pmatrix} 1 & 0 \\ 0 & g \end{pmatrix}$ is an isometric imbedding and i_t^* is bounded. i_t^* maps into $\Omega^{q,p,d}(M)$ because $|di^*\varphi|_x = |i^*d\varphi|_x \leq C \cdot |d\varphi|_x$. Denote $X_0 = \frac{\partial}{\partial t}$ and for $\varphi \in \Omega^{q+1,p,d}(I \times M)$ $\varphi_0(X_1, \ldots, X_q) := \varphi(X_0, X_1, \ldots, X_q)$. Then $\varphi_0 \in \Omega^{q,p,d}(I \times M)$, $|\varphi_0|_{(t,x)} \leq |\varphi|_{(t,x)}$, and we define

$$K\varphi(X_1, \ldots, X_n) := \int_0^1 i_t^* \varphi_0(X_1, \ldots, X_q) dt.$$

Thus K is bounded too. The equation $dK + Kd = i_1^* - t_0^*$ is standard. Replacing $\Omega^{q,p,d}$ by $^b\Omega^{q,d}$ gives the same conclusions.
□

Lemma 1.9 *Let $f, g : M \longrightarrow N$ be smooth mappings, $F : I \times M \longrightarrow N$ a smooth homotopy, $f^*, g^* : \Omega^{q,p,d}(N) \longrightarrow \Omega^{q,p,d}(M)$, $F^* : \Omega^{q,p,d}(N) \longrightarrow \Omega^{q,pd}(I \times M)$ resp. $f^*, g^* : {}^b\Omega^{q,d}(N) \longrightarrow {}^b\Omega^{q,d}(M)$, $F^* : {}^b\Omega^{q,d}(N) \longrightarrow {}^b\Omega^{q,d}(I \times M)$ bounded and $\varphi \in \Omega^{q,p,d}(N)$ resp. $\varphi \in {}^b\Omega^{q,d}(N)$ closed, i. e. $\varphi \in Z^{q,p}(N)$ resp. $\varphi \in {}^bZ^q(N)$. Then there holds $(g^* - f^*)\varphi \in B^{q,p}(M)$ resp. $(g^* - f^*)\varphi \in {}^bB^q(M)$.*

Proof. We consider the case $\Omega^{q,p,d}$. According to our assumption, we have $K\varphi := KF^*\varphi \in \Omega^{q-1,p,d}(M)$ and $(g^* - f^*)\varphi = ((F \circ t_1)^* - (F \circ i_0)^*)\varphi = (i_1^*F^* - i_0^*F^*)\varphi = (i_1^* - i_0^*)F^*\varphi = (dK + Kd)F^*\varphi = dKF^*\varphi = dK\varphi$. The case of bounded forms will be treated by the same equation. $\square$

Now we are able to prove one of our main theorems.

Theorem 1.10 *Let $Q : M_N(C) \longrightarrow C$ be an invariant polynomial, $r \geq 1$, $p = 1$ or 2. Then each component U of ${}^{b,1}C^{p,1}(B_0, f, p)$ determines uniquely a cohomology class $[Q_r(\Omega^U)] \in H^{p,2r}(M)$ resp. $[Q_r(\Omega^U)] \in {}^bH^{2r}(M)$.*

Proof. Assume $\omega_0, \omega_1 \in U$. Then, according to theorem 1.2, d) $\eta := \omega_1 - \omega_0 \in {}^{b,1}\Omega^{1,p,1}(\mathfrak{G}_E, \omega_0)$ and $\omega_t = \omega_0 + t\eta$, $-\delta < t < 1+\delta$, is contained in U. We have to show $[Q_r(\Omega^{\omega_0})] = [Q_r(\Omega^{\omega_1})]$. Consider

$$\Omega_t := \Omega^\omega, \quad \Omega_t = \Omega_0 + td^{\omega_0}\eta + \frac{1}{2}t^2[\eta, \eta]. \tag{1.10}$$

For all $t \in]-\delta, \delta+1[$ is $\int |\Omega_t|^p \, \text{dvol} < \infty$ and $|\Omega_t|_x$ is bounded at M. This follows from (1.10) and the assumption $\omega_0, \omega_1 \in U$. If $\bar{p} :]-\delta, 1+\delta[\times M \longrightarrow M$ denotes the projection $(t, x) \longrightarrow x$, $P' = \bar{p}^*P$ resp. $E' = \bar{p}^*E$ the liftings of the bundles to $]-\delta, 1+\delta[\times M$, p (which covers $\bar{p}$) the associated mapping of the bundle spaces, then $p^*\omega_0, p^*\omega_1$ are connections for the lifted bundles.

$tp^*\omega_1 + (1-t)p^*\omega_0 = p^*\omega_0 + tp^*\eta$ is again a connection ω'. According to (1.10), there holds $\Omega^{\omega'} = p^*\Omega_0 + t\partial^{p^*\omega_0}p^*\eta + \frac{t^2}{2}[p^*\eta, p^*\eta]$. p^*

is bounded. Thus $\Omega^{\omega'}$ is surely p–integrable and bounded if this holds for $d^{p^*\omega_0}p^*\eta$. But this follows from the equation $(p^*\omega_0)_{ij} = p^*(\omega_{0,ij})$ for the connection matrix, $\eta \in {}^{b,1}\Omega^{1,p,d}(\mathfrak{G}_E, \omega_0)$ and from the boundedness of p^*. ω', $\Omega^{\omega'}$ define well determined p–integrable resp. bounded cocycles at $]-\delta, 1+\delta[\times M$. Let i_t again be the mapping $x \longrightarrow (t,x)$. Then $i_0^*(E', \omega')$ resp. $i_1^*(E', \omega')$ can be identified with (E, ω_0) resp. (E, ω_1). i_t, $0 \leq t \leq 1$, is a smooth bounded homotopy between i_0 and i_1. According to 1.9, $i_0^*\Omega_r(\Omega')$ and $i_1^*\Omega_r(\Omega')$ are in $H^{2r,p}$ resp. ${}^bH^{2r}$ cohomologous, i. e. $Q_r(\Omega_0)$ and $Q_r(\Omega_1)$ are in $H^{2r,p}$ resp. ${}^bH^{2r}$ cohomologous. □

Definition. *For a component U of ${}^{b,1}\mathcal{C}^{p,1}(B_0, f, p)$ we define the r-th Chern class $c_r(PU, p)$ by*

$$c_r(E, U, p) = c_r(P, U, p) := \frac{1}{(2\pi)^r}[\sigma_r(\Omega^U)].$$

Then we have $c_r \in H^{2r,p}$, $c_r \in {}^bH^{2r}$.

Remark 1.11 For $\omega_0, \omega_1 \in {}^{b,1}\mathcal{C}^{p,1}(B_0, f, p)$ the cocycles $\sigma_r(\Omega^{\omega_0}/(2\pi i)^r$, $\sigma_r(\Omega^{\omega_1}/(2\pi i)^r$ are contained in the Chern class $c_r(E)$ and therefore they are cohomologous, but they do not need to be cohomologous in $H^{2r,p}$. Take for example an $\omega \in \Omega^{p,1}(B_0, f, p)$ and apply a gauge transformation g with $\omega - g^*\omega \notin {}^{b,1}\Omega^{p,1,d}(\mathfrak{G}_E, \omega)$. There holds $|\Omega^\omega|_x = |\Omega^{g^*\omega}|_x$. An explicit example is given by $M = R^2$, $P = M \times U_N$, ω the canonical flat connection, the gauge transformation g at the point (x,y) given by $e^{i(x^2+y^2)} \cdot \mathrm{id}$, where id denotes the unit matrix. Then $|\omega - g^*\omega|_{(x,y)} = |g^{-1}dg|_{(x,y)} = |i(x\,dx - y\,dy) \cdot \mathrm{id}|_{(x,y)} = |N(x^2+y^2)|^{\frac{1}{2}}$ is neither bounded nor p–integrable. For this reason our above approach seems to be adequate to the general situation on noncompact Riemannian manifolds. □

Definition. *For $\varrho : G \longrightarrow O_N$, $E = P \times_G \mathbb{R}^N$ denote by E^c or P^c the complexification of E or P, respectively. Any connection*

ω on E resp. P extends in a canonical manner to a connection on E^c resp. P^c and we have an inclusion of the components U of $^{b,1}\mathcal{C}^{p,1}(P, B_0, f, p)$ into the components U^c of $^{b,1}\mathcal{C}^{p,1}(P^C, B_0, f, p)$. Then we define the k-th Pontrjagin class $p_k(P, U, p)$ by

$$p_k(P, U, p) = p_k(E, U, p) := (-1)^k c_{2k}(P^c, U^c, p).$$

Let Pf be the Pfaff polynomial for skew symmetric $2N$-matrices, $\varrho : G \longrightarrow SO_{2N}$, $E = P \times_G \mathbb{R}^{2N}$. Then we call for a component U of $^{b,1}\mathcal{C}^{p,1}(P^C, B_0, f, p)$

$$e(E, U, p) := \frac{1}{(2\pi)^N} Pf(\Omega^U)$$

the Euler class of U. There holds $e \in H^{2N,p}(M)$, $e \in {}^b H^{2N}(M)$.

Now come in characteristic numbers. Consider $\varrho : G \longrightarrow U_N$, let be $\dim M = 2k$ and Q an invariant polynomial, $\omega \in {}^{b,1}\mathcal{C}^{p,1}(P^C, B_0, f, p)$, $Q(\Omega^\omega) = a + Q_1(\Omega) + \cdots + Q_k(\Omega)$. Then $Q_{i_1\ldots i_k} := Q_{i_1} \wedge \cdots \wedge Q_{i_k}$ with $i_1 + \cdots + i_k = k$ defines a characteristic $2k$-form and a characteristic number $\int Q_{i_1\ldots i_k} = Q_{i_1\ldots i_k}(P, \omega)(M)$ if the latter integral exists. In particular we consider classes $c_{i_1\ldots i_k} := c_{i_1} \wedge \cdots \wedge c_{i_k}$ and have to ensure the existence of the corresponding integral.

Lemma 1.12 a) If $k = 1$ and $\omega \in {}^{b,1}\mathcal{C}^{1,1}(B_0, f, 1)$, then $\int_M c_1$ converges.
b) If $k > 1$ and $\omega \in {}^{b,1}\mathcal{C}^{1,1}(B_0, f, 1)$ or $\omega \in {}^{b,1}\mathcal{C}^{2,1}(B_0, f, 2)$ then $\int_M c_{i_1\ldots i_k}$ converges.

Proof. a) is clear. We have to prove b). At each $x \in M$, $c_{i_1\ldots i_k}$ is a sum of monomials $a \cdot \Omega_{i_1 j_1} \wedge \cdots \wedge \Omega_{i_k j_k}$. There holds according to lemma 1.5 $|\Omega_{i_1 j_1} \wedge \cdots \wedge \Omega_{i_k j_k}|_x \le D_1 \cdot |\Omega|_x$ resp. $\le D_2 \cdot |\Omega|_x^2$ if $p = 1$ resp. $p = 2$. $\square$

Corollary 1.13 *Under the assumption 1.12, for any invariant polynomial Q converges $\int_M Q_{i_1...i_k}$.*

This follows from 1.4 and the proof of 1.12. □

Lemma 1.12 b) is also valid in the case $\varrho : G \longrightarrow O_N$, $\dim M = 4k$ for $p_{i_1...i_k}$, $i_1 + \cdots + i_k = k$, $k \geq 1$, resp. in the case $\varrho : G \longrightarrow SO_{2N}$, $\dim M = N$ for the Euler form $e(E, \omega, 1, g)$ 1.12 a) for $N = 1$, 1.12 b) for $N > 1$).

The above characteristic numbers are defined until now only for a chosen connection ω. One would like that the charakteristic numbers are constant at least at the components of $^{b,1}\mathcal{C}^{p,1}(B_0, f, p)$. This is in fact the case for $p = 1$.

Theorem 1.14 *The characteristic numbers are constant at the components of $^{b,1}\mathcal{C}^{1,1}(B_0, f, 1)$.*

Proof. If ω, ω' are contained in the same component U, then according to 1.10, $Q_{i_1...i_k}(\omega)$ and $Q_{i_1...i_k}(\omega')$ define the same cohomology clas in $H^{2k,1}(M)$ resp. $H^{4k,1}(M)$, i. e. there exists an absolutely integrable φ with $d\varphi = Q_{i_1...i_k}(\omega) - Q_{i_1...i_k}(\omega')$. A fundamental result of Gaffney then says $\int_M d\varphi = 0$ for (M, g) complete and $d\varphi$ itself absolutely integrable. □

Thus one gets characteristic numbers $Q_{i_1...i_k}(P, U)(M)$.

Remark 1.15 For $\omega, \omega' \in {}^{b,1}\mathcal{C}^{2,1}(B_0, f, 2)$ and $\deg(Q_{i_1...i_k}) \geq 4$ the characteristic numbers $Q_{i_1...i_k}(\omega)(M), Q_{i_1...i_k}(\omega')(M)$ are defined. If $Q_{i_1...i_k}(\omega), Q_{i_1...i_k}(\omega')$ define the same cohomology class in $H^{2k,1}(M^{2k})$ resp. $H^{4k,1}$ resp. $H^{2N,1}$, then the characteristic numbers coincide. □

A very special but interesting case in our considerations is the case $\text{vol}(M) < \infty$. Consider $^{b,1}\mathcal{C}(B_0)$. It is defined by means of $\mathfrak{B} = \{V_\delta\}_{\delta > 0}$, where $V_\delta = \{(\nabla, \nabla') \in \mathcal{C}(B_0)^2 \,|\, {}^{b,1}|\nabla - \nabla'|_\nabla < \delta\}$.

Theorem 1.16 *If $\text{vol}(M) < \infty$ then characteristic numbers are constant on each component of $^{b,1}\mathcal{C}(B_0)$.*

Proof. According to 1.10 each component U of $^{b,1}\mathcal{C}(B_0)$ determines uniquely a cohomology class $[Q_{i_1...i_k}]$ $^bH^{2k}(M^{2k})$ or $^bH^{4k}$ or $^bH^{2N}$ respectively. Taking two cocycles of this class, there exists a bounded C^1–form φ such that there difference equals to $d\varphi$. φ, $d\varphi$ are bounded, vol$(M) < \infty$, thus φ, $d\varphi$ are absolutely integrable and the theorem of Gaffney gives the desired result. □

Remark 1.17 vol$(M) < \infty$ implies $^{b,1}\mathcal{C}(B_0) = {}^{b,1}\mathcal{C}^{p,1}(B_0, f, p)$. Thus the conclusion of 1.16 also holds for the components of $^{b,1}\mathcal{C}^{p,1}(B_0, f, p)$. □

We call the quasi isometry class of g the uniform structure $US(g)$ generated by g. For all metrics of $US(g)$ the cohomology spaces $H^{*,p}(M^n, g')$ coincide. The same holds for $^bH^*(M^n, g')$. This leads immediately to

Theorem 1.18 *The cohomology classes $Q_{i_1...i_k}(\omega)$ resp. characteristic numbers $Q_{i_1...i_k}(\omega)(M)$ in 1.10 respectively 1.14, 1.15 are the same for all metrics $g' \in US(g)$.* □

The situation completely changes if ω itself depends on g. Then it will be wrong in general that for $g' \in US(g)$ $c(\omega(g)) \sim c(\omega(g'))$. The case $\omega = \omega(g)$ is essentially the case $P = $ bundle of orthogonal frames of (M^n, g), $\nabla = $ Levi–Civita connection ∇^g. Therefore we briefly describe the metrics which come into question and describe their admitted variation (for fixed M). Let

$$\mathcal{M}(B_0, p, f) = \{g \mid g \text{ complete, satisfies } (B_0) \text{ and} \int |R^g|_x^p \, \mathrm{dvol}_x(g) < \infty\},$$

$$^{b,2}|g - g'|_{g,p,2} = {}^{b,2}|g - g'|_g + |g - g'|_{g,p,2} =$$
$$= {}^b|g - g'|_g + {}^b|\nabla^g - \nabla^{g'}|_g + {}^b|\nabla^g(\nabla^g - \nabla^{g'})| +$$
$$+ \left(\int \left(|g - g'|_{g,x}^p + \sum_{i=0}^{1} |(\nabla^g)^i(\nabla^g - \nabla^{g'})|_{g,x}^p \right) \mathrm{dvol}_x(g) \right)^{\frac{1}{p}}$$

and set
$$V_\delta = \{(g, g') \in \mathcal{M}(B_0, p, f)^2 \mid C(n,\delta)^{-1} g \leq g' \leq C(n,\delta) g$$
$$\text{and } {}^{b,2}|g - g'|_{g,p,2} < \delta\}.$$

Here $C(n,\delta) = 1 + \delta + \delta\sqrt{2n(n-1)}$.

Lemma 1.19 $\mathfrak{B} = \{V_\delta\}_{\delta > 0}$ is a basis for a metrizable uniform structure. $\square$

Denote by ${}^{b,2}\mathcal{M}^{p,2}(B_0, p, f)$ its completion.

Proposition 1.20 a) ${}^{b,2}\mathcal{M}^{p,2}(B_0, p, f)$ is locally arcwise connected.
b) In ${}^{b,2}\mathcal{M}^{p,2}(B_0, p, f)$ coincide components with arccomponents.
c) ${}^{b,2}\mathcal{M}^{p,2}(B_0, p, f)$ has a representation as a topological sum
$$ {}^{b,2}\mathcal{M}^{p,2}(B_0, p, f) = \sum_{i \in I} {}^{b,2}\text{comp}^{p,2}(g_i).$$

d) $\text{comp}(g) = \{g' \in {}^{b,2}\mathcal{M}^{p,2}(B_0, p, f) \mid {}^{b,2}|g - g'|_{g,p,2} < \infty\}$. $\square$

Proposition 1.21 If $g' \in \text{comp}(g)$ then $\nabla^{g'} \in \text{comp}(\nabla^g)$ is the sense of theorem 1.2, d). $\square$

Hence we obtain well defined characteristic classes $C(\nabla^g) = C(g)$ and characteristic numbers $C\ldots(\nabla^g)(M) = C\ldots(g)(M)$ as above. The main important cases are the Euler form $e = E(g)$,
$$\chi(M^n, g) := \int_M E(g)$$
and the signature case
$$\sigma(M^n, g) := \int_M L(g),$$
where $L(g)$ is the Hirzebruch genus.

There arise the following natural questions.
1) How does $E(g)$ depend on g?
2) What is the topological meaning of $\chi(M^n, g)$?
3) Under which conditions does there hold $\chi(M^n, g) = \chi(M^n)$, i.e. the Gauß–Bonnet formula?

The same questions should be put for $\sigma(M^n, g)$, $\sigma(M^n)$. To the first question we have a partial answer.

Proposition 1.22 *If $g' \in {}^{b,2}\mathrm{comp}^{1,2}(g)$ then*
$$\chi(M^n, g) = \chi(M^n, g')$$
and
$$\sigma(M^n, g) = \sigma(M^n, g').$$
□

In the case $g' \notin {}^{b,2}\mathrm{comp}^{1,2}(g)$ we can't say anything. The examples in [3] for $\chi(M^n, g) \neq \chi(M^n, g')$, $\sigma(M^n, g) \neq \sigma(M^n, g')$ are of this kind, i.e. g' does not lie in the component of g.

Concerning the second question, we start with a simple case in dimension two which has been discussed by Cohn–Vossen [21] and Huber [43] and has been endowed with particular short proofs by Rosenberg [64], which we present below for completion.

Theorem 1.23 *Let (M^n, g) be a finitely connected complete non-compact Riemannian surface with curvature K_g.*
a) *If $K \in L_1$ then $\chi(M) \geq \int_M K \, \mathrm{dvol}_x(g)$.*
b) *If $\mathrm{vol}(M^2, g) < \infty$ and $K \in L_1$ then*
$$\chi(M) = \int_M K \, \mathrm{dvol}_x(g) = \chi(M, g).$$

Proof. M^2 is diffeomorphic to a compact surface with p points deleted. A neighbourhood of each point is diffeomorphic to $S^1 \times R_+$ and the metric can be put in the form

$g_{11}(\theta,t)d\theta^2 + dt^2$. Set $M_k = M \setminus \bigcup_1^p S^1 \times]k, \infty[$. The Gauß–Bonnet theorem for surfaces with boundary yields $\chi(M_k) = \int_{M_k} K \, \mathrm{dvol}_x(g) + \int_{\partial M_k} \omega_{12}$, where ω_{12} is the connection 1–form associated to an orthonormal frame on M. $\chi(M) = \chi(M_k)$, hence one has to show $\lim_{k\to\infty} \int_{\partial M_k} \omega_{12} \geq 0$ for a) and $\lim_{k\to\infty} \int_{\partial M_k} \omega_{12} = 0$ for b). W.r.t. the orthonormal frame $\theta^1 = \sqrt{g_{11}} d\theta$ and $\theta^2 = dt$ the first structure equation $d\theta^1 = \omega_{12} \cap \theta^2$ gives $\omega_{12} = \frac{d}{dt}(\sqrt{g_{11}}) d\theta$ and the second one gives $K \, \mathrm{dvol}_x(g) = \Omega_{12} = d\omega_{12} = \frac{d^2}{dt^2}(\sqrt{g_{11}}) d\theta dt$. $\int_M K \, \mathrm{dvol}_x(g) < \infty$ implies $\lim_{k\to\infty} \int_{\partial M_k} \frac{d^2}{dt^2} \sqrt{g_{11}} d\theta = 0$ or $\lim_{k\to\infty} \int_{\partial M_k} \frac{d^2}{dt^2} \sqrt{g_{11}} d\theta = const = C$. In the case b), $\mathrm{vol}(M, g) < \infty$, i.e. $\int_M \sqrt{g_{11}} d\theta dt < \infty$ which implies $\lim_{k\to\infty} \int_{\partial M_k} \sqrt{g_{11}} d\theta = 0$, hence $\lim_{k\to\infty} \int_{\partial M_k} \omega_{12} = \lim_{k\to\infty} \frac{d}{dt} \int \sqrt{g_{11}} d\theta = C$, $C = 0$. In the case a), $\int_{\partial M_k} \sqrt{g_{11}} d\theta \sim C \cdot k + D$ as $k \to \infty$. $C < 0$ would imply $\int_{\partial M_k} \sqrt{g_{11}} d\theta = 0$ for k sufficiently large. But this is impossible for a positive integrand. □

In the case for arbitrary n, there are many approaches to study the equation $\chi(M, g) = \chi(M)$. To have $\chi(M)$ defined, one must require that each homology group over $\mathbb{R}$ is finitely generated. Sufficient for this is that M has finite topological type, i.e. it has only finitely many ends $\varepsilon_1, \ldots, \varepsilon_s$, each of them collared, $U(\varepsilon_i) \cong \partial U_i \times [0, \infty[$. Then M can be given a boundary ∂M to get a compact manifold $\overline{M}$. The case n odd is absolutely trivial.

Proposition 1.24 *Assume (M^{2k+1}, g) is of finite topological type, g arbitrary. Then*

$$\chi(M) = \int_M E(g) = \chi(M, g) \text{ if and only if } \chi(\partial M) = 0.$$

Proof. For $n = 2k + 1$, the Euler form $E(g)$ vanishes since the

Pfaffian of an odd dimensional skew symmetric matrix is zero, $\int E(g) = \chi(M,g) = 0$. On the other hand, $0 = \chi(\overline{M} \underset{\partial M}{\cup} \overline{M}) = 2\chi(\overline{M}) - \chi(\partial M) = 2\chi(M) - \chi(\partial M)$. □

The more interesting case are even dimensional manifolds. We recall some definitions.

For a local orthonormal frame $\theta^1, \ldots, \theta^n$ the connection 1–forms ω_{ij} satisfy the equations

$$d\theta^i = \sum_j \omega_{ij} \wedge \theta^j \text{ and } \omega_{ji} = -\omega_{ij}.$$

They are related with the curvature 2–forms Ω_{ij} by

$$\Omega_{ij} = d\omega_{ij} - \sum_k \omega_{ik} \wedge \omega_{kj}.$$

Denote by $S(M)$ the tangent sphere bundle which is a $(2n-1)$-dimensional manifold. For a point $(x,\xi) \in S(M)$ let $\theta^1, \ldots, \theta^n$ be a frame such that θ^1 is dual to ξ. We put

$$II(g) := \sum_{0 \leq k < n} c_k \sum_\alpha \text{sign}(\alpha) \Omega_{\alpha(2)\alpha(3)} \wedge \cdots \wedge \Omega_{\alpha(2k)\alpha(2k+1)}$$
$$\wedge \omega_{\alpha(2k+2)1} \wedge \cdots \wedge \omega_{\alpha(n)1}, \qquad (1.11)$$

where we will not specify the c_k and $\sum_\alpha$ means the sum over all permutations α of $\{2, \ldots, n\}$. $II(g)$ can be understand as pull back on an $(n-1)$–form on M to $S(M)$ by means of $pr : S(M) \longrightarrow M$. If $X \in TS(M)$ at (x,ξ), $X = X_1 + X_2$ with X_1 tangent to M and X_2 tangent to S_x^{n-1} then $\Omega_{ij}(X) = \Omega_{ij}(X_1)$ and similarly for $\omega_{i1}(X)$. If M is compact with boundary ∂M and ϱ is the section of $S(M)$ over ∂M given by the outward normal vector, then $\varrho^* \Omega_{ij}(X) = \Omega_{ij}(X_1)$, the same for ω_{i1}. Then, according to Chern,

$$\chi(M) = \int_M E(g) + \int_{\partial M} \varrho^* II(g) = \int_M E(g) + \int_{\partial M} II(g). \quad (1.12)$$

Assume now that $(M^n, g) = (M^{2m}, g)$ is even–dimensional and of finite topological type. By gradient flow of an appropriate Morse function we can introduce coordinates $(x_1, \ldots, x_{n-1}, x_n = r)$ at each end such that $0 \leq r < \infty$, $g_{in} = 0$, $1 \leq i \leq n-1$, $g_{nn} = 1$. Let as above M_k be characterized by $x_n = r \leq k$. Then

$$\chi(M) = \chi(M_k) = \int_{M_k} E(g) + \int_{\partial M_k} II(g). \qquad (1.13)$$

At each end ε $TM|_\varepsilon$ splits as $TM = W \oplus \mathbb{R}$. Suppose additionally that W splits as

$$W = W_2 \oplus \cdots \oplus W_{r_\varepsilon}, \quad r_\varepsilon \geq 2, \quad [W_i, W_j] = 0 \quad \text{if } i \neq j, \qquad (1.14)$$

and that with respect to this splitting g has the form

$$g = f_2^2(r) g_2 \oplus \cdots \oplus f_{r_\varepsilon}^2(r) g_{r_\varepsilon} + dr^2. \qquad (1.15)$$

Then S. Rosenberg calculated in [65] the expression (1.11) at each end can could show if $f_j(r) \xrightarrow[r \to \infty]{} 0$, $f_j'(r) \xrightarrow[r \to \infty]{} 0$, then $\int_{M_k} E(g) \longrightarrow \int_M E(g)$ and $\int_{\partial M_k} II(g) \longrightarrow 0$. We will not repeat the really simple calculations but state Rosenberg's

Theorem 1.25 *Let (M^n, g) be open, complete and of finite topological type. Assume that in an open neighbourhood of each end ε M splits as a product manifold $N_2 \times \cdots \times N_{r_\varepsilon} \times \mathbb{R}$ with the metric $f_2^2(r) g_2 \oplus \cdots \oplus f_{r_\varepsilon}^2(r) g_{r_\varepsilon} + dr^2$, where g_j is a metric on N_j. If $f_j(r) \xrightarrow[r \to \infty]{} 0$ and $f_j'(r) \xrightarrow[r \to \infty]{} 0$, then $\chi(M) = \int_M E(g) \equiv \chi(M, g)$. In particular, any evendimensional manifold of finite topological type admits complete warped product metrics satisfying Gauß–Bonnet (setting $N_2 = \partial M$).* □

Corollary 1.26 *Assume the hypothesis of 1.25 and additionally $g \in {}^{b,2}\mathcal{M}^{1,2}(B_0, f, 1)$. If $g' \in {}^{b,2}\text{comp}^{1,2}(g)$ then $\chi(M) = \int_M E(g') \equiv \chi(M, g')$.* □

Remark 1.27 We see in 1.26 a considerable improvement of 1.25 since now the admitted class of metrics is much larger. □

If one gives up the integrability of the Ws in (1.14), i.e. the product structure of the εs, then one must strengthen the conditions to the f_j. This has been done by Rosenberg too.

Theorem 1.28 *Let* (M^n, g) *be open, complete and of finite topological type. Assume that in an open neighbourhood of each end* ε, $TM|_\varepsilon = W_2 \oplus \cdots \oplus W_{r_\varepsilon} \oplus \mathbb{R}$ *and the metric is of the form* $f_2^2(r)g_2 \oplus \cdots \oplus f_{r_\varepsilon}^2(r)g_{r_\varepsilon} + dr^2$ *with* g_i *a metric on* W_i. *If* $f_i(r) \underset{r \to \infty}{\longrightarrow} 0$, $f_i'(r) \underset{r \to \infty}{\longrightarrow} 0$ *and* $f_j f_i^{-1}$ *and* $(f_j f_i^{-1})'$ *are bounded for all* r, i, j *then*

$$\chi(M) = \int E(g).$$

□

Example. Let $M \backslash G/K$ be an arithmetic quotient of an evendimensional split rank–one symmetric space. Then at each component ∂M_i of ∂M, ∂M is the total space of a fibration over a torus T_1 with a torus T_2 as fiber. We have $TM|_{V \times \mathbb{R}} = W_1 \oplus W_2 \oplus \mathbb{R}$ for open $V \subset \partial M$ where the fibration restricted to V is trivial. W_i is the tangent space to the torus T_i. But in general the G–invariant metric g does not respect this splitting. Donnelly has shown in [24] that each end ε has the structure $N \times \mathbb{R}$, N at most two–step nilpotent. The Lie algebra $\mathfrak{n}$ of N splits as $\mathfrak{n} = V_2 \oplus V_3$ of root spaces, $V_3 = Z(\mathfrak{n})$, and the invariant metric at the identity of N has the form

$$e^{-2r}g_2 + e^{-4r}g_3 + dr^2, \tag{1.16}$$

where g_2 is a metric on V_2, g_3 a metric on V_3. $[\mathfrak{n}, \mathfrak{n}] \subset Z(\mathfrak{n})$ and the G–invariant distribution V_2 is not integrable. Hence theorem 1.25 is not applicable in general. In the hyperbolic case $G/K = SO(n,1)/SO(n)$, one has $V_2 = \mathfrak{n}$, which yields Gauß–Bonnet. □

Corollary 1.29 *Assume the hyptheses of theorem 1.28 and additionally* $g \in {}^{b,2}\mathcal{M}^{1,2}(B_0, f, 1)$. *If* $g' \in {}^{b,2}\text{comp}^{1,2}(g)$ *then* $\chi(M) = \int E(g') \equiv \chi(M, g')$. □

There is another Gauß–Bonnet case which does not fall under 1.2 – 1.29.

Proposition 1.30 *Let* (M^{2m}, g) *be open, complete, oriented, of finite topological type and the metric at ∞ constant with respect to r, i.e. there exists an $r_0 \geq 0$ such that $g(r_1, x) = g(r_2, x)$ for all $x \in \partial M$ and $r_1, r_2 > r_0$. Then*

$$\chi(M) = \int_M E(g) \equiv \chi(M, g).$$

Proof. Let $k > r_0 + \delta$. Then $M_k \cup M_k$ yields a smooth closed manifold. Hence

$$\chi(M_k \cup M_k) = \int_{M_k \cup M_k} E(g_{M_k \cup M_k}) = 2 \int_{M_k} E(g_{M_k}),$$

$$\chi(M_k \cup M_k) = 2\chi(M_k) - \chi(\partial M_k) = 2\chi(M)$$

$$\chi(M) = \int_{M_k} E(g_{M_k}). \qquad (1.17)$$

Forming $\lim_{k \to \infty}$ in (1.17) gives the desired result. □

A special case of 1.28 would be a metric cylinder at infinity, $g|_{U(\infty)} = g_{\partial M} \otimes +dr^2$. This is simultaneously a warped product with warping function $f(r) = 1$. $f(r) = 1$ does not satisfy $f(r) \xrightarrow[r \to \infty]{} 0$, 1.25 is not applicable. Clearly, such an (M^{2m}, g) satifies (B_0) but either $\int_{U(\infty)} |R|^p \, \text{dvol}_x(g) = 0$ or $\int_{U(\infty)} |R|^p \, \text{dvol}_x(g) = \infty$, similarly either $\int_{U(\infty)} |E(g)| \, \text{dvol}_x(g) = 0$ or $\int_{U(\infty)} |E(g)| \, \text{dvol}_x(g) = \infty$. In the second case $\int E(g)$ exists but $|E(g)| \notin L_p, p \geq 1$.

Another class of examples which submits very useful insights are surfaces of revolution. We state from [65] without proof

Proposition 1.31 Let $f :]0,\infty[\longrightarrow \mathbb{R}$ be smooth, $f(0) = f'(0) = 0$ and $(M^2 = \{z = f(x^2+y^2)\}$, induced metric from $\mathbb{R}^3)$ be the associated surface of revolution. Then

$$\chi(M) = \frac{1}{2\pi}\int_M K\,\mathrm{dvol}_x(g) = \chi(M,g) \qquad (1.18)$$

if and only if

$$r^{\frac{1}{2}}f'(r) \xrightarrow[r\to\infty]{} \pm\infty.$$

□

Hence, if f is for all $r > 0$ strongly convex or concave, (1.18) holds. In both cases M has for $r > 0$ positive curvature and infinite volume. On the other hand, we have 1.15 in the case of 1.23 b) in the finite volume case, i.e. one can have $\chi(M) = \chi(M,g)$ as in the finite volume case. For this reason we should find additional conditions which assure in the finite volume case or the infinite volume case, respectively, that
1) $\chi(M,g)$ is a (proper) homotopy invariant,
2) $\chi(M,g) = \chi(M)$ if M has finite topological type.
We start with $\mathrm{vol}(M^n,g) < \infty$ and $|K| \leq 1$ where the letter (after rescaling) is equivalent to (B_0). Then

$$\chi(M,g) = \int_M E(g)$$

is well defined and for $g' \in {}^{b,2}\mathrm{comp}^{1,2}(g)$

$$\chi(M,g) = \chi(M,g'). \qquad (1.19)$$

Lemma 1.32 Let (M^n,g) be complete, $\mathrm{vol}(M,g) < \infty$ and $|K| \leq 1$. Then M^n admits an exhaustion by compact manifolds with smooth boundary, $M_1^n \subset M_2^n \subset \cdots$, $\bigcup_k M_k^n = M$, such that $\mathrm{vol}(\partial M_k^n) \longrightarrow 0$ and for which the second fundamental forms $II(\partial M_k^n)$ are uniformly bounded.

This is just a corollary of theorem 1.33 below. □

If we take such an exhaustion as just described then

$$\chi(M_k^n) = \chi(M_k^n, g) + \int_{\partial M_k^n} II(\partial M_k^n). \qquad (1.20)$$

$\int_{\partial M_k^n} II(\partial M, g) \xrightarrow[k \to \infty]{} 0$, $\chi(M_k^n) \in \mathbb{Z}$, hence for k sufficiently large $\chi(M_k^n, g) \in \mathbb{Z}$, but we are far from a certain convergence of $(\chi(M_k^n, g))_k$ and don't know anything about the topological properties of such a limit if it exists. To obtain more insight and definite results we follow [3] and consider the following additional hypothesis.

For some neighbourhood $U(\infty) \subset M$, some profinite or normal covering space $\tilde{U}(\infty)$ has the injectivity radius at least (say) 1 for the pull back metric,

$$r_{\text{inj}}(\tilde{U}_\infty)) \geq 1. \qquad (1.21)$$

Together with $|K| \leq 1$ on $\tilde{U}(\infty)$ we write $\text{geo}_\infty(M) \leq 1$. If $U = M$ then we denote $\text{geo}(\tilde{M}) \leq 1$. In any case we assume in this hypothesis that $\tilde{U}$ or $\tilde{M}$ are profinite or normal coverings.

Here $\tilde{M} \longrightarrow M$ is profinite if there exists a decreasing sequence $\{\Gamma_j\}_j$ of subgroups of finite index, $\Gamma_j \subset \pi_1(M)$, such that $\bigcap \Gamma_j = \pi_1(\tilde{M})$.

The key for everything is the following very general theorem which assures the existence of sufficiently "smooth" exhaustions and which yields 1.32 in the case of $\text{vol}(M, g) < \infty$.

Theorem 1.33 *(Neighborhoods of bounded geometry).*
Let (M^n, g) be complete, $X \subset M^n$ a closed subset and $0 < r \leq 1$. Then there is a submanifold U^n with smooth boundary ∂U^n such that for some constant $c(n)$ depending only on n

    a) $X \subset U \subset T_r(x) = r-$ tubular neighbourhood of X,
    b) $\text{vol}(\partial U) \leq c(n) \cdot \text{vol}(T_r(X) \setminus X) \cdot r^{-1}$,   (1.22)
    c) $|II(\partial U)| \leq c(n) \cdot r^{-1}$.   (1.23)

We refer to [17] for the proof. □

Now we will discuss $\chi(M,g)$ in the profinite or normal case, $\text{geo}(\tilde{M}) \leq 1$. Here we follow [16]. Put for $j : A_1 \subset A_2$ and real coefficients $\beta^i(A_1, A_2) = \dim\{j^*(H^i(A_2)) \subset H^i(A_1)\}$ and $\beta^i(A) = \dim\{j^*(H^i(A, \partial A)) \subset H^i(A)\}$. b^i shall denote the usual Betti number. Then for $A_1 \subset A_2 \subset A_3 \subset A_4$ and $A \subset Y$ a finite closed and $f : Y \longrightarrow Z$, $g : Z \longrightarrow Y$ simplicial, determining a homotopy equivalence,

$$\beta^i(A_1) \subseteq \beta^i(A_2) \leq \beta^i(A_2, A_4) \leq \beta^i(A_3, A_4) \quad (1.24)$$

and

$$\beta^i(A, Y) \leq \beta^i(f(A), Z) \leq \beta^i(g \circ f(A), Y). \quad (1.25)$$

Put for $p : \tilde{Y}^n \longrightarrow Y^n$ profinite with $\text{ind}(\Gamma_j) = d_j$ and corresponding covering spaces $p_j : \tilde{Y}^n_j \longrightarrow Y^n$

$$\sup \tilde{\chi}(Y^n) := \varlimsup_{A\to\infty} \varlimsup_{j\to\infty} \sum_{i=1}^{n} (-1)^i \frac{1}{d_j} \beta^i(P_j^{-1}(A), \tilde{Y}^n_j) \leq \infty \quad (1.26)$$

and define $\inf \tilde{\chi}(Y^n)$ similarly. $A \longrightarrow \infty$ is defined by partial ordering of finite subcomplexes induced by inclusion. Using (1.24) and a diagonal argument, there are subsequences $S = \tilde{Y}^n_{j(e)}$ s.t.

$$\infty \geq \tilde{\beta}^i(Y^n, S) := \varlimsup_{A\to\infty} \varlimsup_{e\to\infty} \frac{1}{d_{j(e)}} \beta^i(P_{j(e)}^{-1}(A), \tilde{Y}^n_{j(e)})$$
$$= \varliminf_{A\to\infty} \varliminf_{e\to\infty} \frac{1}{d_{j(e)}} \beta^i(P_{j(e)}^{-1}(A), \tilde{Y}^n_{j(e)}) \quad (1.27)$$

exists. From (1.25) we infer immediately that $\tilde{\beta}^i(Y^n, S)$ is a homotopy invariant. Suppose $\tilde{\beta}^i(Y^n, S) < \infty$, $i = 0, \ldots, n$ and $\sup \tilde{\chi}(Y^n) = \inf \tilde{\chi}(Y^n)$, then the latter number is also a homotopy invariant.

Theorem 1.34 *Suppose (M^n, g) complete, $\text{vol}(M^n, g) < \infty$, $\tilde{M}$ either profinite or normal and $\text{geo}(\tilde{M}) \leq 1$.*
a) *Then $\chi(M^n, g)$ is a proper homotopy invariant,*

b) *in the case $\tilde{M}$ profinite*

$$\chi(M,g) = \sup \tilde{\chi}(M) = \inf \tilde{\chi}(M),$$

c) *if additionally M has finite topological type,*

$$\chi(M,g) = \chi(M).$$

Proof. Assume $\tilde{M} \longrightarrow M$ profinite, let $M_1 \subset M_2 \subset \cdots$, $\bigcup_k M_k = M$ be an exhaustion of M by compact submanifolds with boundary and denote $M_k - R = \{x \in M_k | \text{dist}(x, \partial M_k) = R\}$. For j sufficiently large, theorem 1.33 is applicable and we apply it to $p_j^{-1}(M_{k-1})$, $p_j^{-1}(M_k)$ with $\varepsilon = \frac{1}{2}$. This yields submanifolds $A_{jk} \subset p_j^{-1}(M_k) \subset B_{jk}$. Given $\varepsilon > 0$ arbitrary, there exist k_0, $N(k)$ such that for $k > k_0$, $j > N(k)$

$$\left| \chi(M^n, g) - \frac{1}{d_j} \chi(B_{jk}) \right| \leq \left| \chi(M^n, g) - \frac{1}{d_j} \int_{B_{jk}} E(g) \right|$$

$$+ \left| \frac{1}{d_j} \int_{B_{jk}} E(g) - \frac{1}{d_j} \chi(B_{jk}) \right| < \varepsilon.$$

(1.28)

We see this immediately from (1.12) and (1.22), (1.23): $\chi(M^n, g) = \chi(M_k^n, g) + \chi(M^n \setminus M_k^n, g)$, here $|\chi(M^n \setminus M_k^n, g)|$ becomes arbitrarily small for k sufficiently large.

$$\left| \chi(M^n, g) - \frac{1}{d_j} \int_{B_{jk}} E(g) \right| \leq |\chi(M^n, g) - \chi(M_k^n, g)|$$

$$+ \left| \chi(M_k^n, g) - \frac{1}{d_j} \int_{B_{jk}} E(g) \right|,$$

$$\left|\chi(M_k^n,g) - \frac{1}{d_j}\int_{B_{jk}} E(g)\right| \leq \left|\chi(M_k^n,g) - \frac{1}{d_j}\int_{p_j^{-1}(M_k^n)} E(g)\right|$$

$$+ \left|\frac{1}{d_j}\int_{p_j^{-1}(M_k^n)} E(g) - \frac{1}{d_j}\int_{B_{jk}} E(g)\right|$$

$$= \left|\frac{1}{d_j}\int_{B_{jk}\setminus p_j^{-1}(M_k^n)} E(g)\right|,$$

but this becomes arbitrarily small for j and k sufficiently large. Finally

$$\left|\frac{1}{d_j}\int_{B_{jk}} E(g) - \frac{1}{d_j}\chi(B_{jk})\right| = \left|\frac{1}{d_j}\int_{\partial B_{jk}} II(\partial B_{jk})\right| \xrightarrow[j,k\to\infty]{} 0$$

according to (1.22). (1.28) is proven.
We obtain from (1.24)

$$\beta^i(A_{jk}) \leq \beta^i(p_j^{-1}(M_k)) \leq \beta^i(p_j^{-1}(M_k),\tilde{M}_j) \leq b^i(B_{jk}) \quad (1.29)$$

and from the exact cohomology sequence of the pair $(B_{jk},\overline{B_{jk}\setminus A_{jk}})$ together with the excision property

$$|\beta^i(A_{jk}) - b^i(B_{jk})| \leq b^{i-1}(\overline{B_{jk}\setminus A_{jk}}) + b^i(\overline{B_{jk}\setminus A_{jk}}):$$

$$\cdots \longrightarrow H^{i-1}(\overline{B_{jk}\setminus A_{jk}}) \longrightarrow H^i(B_{jk},\overline{B_{jk}\setminus A_{jk}}) \cong H^i(A_{jk},\partial A_{jk})$$
$$\longrightarrow H^i(B_{jk}) \longrightarrow H^i(\overline{B_{jk}\setminus A_{jk}}) \longrightarrow \cdots$$

The manifold $\overline{B_{jk}\setminus A_{jk}}$ satisfies (B_0), (I) for $j > N(k)$ and for k sufficiently large,

$$\mathrm{vol}(\overline{B_{jk}\setminus A_{jk}}) \leq d_j\varepsilon. \quad (1.30)$$

According to a theorem of Gromov,

$$\sum_i \beta^i(\overline{B_{jk} \setminus A_{jk}}) \leq c(n) \cdot \text{vol}(\overline{B_{jk} \setminus A_{jk}}). \tag{1.31}$$

We infer from (1.29) – (1.32) that we can replace in (1.28) $\chi(B_{jk})$ by $\chi(p_j^{-1} M_k, \tilde{M}_j)$, hence

$$\left| \chi(M^n, g) - \frac{1}{d_j} \chi(p_k^{-1}(M_k), \tilde{M}_j) \right|$$

becomes arbitrarily small, any proper homotopy equivalence preserves a subsequence of $\left(\frac{1}{d_j} \chi(p_j^{-1}(M_k), \tilde{M}_j) \right)_{j,k}$, $\chi(M^n, g)$ is a proper homotopy invariant. By the same argument we conclude in the profinite case assertion b). If M has finite topological type then for k sufficiently large $\beta^i(p_j^{-1}(M_k), \tilde{M}_j) = \beta^i(\tilde{M}_j)$ and

$$\chi(p_j^{-1}(M_k), \tilde{M}_j) = \chi(\tilde{M}_j) \cdot \frac{1}{d_j} \chi(\tilde{M}_j) = \chi(M_j) = \chi(M)$$

yields assertion c). □

The case of a normal covering $\tilde{M} \longrightarrow M$ will be discussed in theorem 1.38.

The second characteristic number of particular importance is given by $\sigma(M, g) = \int_M L(M, g)$, where $L(M, g)$ is the Hirzebruch genus. For closed M it is the topological index of the signature operator, i.e. it coincides with the topological signature. For simple open manifolds this equality does not longer hold in general. Nevertheless, we could ask for $\sigma(M, g)$ the same questions as for $\chi(M, g)$, the question for the invariance properties and the topological significance of $\sigma(M, g)$. Concerning the invariance, a first answer is given by proposition 1.22.

But we consider also other variations of g. A key role plays again the formula for the compact case with boundary, $\partial M = N$,

$$\sigma(M, g) + \eta(N, g) + \int_N II_\sigma(N, g) = \sigma(M), \tag{1.32}$$

where $II_\sigma(N,g)$ essentially involves the second fundamental form and $\eta(N,g)$ is the eta invariant. If M^n is open and $M_1 \subset M_2 \subset \cdots$, $\bigcup_k M_k = M$, an appropriate exhaustion such that $\int II_\sigma(\partial M_k) \longrightarrow 0$ and $\eta(\partial M_k) \longrightarrow 0$ then we would have in fact $\sigma(M_k, g) \longrightarrow \sigma(M)$. Hence we should ask for conditions which assure $\eta(\partial M_k) \longrightarrow 0$. There is a clear (and for our case complete) answer.

Theorem 1.35 *Let (N^{4l-1}, g) be compact satisfying* $\mathrm{geo}(N) \leq 1$. *Then there is a constant $c = c(4l-1)$ such that*

$$|\eta(N^{4l-1})| \leq c(4l-1) \cdot \mathrm{vol}(N^{4l-1}, g). \qquad (1.33)$$

We refer to [16], [27] for the proof. □

Now we define $\sup \tilde\sigma(M)$, $\inf \tilde\sigma(M)$ quite analogous to the Euler characteristic as follows. Let M^{4l} be complete, $\tilde M^{4l} \longrightarrow M$ profinite and $M_k^{4l} \subset M^{4l}$ a compact submanifold with boundary. Put

$$\sup \tilde\sigma(M_k) := \limsup_j \frac{1}{d_j} \sigma(P_j^{-1}(M_k)),$$

$$\sup \tilde\sigma(M) := \limsup_{M_k} \sup \tilde\sigma(M_k))$$

and similarly $\inf \tilde\sigma(M_k)$, $\inf \tilde\sigma(M)$. Here as always $\sigma(M_k)$ is defined as the signature of the cup product pairing on $j^* H^{2l}(M_k^{4l}, \partial M_k^{4l}) \subset H^{2l}(M_k^{4l})$.

Theorem 1.36 *Let (M^{4l}, g) be complete, $\mathrm{vol}(M, g) < \infty$ and suppose $\tilde M$ either profinite or normal and $\mathrm{geo}(\tilde M) \leq 1$. Then there holds*

a) *Assume $\tilde M$ normal. Then $\sigma(M, g)$ is a proper homotopy invariant of M.*

b) *In the case $\tilde M \longrightarrow M$ profinite, for any exhaustion $M_1 \subset M_2 \subset \cdots$, $\bigcup_k M_k = M$, by compact manifolds,*

$$\sigma(M, g) = \sup \tilde\sigma(M) = \inf \tilde\sigma(M).$$

c) *If, additionally, M has finite topological type,*

$$\sigma(M,g) = \lim_{j\to\infty} \frac{1}{d_j}\sigma(\tilde{M}_j).$$

Proof. In the normal case $\tilde{M} \longrightarrow M$ below a) follows from theorem 1.38. The proof of b) is quite analogous to that of theorem 1.34 b), using a chopping of M according to theorem 1.33, (1.32) and theorem 1.35. c) then follows from b) and the fact that for sufficiently large k, $\frac{1}{d_j}\sigma(p_j^{-1}(M_k)) = \frac{1}{d_j}\sigma(\tilde{M}_j)$. □

We now turn to the normal case $\tilde{M} \longrightarrow M$, being even more explicit than in the profinite case. The first key here is the extension of Atiyah's L_2–index theorem for normal coverings $\tilde{M} \longrightarrow M$ of closed M to normal coverings $\tilde{M} \longrightarrow M$, $M = \tilde{M}/\Gamma$, $r_{\text{inj}}(\tilde{M}) \geq 1$, (M^n, g) complete, $\text{vol}(M^n, g) < \infty$, $|K| \leq 1$. We denote by $\mathcal{H}^{q,2}(\tilde{M})$ the space of L_2–harmonic q–forms, by $P_{\mathcal{H}^{q,2}} : L_2(\Lambda^q T^*A) = \Omega^{q,2} \longrightarrow \mathcal{H}^{q,2}$ the orthogonal projection. $P_{\mathcal{H}}$ has Schwartz kernel $\tilde{h}^q(x,y)$ which is a symmetric C^∞ double form whose pointwise norm satisfies

$$|h^q(x,y)| \leq c(n). \qquad (1.34)$$

(1.34) comes from $\text{geo}(\tilde{M}) \leq 1$ and the elliptic estimate for the Laplacian. $\tilde{h}^q(x,y)$ is invariant under the isomtries Γ, hence the pointwise trace $\text{tr}\tilde{h}^q(x,x)$ can be understood as function on M and we put as usual

$$\tilde{b}^{q,2}(M) := \text{tr}_\Gamma P_{\mathcal{H}^{q,2}(\tilde{M})} = \int_M \text{tr}\tilde{h}^q(x,x)\,\text{dvol}_x(g) < \infty.$$

$\tilde{b}^{q,2}(M)$ is just the von Neumann dimension $\dim_\Gamma \overline{H}^{q,2}(\tilde{M})$ of the Γ–module $\overline{H}^{q,2}(\tilde{M})$. We define the L_2–Euler characteristic and L_2–signature by

$$\tilde{\chi}_{(2)}(M) := \sum_{q=0}^{n}(-1)^q \tilde{b}^{q,2}(M)$$

and
$$\tilde{\sigma}_{(2)}(M) := \mathrm{tr}_\Gamma(*P_{\mathcal{H}^{2k,2}(\tilde{M}^{4k})}).$$

Now we state the L_2–index theorem for open manifolds with finite volume and bounded curvature.

Theorem 1.37 *Suppose* (M, g) *complete with* $\mathrm{vol}(M^n, g) < \infty$, $|K| \leq 1$ *and* $\tilde{M} \longrightarrow M$ *normal with* $\mathrm{geo}(\tilde{M}) \leq 1$. *Then*

$$\chi(M, g) = \tilde{\chi}_{(2)}(M) \tag{1.35}$$

and

$$\sigma(M, g) = \tilde{\sigma}_{(2)}(M). \tag{1.36}$$

We refer to [15], [27] for the proof. □

We recall the existence of good chopping sequences $M_1 \subset M_2 \subset \cdots$, $\bigcup_1^\infty M_k = M$, $\mathrm{vol}(\partial M_k) \longrightarrow 0$, $|II(\partial M_k)| \leq c$, $|\tilde{h}_k^q(x,y)| \leq c(n)|$, where $\tilde{h}_k^q$ denotes the kernel corresponding to projection on the harmonic q–forms for $p^{-1}(M_k) \subset \tilde{M}$. Then we obtain

$$\lim_{k \to \infty} \tilde{b}^{q,2}(\partial M_k) = 0 \tag{1.37}$$

and

$$\lim_{k \to \infty} \tilde{b}^{q,2}(M \setminus M_k, \partial(M \setminus M_k)) = 0. \tag{1.38}$$

Define $\tilde{\beta}^{q,2}(B)$ by

$$\tilde{\beta}^{q,2}(B) := \dim_\Gamma \mathrm{im}\, (H^{q,2}(p^{-1}(B), p^{-1}(\partial B)) \subset \overline{H}^{q,2}(p^{-1}(B))) \tag{1.39}$$

and for $A \subset B$

$$\tilde{\beta}^{q,2}(A, B) := \dim_\Gamma \mathrm{im}\, (\overline{H}^{q,2}(p^{-1}(B)) \subset \overline{H}^{q,2}(p^{-1}(A))). \tag{1.40}$$

It follows from the properties of $\dim_\Gamma$ that

$$\tilde{\beta}^{q,2}(A) \leq \tilde{\beta}^{q,2}(B) \tag{1.41}$$

and

$$\tilde{\beta}^{q,2}(A) \leq \tilde{\beta}^{q,2}(A, B) \leq \tilde{b}^{2,q}(A). \tag{1.42}$$

We remark that (1.41) and (1.42) are the adequate reformulation of (1.24), (1.29) in the language of $\dim_\Gamma$. We established in theorem 1.37 the equations $\chi(M,g) = \tilde{\chi}_{(2)}(M)$, $\sigma(M,g) = \tilde{\sigma}_2(M)$. Now we discuss the invariance properties of the right hand sides. This is the content of

Theorem 1.38 *Let (M^n, g) be complete, $|K| \leq 1$, $\mathrm{vol}(M,g) < \infty$ and assume for some normal covering $\mathrm{geo}(\tilde{M}) \leq 1$.*
a) *If $M_1 \subset M_2 \subset \cdots$, $\bigcup M_k = M$ is an exhaustion then*

$$\lim_{k\to\infty} \tilde{\beta}^{q,2}(M_k) = \lim_{k\to\infty} \lim_{l\to\infty} \tilde{\beta}^{q,2}(M_k, M_l) = \tilde{b}^{q,2}(M). \qquad (1.43)$$

This implies the homotopy invariance of the $\tilde{b}^{q,2}(M)$.
b) *$\chi(M,g)$ resp. $\sigma(M,g)$ is a homotopy invariant resp. proper homotopy invariant of M.*
c) *If M has the topological type of some $M_k \subset M$, then*

$$\tilde{b}^{q,2}(M_k) = \tilde{b}^{q,2}(M) \qquad (1.44)$$

and

$$\chi(M,g) = \chi(M_k). \qquad (1.45)$$

Proof. b) follows immediately from theorem 1.37 and a). For c) suppose that M has finite topological type. Then there exists an exhaustion $M_1 \subset M_2 \subset \cdots$ s.t. each inclusion $M_k \longrightarrow M$ is a homotopy equivalence. This implies

$$\tilde{\beta}^{q,2}(M_k, M_k) = \tilde{b}^{q,2}(M_k)$$

and we obtain (1.44) from (1.43) and moreover $\chi(M,g) = \chi(M_k)$. Hence there remains to show a). For this we must refer to [15]. $\square$

We apply these results on characteristic numbers to 4–manifolds. Let (M^4, g) be open, complete and oriented, $* : \Lambda^2 M \longrightarrow \Lambda^2 M$ the Hodge operator, $*^2 = 1$, $\Lambda^2 = \Lambda^2_+ \oplus \Lambda^2_-$. The special orthogonal group acts on the space of algebraic curvature tensors $\mathcal{C}^2_b$

(cf. [57]). Let $C_b^2 = \mathcal{U} + \mathcal{S} + \mathcal{W}$ be the corresponding (fiberwise) decomposition into irreducible subspaces. Then this induces for the curvature tensor $R = R^g$ a decomposition $R = U + S + W$. For $R = R^g = R_+ + R_-$, we denote by $\text{Ric} = \text{Ric}^g$ the Ricci tensor, by $\tau = \tau^g$ the scalar curvature, by $K = K^g$ the sectional curvature and by $W = W^g = W_+ + W_-$ the Weyl tensor. There are decompositions for the pointwise norms $|\ |_x$ as follows

$$\begin{aligned}
|R|^2 &= |R_+|^2 + |R_-|^2 = |U|^2 + |S|^2 + |W|^2 \\
&= 4|W_+|^2 + |W_-|^2 + 2|\text{Ric}|^2 - \frac{1}{3}\tau^2, \quad (1.46) \\
|\text{Ric}|^2 &= 6|U|^2 + 2|S|^2, \quad (1.47) \\
\tau^2 &= 24|U|^2. \quad (1.48)
\end{aligned}$$

We obtain still other decompositions if we consider the curvature operator R as acting from $\Lambda^2 = \Lambda_+^2 \oplus \Lambda_-^2$ to $\Lambda_+^2 \oplus \Lambda_-^2$, for an orthonormal basis e_1, e_2, e_3, e_4

$$R(e_i \wedge e_j) = \frac{1}{2}\sum R_{ijkl} e_k \wedge e_l = \Omega_{ij},$$

$\Omega = (\Omega_{ij}) =$ matrix of curvature forms, $\Omega_{ij}(e_k, e_l) = R_{ijkl}$. We can write R with respect to the orthogonal basis $e_1 \wedge e_2 + e_3 \wedge e_4$, $e_1 \wedge e_4 + e_2 \wedge e_3$, $e_1 \wedge e_3 + e_2 \wedge e_4$ in Λ_+^2, $e_1 \wedge e_3 + e_2 \wedge e_4$, $e_1 \wedge e_2 - e_3 \wedge e_4$, $e_1 \wedge e_4 - e_2 \wedge e_3$ in Λ_-^2, as

$$R = \begin{pmatrix} A & B \\ C & D \end{pmatrix}$$

with $A = A^*$, $C = B^*$, $D = D^*$, $\text{tr} A = \text{tr} D = \frac{\tau}{4}$, $B = \text{Ric} - \frac{1}{4}\tau g$ and $\begin{pmatrix} A & 0 \\ 0 & D \end{pmatrix} - \frac{\tau}{12} = W$, $W^+ = A - \frac{\tau}{12}$, $W^- = D - \frac{\tau}{12}$. We obtain for the first Pontrjagin form p_1

$$\begin{aligned}
p_1 &= -\frac{1}{8\pi^2}\text{tr}(R \wedge R) = -\frac{1}{8\pi^2}\text{tr}(A \wedge A) + \text{tr}(D \wedge D) \\
&= -\frac{1}{8\pi^2}(-2)(|W_+|^2 - |W_-|^2)\,\text{dvol} \\
&= \frac{1}{4\pi^2}(|W_+|^2 - |W_-|^2)\,\text{dvol} \\
&= \frac{1}{12\pi^2}(|R_+|^2 - |R_-|^2)\,\text{dvol}
\end{aligned}$$

and for $\sigma(M^4, g) = \int L(g) = \int \frac{1}{3} p_1 = \frac{1}{12\pi^2} \int (|W_+|^2 - |W_-|^2)$ dvol. Assuming $g \in {}^{b,2}\mathcal{M}^{1,2}(B_0, 1, f)$, $\sigma(M^4, g)$ is well defined. The Euler form $E(g)$ has the representation

$$\begin{aligned} E(g) &= \frac{1}{8\pi^2} \operatorname{tr}(*R)^2 \, \text{dvol} \\ &= \frac{1}{8\pi^2} (|U|^2 - |S|^2 + |W|^2) \, \text{dvol} \\ &= \frac{1}{8\pi^2} \operatorname{tr}(A^2 - 2BB^* + D^2) \, \text{dvol} \\ &= \frac{1}{32\pi^2} (|R|^2 - 4|\text{Ric}\,|^2 + \tau^2) \, \text{dvol}. \end{aligned}$$

For $g \in {}^{b,2}\mathcal{M}^{1,2}(B_0, 1, f)$, $\int E(g) = \chi(M, g)$ is well defined. Hence we obtain

Proposition 1.39 *Let (M^4, g) be open, complete, oriented and $g \in {}^{b,2}\mathcal{M}^{1,2}(B_0, 1, f)$. Then $\sigma(M, g)$ and $\chi(M, g)$ are well defined and an invariant of* comp(g). □

Remark 1.40 According to (1.46) – (1.48), $\int |R^g|^2$ dvol $< \infty$ would be sufficient for the existence of $\sigma(M, g)$ and $\chi(M, g)$. But this condition would not establish a uniform strucutre, we would not have components and invariance properties (where we used in particular Gaffney's theorem). Moreover, we need the bounded curvature property for the connection with the theorems 1.37, 1.38. □

We obtain from proposition 1.39 and its proof the simple

Corollary 1.41 *If (M^4, g) is additionally Einstein then $\chi(M, g) \geq 0$ and $|\sigma(M^4, g)| \leq \frac{2}{3} \chi(M^4, g)$. Moreover, $\chi(M^4, g) = 0$ if and only if (M^4, g) is flat.*

Proof. If (M^4, g) is Einstein then $S \equiv 0$, $B \equiv 0$ and $\frac{1}{12\pi^2}(|W_+|^2 - |W_-|^2) \leq \frac{2}{3} \frac{1}{8\pi^2}(|U|^2 + |W_+|^2 - |W_-|^2)$. Hence $\sigma(M^4, g) \leq \frac{2}{3} \chi(M, g)$. Changing the orientation replaces

$\sigma(M^4,g)$ by $-\sigma(M^4,g)$ and we get altogether $|\sigma(M^4,g)| \leq \frac{2}{3}\chi(M^4,g)$. □

The same estimate holds for $\frac{2}{3}$-pinched Ricci curvature.

Proposition 1.42 *Suppose the hypotheses of 1.39 and additionally that the Ricci curvature of (M^4,g) is negative and $\frac{2}{3}$-pinched, i.e. there exists $A > 0$ s.t.*

$$-Ag \leq \mathrm{Ric} \leq -\frac{2}{3}Ag. \quad (1.49)$$

Then there holds for all $g' \in \mathrm{comp}(g) \subset {}^{b,2}\mathcal{M}^{1,2}(B_0,1,f)$

$$|\sigma(M^4,g')| \leq \frac{2}{3}\chi(M^4,g'). \quad (1.50)$$

Proof. We have

$$|\sigma(M^4,g)| = \left|\int L(g)\right| \leq \int |L(g)|\,\mathrm{dvol}$$
$$= \frac{1}{12\pi^2}\int (|W_+|^2 + |W_-|^2)\,\mathrm{dvol}$$

and

$$\chi(M^4,g) = \int E(g) = \frac{1}{8\pi^2}\int (|U|^2 - |S|^2 + |W|^2)\,\mathrm{dvol}.$$

Sufficient for (1.50) would be $|S|^2 \leq |U|^2$ and sufficient for this is (1.49) as pointed out by [57].

□

Examples 1.43 1) Examples for 1.39 with infinite volume are e.g. manifolds M^4 of the smooth type $M^4 = M_0^4 \cup \partial M_0^4 \times [0,\infty[$ where the curvature at the cylinder $\partial M_0^4 \times [0,\infty[$ is bounded and asymptotically flat in the sense $\int_{\partial M_0^4 \times [0,\infty[} |R|\,\mathrm{dvol} < \infty$. This can be easily realized by warped product metrics.

2) Examples for 1.39, 1.41, 1.42 with finite volume are given by hyperbolic 4–manifolds of finite volume.

3) Generalizations of these examples are given by variation of g inside comp(g). □

Theorem 1.44 *Let (M^4, g) be open, complete, $\mathrm{vol}(M^4, g) < \infty$, $|K| \leq 1$ and suppose that (M^4, g) admits a normal covering $(\tilde{M}, g)$ satisfying $\mathrm{geo}(\tilde{M}) \leq 1$.*

a) *If $\chi(M^4, g) < 0$ then M^4 does not admit a complete Einstein metric g' satisfying $\mathrm{vol}(M^4, g') < \infty$, $|K_{g'}| \leq 1$, $\mathrm{geo}(\tilde{M^4}, g') \leq 1$ for some normal covering.*

b) *If $\chi(M^4, g) > 0$ and $|\sigma(M^4)| > \frac{2}{3}\chi(M^4, g)$ then M^4 does not admit a complete Einstein metric g', s.t. $\mathrm{vol}(M^4, g') < \infty$, $|K_{g'}| \leq 1$, $\mathrm{geo}(\tilde{M^4}, g') \leq 1$. Moreover, there does not exist a complete metric g' satisfying*

$$-Ag' \leq \mathrm{Ric}\,(g') \leq -\frac{2}{3}Ag'$$

and $|K_{g'}| \leq 1$, $\mathrm{vol}(M^4, g') < \infty$ and $\mathrm{geo}(\tilde{M^4}, g') \leq 1$ for some normal covering.

Proof. a) Suppose the existence of an Einstein metric g' with the required properties. Then $\chi(M^4, g)$, $\chi(M^4, g')$ are well defined. $\chi(M^4, g) = \chi(M^4, g')$, according to theorem 1.38 b). But this contradicts $\chi(M^4, g') = \frac{1}{8\pi^2} \int (|U|^2 + |W|^2)\,\mathrm{dvol} \geq 0$. b) and c): Quite analogously we derive by means of theorem 1.38 b), corollary 1.41 and proposition 1.42 a contradiction. □

Until now we defined characteristic numbers in the following cases

1) $R \in L_1$ and bounded, $\mathrm{vol}(M)$ arbitrary,
2) R bounded, $\mathrm{vol}(M) < \infty$.

There remains the case R bounded, $\mathrm{vol}(M) = \infty$. It is clear that in this case we will not get characteristic numbers by integration. (M^n, g) is called closed at infinity if for any $\varphi \in C(M)$, $0 <$

$A^{-1} < \varphi < A$, $A > 0$ some constant, the form $\varphi \cdot \mathrm{dvol}$ generates a nontrivial cohomology class in ${}^b H^n(M^n, g)$. A fundamental class for M is a positive continuous linear function $\mathfrak{m} : {}^b\Omega^n(M) \longrightarrow \mathbb{R}$ such that $\langle \mathfrak{m}, \mathrm{dvol} \rangle \neq 0$ and $\langle \mathfrak{m}, d\psi \rangle = 0$.

Proposition 1.45 *M has a fundamental class if and only if M is closed at infinity.*

Proof. Denote $\mathcal{L}(\mathrm{dvol})$ for the linear hull of dvol, let $0 \notin [\mathrm{dvol}] \in {}^b\overline{H}^n(M)$ and set $\langle \mathfrak{m}, \mathrm{dvol} \rangle = 1$, $\mathfrak{m}|_{{}^b\overline{B}^n} \equiv 0$. Then we obtain by linear extension $\mathfrak{m}$ on $\mathcal{L}(\mathrm{dvol}) \oplus {}^b\overline{B}^n$ as positive continuous linear functional. The Hahn–Banach theorem for the extension of such functionals yields the desired $\mathfrak{m}$. The other direction is absolutely trivial. □

Define the penumbra for $K \subset M$.

$$\mathrm{Pen}^+(K, r) = \mathrm{CL}(\bigcup_{x \in K} B_r(x)),$$
$$\mathrm{Pen}^-(K, r) = \mathrm{CL}(M \setminus \mathrm{Pen}^+(M \setminus K, r)).$$

We call an exhaustion $M_1 \subset M_2 \subset \cdots$, $\bigcup_i M_i = M$, by compact submanifolds a regular exhaustion if for each $r \geq 0$

$$\lim_{i \to \infty} \mathrm{vol}(\mathrm{Pen}^+(M_i, r))/\mathrm{vol}(\mathrm{Pen}^-(M_i, r)) = 1.$$

It is clear that then automatically

$$\lim_{i \to \infty} \mathrm{vol}(\mathrm{Pen}^+(M_i, r))/\mathrm{vol}(M_i) = 1,$$
$$\lim_{i \to \infty} \mathrm{vol}(M_i)/\mathrm{vol}(\mathrm{Pen}^-(M_i, r)) = 1.$$

Examples 1.46 1) $(M^n, g) = (\mathbb{R}^n, g_{standard})$ admits a regular exhaustion.

2) Any (M^n, g) with subexponential growth admits a regular exhaustion.

3) The hyperbolic space admits no regular exhaustion. □

Let $\{M_i\}_{i\geq 1}$ be a regular exhaustion and set for $\omega \in {}^b\Omega^n$

$$\langle \mathfrak{m}_i, \omega \rangle := \frac{1}{\text{vol}(M_i)} \int_{M_i} \omega.$$

Then $|\langle \mathfrak{m}_i, \omega \rangle| \leq \sup_x |\omega|_x = {}^b|\omega|$, i.e. $|\mathfrak{m}_i| \leq 1$, the $\mathfrak{m}_i$ belong to the unit ball in $({}^b\Omega^n)^*$. This unit ball is compact in the weak star topology, according to the Banach–Alaoglu theorem, hence the sequence $\{\mathfrak{m}_i\}_i$ has a weak star limit point $\mathfrak{m}$. $\mathfrak{m}$ is then called associated to the regular exhaustion $\{M_i\}_i$.

Proposition 1.47 *Let $\mathfrak{m}$ be associated to a regular exhaustion $\{M_i\}_i$. Then $\mathfrak{m}$ is a fundamental class for M.*

Proof. There remains only to show $\langle \mathfrak{m}, d\psi \rangle = 0$. Let $\Phi_i \in C^\infty(M)$ such that $0 \leq \Phi_i(x) \leq 1$, $\Phi_i = 1$ on M_i, $\Phi_i = 0$ outside $\text{Pen}^+(M_i, 1)$, $|\nabla \Phi_i| \leq 2$. We obtain for $\omega \in {}^b\Omega^n$

$$\left| \int_{M_i} \omega - \int_M \Phi_i \omega \right| \leq (\text{vol}(\text{Pen}^+(M_i, 1)) - \text{vol}(M_i)){}^b|\omega|,$$

hence

$$\lim_{i \to \infty} \frac{1}{\text{vol}(M_i)} \left(\int_{M_i} \omega - \int_M \Phi_i \omega \right) = 0.$$

Therefore we would be done if we could show

$$\lim_{i \to \infty} \frac{1}{\text{vol}(M_i)} \int_M \Phi_i d\psi = 0.$$

Integration by parts yields

$$\int \Phi_i d\psi = -\int d\Phi_i \wedge \psi,$$

$$\left| \int \Phi_i d\psi \right| = \left| \int d\Phi_i \wedge \psi \right| \leq 2(\text{vol}(\text{Pen}^+(M_i, 1)) - \text{vol}(M_i)){}^b|\psi|,$$

which implies the assertion. □

Define for $\omega \in {}^{b,1}\mathcal{C}_p(B_0)$, $[Q_{i_1...i_k}(\omega)] \in {}^bH^n(M)$ a (bounded) characteristic class and a regular exhaustion $\{M_i\}_i$ with associated fundamental class $\mathfrak{m}$ the characteristic number

$$Q_{i_1...i_k}(P,\text{comp}(\omega))[\mathfrak{m}] := \langle \mathfrak{m}, [Q_{i_1...i_k}] \rangle := \lim_{i\to\infty} \frac{1}{\text{vol}(M_i)}(Q_{i_1...i_k}).$$

Then, according to proposition 1.47, $Q_{i_1...i_k}(P,\text{comp}(\omega))[\mathfrak{m}]$ is well defined. In particular we obtain in this case avarage Euler numbers, avarage signatures, which are special cases of Roe's (avarage) topological index (cf. [60]). Average characteristic numbers are also considered in [44], [45], [42]. Some simple geometric examples are calculated in [44].

In all cases discussed until now, we restricted to the case of connections (or metrics) with finite p–action or bounded curvature or both. The next proposition shows that this is in fact a restriction.

Proposition 1.48 *Let (M^n, g) be open, complete, satisfying (I), G a compact Lie group, $P = P(M,G)$ a G-principal fibre bundle, $\varrho : G \longrightarrow U(N)$ resp. $O(N)$ a faithful representation, E the associated vector bundle, $p \leq 1$. Then there exist G-connections ω such that their p-action is infinite or the curvature is unbounded or both, respectively.*

Proof. Consider the closed unit ball $\overline{B}_1(0) \subset \mathbb{R}^n$ and set up in $\overline{B}_1(0)$ constant 1-forms ω_{ij}, $\omega_{ij} = -\overline{\omega}_{ji}$ or $\omega_{ij} = -\omega_{ji}$, $1 \leq i,j \leq N$, respectively, such that some $\Omega_{ij} = d\omega_{ij} - \sum_k \omega_{ik} \wedge \omega_{kj}$ are $\neq 0$. Now consider an infinite sequence $U_\nu = U_{\varepsilon_\nu}(x_\nu)$ of closed geodesic balls with pairwise distance $\geq d > 0$, introduce in each geodesic ball normal coordinates $u^1, \ldots, u^n$, $\sum_i (u^i)^2 \leq \varepsilon_\nu$, choose over U_ν orthonormal bases $e_{1,\nu}, \ldots e_{N,\nu}$ and define with respect to these bases local connection matrices $\omega'_{ij,\nu}$ by $\omega'_{ij,\nu}(u_1,\ldots,u_n) := \omega_{ij}$. If $\int_{U_\nu} |\Omega'_{ij,\nu}|^p_x \, d\text{vol}_x(g) = a_\nu \neq 0$, set

$\omega''_{ij,\nu} = (a_\nu + \frac{1}{a_\nu})^{\frac{1}{2}p} \omega'_{ij,\nu}$. This connection over $\bigcup_\nu U_\nu$ is smoothly extendable over the whole of M and gives a connection with $\int_M |\Omega''|^p_x \, \mathrm{dvol}_x(g) \geq \sum_\nu \int_{U_\nu} |\Omega''|^p_x \, \mathrm{dvol}_x(g) \geq \sum_\nu 1 = \infty$. Setting $\omega''_{ij,\nu} = \nu \cdot (a_\nu + \frac{1}{a_\nu})^{\frac{1}{2}p} \cdot \omega'_{ij,\nu}$ yields examples for the other cases. □

The conditions of finite p–action or boundedness can be reformulated in the language of classifying spaces and classifying mappings.

We start with $G = U(N)$. Let $V_{N,k} \xrightarrow{U(k)} G_{N,k}$ be the Stiefel bundle over the complex Grassmann manifold $G_{N,k}$ of all k–subspaces $\subset \mathbb{C}^N$ and S the matrix valued function on $V_{N,k}$ defined by $S(v_1, \ldots, v_k) = a_{ij} := (b_{ij})^t$, where $v_1, \ldots, v_k$ is a unitary k–frame, $e_1, \ldots, e_N$ the standard base in $\mathbb{C}^N$ and $v_i = \sum_{j=1}^N b_{ij} e_j$.

Proposition 1.49 a) $\gamma_U = S^* dS$ is a $U(N)$–invariant connection form at $V_{N,k}$.
b) Let be $m = (n+1)(2n+1)k^3$. If P is a $U(k)$–principal fibre bundle over a manifold of dimension $\leq n$ and ω a connection form for P, then there exists a smooth bundle morphism $f_P : P \longrightarrow V_{m,k} = P_{n,U(k)}$ such that $f_P^* \gamma = \omega$.

We refer to [55], p. 564, 568 for the proof. □

γ_0 is called a n–universal connection for $U(k)$. In a similar manner one defines on the real Stiefel bundle $V^r_{m,k} \xrightarrow{O(k)} G^r_{m,r}$ an n–universal $O(k)$–connection γ^r_0.

For an arbitrary compact Lie group G one constructs by means of a faithful representation $G \longrightarrow O(k)$ an n–universal connection γ_G on the n–universal bundle $P_{n,G} \longrightarrow B_{n,G}$ (cf. [20], p. 570).

According to proposition 1.49, we refine the bundle concept and consider instead of a bundle P pairs (P, f_P), $f_P : P \longrightarrow P_{n,G}$ a

C^1–classifying bundle map.

(P, f_P) is called a (p, f)–bundle if $f_p^* \gamma_G \in C^1 C_p(f, p) = \{\omega$ a C^1-connection$| \int |\Omega^\omega|_x^p \, \mathrm{dvol}_x(g) < \infty\}$, i.e. $\int |\Omega^{f_p^* \gamma_G}|_x^p \, \mathrm{dvol}_x(g) < \infty$. In the same manner we define (P, f_P) to be a b–bundle if $f_p^* \gamma_G \in C^1 C_p(B_0)$, i.e. $^b|\Omega^{f_p^* \gamma_G}| < \infty$.

The for the applications most interesting case is the case assuming (B_0) and finite p–action.

Hence we assume (B_0) for (M^n, g). (P, f_P) is a (b, p, f)–bundle, if $f_p^* \gamma_G \in {}^{b,1} C_p^{p,1}(B_0, f, p)$. Two (b, p, f)–bundles (P, f_P), (P, f'_P) are called equivalent if $f_P^* \gamma_G$, $f'^*_P \gamma_G$ are contained in the same component of ${}^{b,1} C_P^{p,1}(B_0, f, p)$.

Assume G to be a subgroup of $U(N)$, $\dim M^n = 2k$. At the level of base spaces we consider classifying maps $f_M : M \longrightarrow B_{n,G}$. A pair (M, f_M) is called a (p, c)–bundle if all classes $f_M^* c_{i_1 \ldots i_k}$, $i_1 + \cdots i_k = k$, are elements of $H^{2k,p}(M)$. (M, f_M) is called a (b, c)–bundle if all classes $f_M^* c_{i_1 \ldots i_k}$ are elements of ${}^b H^{2k}(M)$. (M, f_M) is called a (b, p, c)–bundle if all classes $f_M^* c_{i_1 \ldots i_k}$ are elements of ${}^b H^{2k,p}(M)$. It is clear that a given $f_P : P \longrightarrow P_{n,G}$ uniquely determines $f_M : M \longrightarrow B_{n,G}$.

The case $G \subseteq O(N)$, $\dim M = 4k$, is quite parallel. Then we consider the $p_{i_1 \ldots i_k}$, $i_1 + \cdots i_k = k$ and define (M, f_M) to be a (p, po)–bundle if all classes $f_M^* p_{i_1 \ldots i_k}$, $i_1 + \cdots i_k = k$ are elements of $H^{4k,p}(M)$. Analogously for (b, po)– and (b, p, po)–bundles (M, f_M).

If we replace $p_{i_1 \ldots i_k}$ by the class of Hirzebruch genuss L_k then we get the notion of a (p, L_k)–, (b, L_k)– or (b, p, L_k)–bundle (M, f_M), respectively.

Theorem 1.50 a) *Suppose $G \subset U(N)$, $\dim M = 2k$. (M, g) satisfying (B_0), $p \geq 1$. A (b, p, f)–bundle (P, f_P) defines a unique (b, p)–bundle (M, f_M). If (P, f_P), (P, f'_P) are equivalent then $f_M^* c_{i_1 \ldots i_k} = f'^*_M c_{i_1 \ldots i_k}$ for all $c_{i_1 \ldots i_k}$, $i_1 + \cdots + i_k = k$. If additionally $p = 1$ and (M, g) is complete then even the corresponding characteristic numbers coincide.*

b) *Suppose $G \subseteq O(N)$, $\dim M = 4k$, (M, g) satisfying (B_0), $p \geq 1$. A (b, p, f)–bundle (P, f_P) defines a unique (b, p, po)–*

bundle (M, f_M) which is simultaneously a (b, p, L_k)-bundle. If (P, f_P), (P, f'_P) are equivalent then $f_M^* p_{i_1...i_k} = f'^*_M p_{i_1...i_k}$ and $f_M^* L_k = f'^*_M L_k$. If additionally $p = 1$ and (M, g) is complete then the corresponding characteristic numbers coincide.

The proof follows immediately from the definitions and theorem 1.14. □

Example 1.51 It it possible that $^{b,1}\mathcal{C}_p^{1,1}(B_0, 1, f) = \emptyset$. Let (M^2, g) be an infinitely connected open complete Riemannian manifold with bounded sectional curvature K, $K = K_+ - K_-$. $K_+ = \begin{cases} K, & K \geq 0 \\ 0, & K < 0 \end{cases}$, $K_- = \begin{cases} -K, & K \leq 0 \\ 0, & K > 0 \end{cases}$, Then there holds $\int K_- \, dvol = \infty$ (cf. [43], theorem 13). In particular $\int |K| \, dvol = \infty$ which implies $\int |\Omega^\omega(g)| \, dvol = \infty$. The proof essentially relies on the Gauß–Bonnet theorem (as one would expect) for compact surfaces. But this theorem holds for any metrizable connection in the orthogonal 2–frame bundle $P(M^2, O(2))$ over M^2 ([47], p. 305/306). The sectional curvature K is defined by $\Omega_{1,2} = K \, dvol$. As conclusion we obtain $^{b,1}\mathcal{C}_p(B_0, 1, f) = \emptyset$. □

We conclude this section with some remarks concerning the Novikov conjecture for open manifolds. As very well known, the Novikov conjecture for closed manifolds stimulated many outstanding topologists to prove this and on this road deep results in C^* algebraic topology, C^* K–theory and geometric group theory have been achieved. Hence, the Novikov conjecture has not only its own meaning but even more meaning as a stimulating question.

If M^n is open and we consider the classifying diagram

$$\begin{array}{c} \tilde{M} \\ \downarrow \\ M \xrightarrow{f} B_\pi \end{array}$$

and $a \in H^*(B_\pi)$ then
$$\langle L(M) \cdot f^*a, [M] \rangle$$
will not be defined in general. For this reason, Gromov proposes to consider
$$\sigma_a(M) = \langle L(M) \cdot f^*a, [M] \rangle$$
for $a \in H_c^*(B_\pi)$.

Then the NC for open manifolds would mean the "invariance of $\sigma_a(M)$ under proper homotopy equivalences". Probably much more appropriate would be an approach in the sense of our "open category", i.e.

1) everything is uniformly metrized, we have (I), (B_k), uniform triangulations etc.,

2) maps are bounded and uniformly proper, in particular this holds for homotopy equivalences,

3) one works within functional algebraic topology.

Hence one should consider
$$\langle L(M) \cdot f^*a, [M] \rangle \quad \text{with} \quad L(M) \in L_p, f^*a \in L_q.$$

Of particular meaning would be the cases
$$L(M) \in {}^b H^*(M) \quad \text{and} \quad a \in H^{*,1}(B_\pi) \tag{1.51}$$
or
$$L(M) \in H^{*,2}(M) \quad \text{and} \quad a \in H^{*,2}(B_\pi), \tag{1.52}$$
respectively. If we suppose (M,g) satisfying (B_0) then automatically $L(M) \in {}^b H^*(M)$. (B_0) does not restrict to topological type since any open manifold admits a metric g satisfying even (B_∞) and (I).

In the second case one should additionally assume
$$\inf \sigma_e(\Delta_*(M,g)|_{(\ker \Delta_*)^\perp}) > 0, \tag{1.53}$$
i.e. there is a spectral gap of Δ_* above zero. In this case $H^{*,2} = \mathcal{H}^{*,2} = L_2$-harmonic forms, $C^{*,2}$, $C_{*,2}$ are L_2-complexes and

form an L_2-Poincare complex. Every L_2-(co-)homology class can be represented by an L_2-harmonic (co-)cycle. Bordism of L_2-Poincare complexes can be defined easily.

We proved in [34] that (1.53) is invariant under bounded uniformly proper homotopy equivalences. W.l.o.g., classifying maps can be assumed to be bounded and uniformly proper,

$$M^n \longrightarrow B_\pi = M^n \cup \text{cells}.$$

We present now 3 versions of NC (for open manifolds).

1. Version. In the class of open oriented manifolds (M^n, g), $g \in {}^{b,2}\mathcal{M}^{2,2}(B_0, 2, f)$ with $\inf \sigma_e(\Delta_*(g)|_{(\ker \Delta_*)^\perp}) > 0$ is

$$\langle L(M) f^* a, [M] \rangle, \quad a \in H^{*,2}(B_\pi), \quad f \text{ bounded and}$$
uniformly proper classifying map, invariant under
bounded and uniformly proper homotopy equivalences.
$$\tag{NCO1}$$

Criticism. This version should hold only in very restricted cases. Starting point in the compact case is the equality

$$\sigma(M^{4k}) = \int L_k(M) \tag{1.54}$$

where the l.h.s. is a priori a homotopy invariant and the r.h.s. is a certain characteristic number. The L_2-version of (1.54) is already wrong in simple open cases. Let (M^{4k}, g) be an open manifold with cylindrical ends, i.e. $(M^{4k}, g) = (M'^{4k} \cup \partial M'^{4k} \times [0, \infty[, g)$ with $g|_{\partial M'^{4k} \times [0,\infty[} \cong g|_{\partial M'} + dt^2$. Then it is well known that

$$\sigma(M^{4k}) = \sigma_{L_2}(M^{4k}) = \int L_k(M) - \eta(\partial M'^{4k}),$$

i.e. already the starting point which guarantees the invariance of $L(M)$ in the simplest case is wrong. Hence the first version of NC for open manifolds makes sense only for that classes of open manifolds for which

$$\sigma_{L_2}(M^n) = \int L(M)$$

in the case $n = 4k$ holds.

2. Version of NC, relative version. Fix (M^n, g) and suppose $M_1, M_2 \in \text{gen}^b\text{comp}_{L,iso,rel}(M, g)$

$$M_1 \setminus K_1 \cong M \setminus K$$
$$M_2 \setminus K_2 \cong M \setminus K$$

with a Riemannian collar at ∂K_1, ∂K_2, ∂K. Then we define

$$\sigma(M_i, M) := \int_{K_i} L(M_i) - \int_K L(M)$$

$$\sigma(M_1, M_2) := \sigma(M_1, M) - \sigma(M_2, M)$$

$$= \int_{K_1} L(M_1) - \int_{K_2} L(M_2)$$

$$= \sigma(K_1 \cup K_2)$$

$$= \int_{K_1} L(M_1) - \eta(\partial K_1) - \left(\int_{K_2} L(M_2) - \eta(\partial K_2)\right)$$

$$= \sigma(K_1) - \sigma(K_2).$$

The relative NC becomes

$$\int_{K_1} L(M_1) f_1^* a = \int_{K_2} L(M_2) f_2^* a \qquad \text{(NCO2)}$$

if there exist $\Phi_{12} : M_1 \longrightarrow M_2$, $\Phi_{21} : M_2 \longrightarrow M_1$, bounded, uniformly proper, $\Phi_{21}\Phi_{12} \sim \text{id} M_1$, $\Phi_{12}\Phi_{21} \sim \text{id} M_2$ bounded and u.p. and $\Phi_{21}\Phi_{12} = \text{id}$ outside $\tilde{K}_1 \subset M_1$, $\Phi_{12}\Phi_{21} = \text{id}$ outside $\tilde{K}_2 \subset M_2$ and $f_i : M_i \longrightarrow B_\pi$ are bounded and u.p. classifying maps, $a \in H^*(B_\pi)$.

This relative version has the advantage that we require no conditions for (M^n, g) and NC splits to NC for the generalized Lipschitz components (cf. [27], [33]).

3. Version of NC. Consider (M^n, g) open, oriented with (B_0), $r_{\text{inj}} > 0$, embeddings $N^{4k} \hookrightarrow M^n \times \mathbb{R}^j$ with trivial normal bundle and bounded second fundamental form such that $PD[N] = f^*a$, $a \in H^{n-4k,1}(B_\pi)$, $f : M^n \longrightarrow B_\pi$ bounded and uniformly proper classifying map and such that $\sigma_{L_2}(N^{4k})$ is defined (i.e. $\dim \mathcal{H}^{2k,2}(N) < \infty$).

Then the number $\sigma_a(M) := \sigma_{L_2}(N^{4k})$ is invariant under bounded and uniformly proper homotopy invariants.

(NCO3)

How to attack these conjectures will be the content of a forthcoming investigation.

2 Index theorems for open manifolds

Let (M^n, g) be closed, oriented, $(E, h_E), (F, h_F) \longrightarrow M^n$ smooth vector bundles, $D : C^\infty(E) \longrightarrow C^\infty(F)$ an elliptic differential operator. Then $L_2(E) \supset \mathcal{D}_{\overline{D}} \xrightarrow{\overline{D}} L_2(F)$ is Fredholm, i.e. there exists $P : L_2(F) \longrightarrow L_2(E)$ s.t. $PD - \mathrm{id} = K_1$, $DP - \mathrm{id} = K_2$, K_i integral operators with C^∞ kernel $\mathcal{K}_i$ and hence compact. It follows $\dim \ker D, \dim \operatorname{coker} D < \infty$, $\operatorname{ind}_a D = \dim \ker D - \dim \operatorname{coker} D$ is well defined and there arises the question to calculate $\operatorname{ind}_a D$. The answer is given by the seminal Atiyah–Singer index theorem

Theorem 2.1
$$\operatorname{ind}_a D = \operatorname{ind}_t D,$$
where
$$\operatorname{ind}_t D = \langle \mathrm{ch}\, \sigma(D) \mathcal{T}(M), [M] \rangle.$$
□

Assume now (M^n, g) open, E, F, D as above. K_1, K_2 still exist as operators with a smooth kernel where in good cases one can achieve that the support of $\mathcal{K}_i$ is located near the diagonal.

But there arise several troubles.

1) If K_i bounded is achieved then K_i must not be compact.

2) If K_i would be compact then $\mathrm{ind}_a D$ would be defined.

3) If $\mathrm{ind}_a D$ would be defined then $\mathrm{ind}_t D$ must not be defined.

4) If $\mathrm{ind}_a D$, $\mathrm{ind}_t d$ (as above) would be defined then they must not coincide. There are definite counterexamples.

There are 3 ways out from this difficult situation.

1) One could ask for special conditions in the open case under which an elliptic D is still Fredholm, then try to establish an index formula and finally present applications. These conditions could be conditions on D, on M and E or a combination of both. In [2] the author formulates an abstract (and very natural) condition for the Fredholmness of D and assumes nothing on the geometry. But in all substantial applications this condition can be assured by conditions on the geometry. The other extreme case is that discussed in [22], [50], [48], where the authors consider the L_2–index theorem for locally symmetric spaces. Under relatively restricting conditions concerning the geometry and topology at infinity the Fredholmness and an index theorem are proved in [11] and [12].

2) One could generalize the notion of Fredholmness (using other operator algebras) and then establish a meaningful index theory with applications. The discussion of these both approaches will be the content of this section.

3) Another approach will be relative index theory which is less restrictive concerning the geometrical situation (compared with the absolute case) but its outcome are only statements on the relative index, i.e. how much the analytical properties of D differ from those of D'. This approach will be discussed in detail in chapter V.

4) For open coverings $(\tilde{M}, \tilde{g})$ of closed manifolds (M^n, g) and lifted D there is an approach which goes back to Atiyah, (cf. [4]). This has been further elaborated by Cheeger, Gromov and others. The main point is that all considered (Hilbert–) modules are modules over a von Neumann algebra and one replaces the usual trace by a von Neumann trace. We will not dwell on this approach since there is a well established highly elaborated theory. Moreover special features of openess come not

into. The openess is reflected by the fact that all modules under consideration are modules over the von Neumann algebra $\mathcal{N}(\pi)$, $\pi = \mathrm{Deck}(\tilde{M} \longrightarrow M)$. We refer to the very comprehensive representation [46].

This section is a brief review of absolute index theorems under additional strong assumptions. It shows that these approaches are successful only in special situations. In chapter V we will establish very general relative index theorems.

We start with the first approach and with the question which elliptic operators over open manifolds are Fredholm in the classical sense above. Let (M^n, g) be open, oriented, complete, $(E, h) \longrightarrow (M^n, g)$ be a Hermitean vector bundle with involution $\tau \in \mathrm{End}(E)$, $E = E^+ \oplus E^-$, $D : C^\infty(E) \longrightarrow C^\infty(E)$ an essentially self-adjoint first order elliptic operator satisfying $D\tau + \tau D = 0$. We denote $D^\pm = D|_{C^\infty(E^\pm)}$. Then we can write as usual

$$D = \begin{pmatrix} 0 & D^- \\ D^+ & 0 \end{pmatrix} : \begin{matrix} C^\infty(E^+) \\ \oplus \\ C^\infty(E^-) \end{matrix} \longrightarrow \begin{matrix} C^\infty(E^+) \\ \oplus \\ C^\infty(E^-) \end{matrix}. \quad (2.1)$$

The index $\mathrm{ind}_a D$ is defined as

$$\begin{aligned} \mathrm{ind}_a D &:= \mathrm{ind}_a D^+ := \dim \ker D^+ - \dim \mathrm{coker}\, D^+ \\ &= \dim \ker D^+ - \dim \ker D^- \end{aligned} \quad (2.2)$$

if these numbers would be defined. Denote by $\Omega^{2,i}(E, D)$ the Sobolev space of order i of sections of E with D as generating differential operator. We essentially follow [2].

Proposition 2.2 *The following statements are equivalent*
a) *D is Fredholm.*
b) $\dim \ker D < \infty$ *and there is a constant $c > 0$ such that*

$$|D\varphi|_{L_2} \geq c \cdot |\varphi|_{L_2}, \quad \varphi \in (\ker D)^\perp \cap \Omega^{2,1}(E, D), \quad (2.3)$$

where $(\ker D)^\perp \equiv \mathcal{H}^\perp$ *is the orthogonal complement of* $\mathcal{H} = \ker D$ *in* $L_2(E)$.

c) There exists a bounded non–negative operator $P : \Omega^{2,2}(E, D) \longrightarrow L_2(E)$ and bundle morphism $R \in C^\infty(\operatorname{End} E)$, R positive at infinity (i.e. there exists a compact $K \subset M$ and a $k > 0$ s. t. pointwise on $E|_{M \setminus K}$, $R \geq k$), such that on $\Omega^{2,2}(E, D)$

$$D^2 = P + R. \qquad (2.4)$$

d) There exist a constant $c > 0$ and compact $K \subset M$ such that

$$|D\varphi|_{L_2} \geq c \cdot |\varphi|, \quad \varphi \in \Omega^{2,1}(E, D), \quad \operatorname{supp}(\varphi) \cap K = \emptyset. \qquad (2.5)$$

$\square$

The main task now is to establish a meaningful index theorem. This has been performed in [2].

Theorem 2.3 Let (M^n, g) be open, complete, oriented, $(E, h, \tau) = (E^+ \oplus E^-, h) \longrightarrow (M^n, g)$ a $\mathbb{Z}_2$-graded Hermitean vector bundle and $D : C_c^\infty(E) \longrightarrow C_c^\infty(E)$ first order elliptic, essentially self-adjoint, compatible with the $\mathbb{Z}_2$-grading (i.e. supersymmetric), $D\tau + \tau D = 0$. Let $K \subset M$ be a compact subset such that 2.2 a) for K is satisfied, and let $f \in C^\infty(M, \mathbb{R})$ be such that $f = 0$ on $U(K)$ and $f = 1$ outside a compact subset. Then there exists a volume density ω and a contribution I_ω such that

$$\operatorname{ind}_a \overline{D}^+ = \int_M (\omega(1 - f(x)) \operatorname{dvol}_x(g) + I_\omega, \qquad (2.6)$$

where ω has an expression locally depending on D and I_ω depends on D and f restricted to $\Omega = M \setminus K$. $\square$

Until now the differential form $\omega \operatorname{dvol}_x(g)$ is mystery. One would like to express it by well known canonical terms coming e.g. from the Atiyah–Singer index form $\operatorname{ch} \sigma(D^+) \cup \mathcal{T}(M)$, where $\mathcal{T}(M)$ denotes the Todd genus of M. In fact this can be done.

Index Theorem 2.4 Let (M^n, g) be open, oriented, complete, $(E, h, \tau) \longrightarrow (M^n, g)$ a $\mathbb{Z}_2$-graded Hermitean vector bundle, $D :$

$C_c^\infty(E) \longrightarrow C_c^\infty(E)$ a first order elliptic essentially self-adjoint supersymmetric differential operator, $D\tau + \tau D = 0$, which shall be assumed to be Fredholm. Let $K \subset M$ compact such that 2.2 d) is satisfied. Then

$$\mathrm{ind}_a D^+ = \int_K \mathrm{ch}\ \sigma(D^+) \cup \mathcal{T}(M) + I_\Omega, \qquad (2.7)$$

where $\mathrm{ch}\ \sigma(D^+) \cup \mathcal{T}(M)$ is the Atiyah–Singer index form and I_Ω is a bounded contribution depending only on $D|_\Omega$, $\Omega = M \setminus K$. □

Remarks 2.5 a) As we already mentioned, $\mathbb{Z}_2$–graded Clifford bundles and associated generalized Dirac operators D such that in $D^2 = \Delta^E + \mathcal{R}$, $\mathcal{R} \geq c \cdot \mathrm{id}$, $c > 0$, outside some compact $K \subset M$, yield examples for theorem 2.3. A special case is the Dirac operator over a Riemannian spin manifold with scalar curvature $\geq c > 0$ outside $K \subset M$.

b) Much more general perturbations than compact ones will be considered in section V 1. □

The other case of a very special class of open manifolds are coverings $(\tilde{M}, \tilde{g})$ of a closed manifold (M^n, g). Let $E, F \longrightarrow (M^n, g)$ be Hermitean vector bundles over the closed manifold (M^n, g). $D : C^\infty(E) \longrightarrow C^\infty(F)$ be an elliptic operator, $(\tilde{M}, \tilde{g}) \longrightarrow (M, g)$ a Riemannian covering, $\tilde{D} : C_c^\infty(\tilde{E}) \longrightarrow C_c^\infty(\tilde{F})$ the corresponding lifting and $\Gamma = \mathrm{Deck}\ (\tilde{M}^n, \tilde{g}) \longrightarrow (M^n, g)$. The actions of Γ and $\tilde{D}$ commute. If $P : L_2(\tilde{M}, \tilde{E}) \longrightarrow \mathcal{H}$ is the orthogonal projection onto a closed subspace $\mathcal{H} \subset L_2(\tilde{M}, \tilde{E})$ then one defines the Γ–dimension $\dim_\Gamma \mathcal{H}$ of $\mathcal{H}$ as

$$\dim_\Gamma \mathcal{H} := \mathrm{tr}_\Gamma P,$$

where tr_Γ denotes the von Neumann trace and $\mathrm{tr}_\Gamma P$ can be any real number ≥ 0 or $= \infty$.

If one takes $\mathcal{H} = \mathcal{H}(\tilde{D}) = \ker \tilde{D} \subset L_2(\tilde{E})$, $\mathcal{H}^* = \mathcal{H}(\tilde{D}^*) = \ker(\tilde{D}^*) \subset L_2(\tilde{F})$ then one defines the Γ–index $\mathrm{ind}_\Gamma \tilde{D}$ as

$$\mathrm{ind}_\Gamma \tilde{D} := \dim_\Gamma \mathcal{H}(\tilde{D}) - \dim_\Gamma \mathcal{H}(\tilde{D}^*).$$

Atiyah proves in [4] the following main

Theorem 2.6 *Under the assumptions above there holds*
$$\mathrm{ind}_a D = \mathrm{ind}_\Gamma \tilde{D}.$$
□

It was this theorem which was the orign of the von Neumann analysis as a fastly growing area in geometry, topology and analysis. Moreover, the proof of theorem 2.3 is strongly modeled by that of 2.6. Another very important special case which is related to the case above of coverings are locally symmetric spaces of finite volume. There is a vast number of profound contributions, e. g. [7], [22], [48], [50], [51]. We do not intend here to give a complete overview for reasons of space. But we will sketch the main features and main achievements of these approaches.

Let G be semisimple, noncompact, with finite center, $K \subset G$ maximal compact, $\tilde{X} = G/K$ a symmetric space of noncompact type, $\Gamma \subset G$ discrete, torsion free and $\mathrm{vol}(\Gamma\backslash G) < \infty$. Then $X = \Gamma\backslash \tilde{X} = \Gamma\backslash G/K$ is a locally symmetric space of finite volume. If V_E, V_F are unitary K–modules then we obtain homogeneous vector bundles $\tilde{E} = G/K \times_K V_E \longrightarrow G/K = \tilde{X}$, $\tilde{F} = G/K \times_K V_F \longrightarrow G/K = \tilde{X}$, over $\tilde{X}$ and corresponding bundles $E, F \longrightarrow X$ over X. A G–invariant elliptic differential operator $\tilde{D} : C^\infty(\tilde{E}) \longrightarrow C^\infty(\tilde{F})$ descends to an elliptic operator $D : C^\infty(E) \longrightarrow C^\infty(F)$. There arise the following natural questions: to describe the $\tilde{D}$ in question, to establish a formula for the analytical index, to calculate the index via a topological index and an index theorem. We indicate (partial) answers given by Barbasch, Connes, Moscovici and Müller.

Denote by $R(k)$ the right regular representation $R(k)f(g) = f(gk)$, $\tau_E : K \longrightarrow U(V_E)$. Then $k \longrightarrow R(k) \otimes \tau_E(k)$ acts on $C^\infty(G) \otimes V_E$. We identify $C^\infty(\tilde{E})$ with $(C^\infty(G) \otimes V_E)^K$, similarly $L_2(\tilde{E})$ with $(L_2(G) \otimes V_E)^K$. If $\mathfrak{G}$ is the Lie algebra of G, $\mathfrak{G}_c$ its complexification, $\mathfrak{U}(\mathfrak{G})$ the universal enveloping algebra of $\mathfrak{G}$, $\tau_E : K \longrightarrow U(V_E)$, $\tau_F : K \longrightarrow U(V_F)$ are unitary representations then $(\mathfrak{U}(\mathfrak{G}) \otimes \mathrm{Hom}\,(V_E, V_F))^K$ shall denote the subspace of all elements in $\mathfrak{U}(\mathfrak{G}) \otimes \mathrm{Hom}\,(V_E, V_F)$ which are fixed under $k \longrightarrow Ad_G(k) \otimes \tau_E(k^{-1})^t \otimes \tau_F(k)$. Let $d = \sum_i X_i \otimes A_i \in (\mathfrak{U}(\mathfrak{G}) \otimes \mathrm{Hom}\,(V_E, V_F))^K$. Then $\tilde{D} = \sum_i R(X_i) \otimes A_i$ defines a differential operator $\tilde{D} : C^\infty(\tilde{E}) \longrightarrow C^\infty(\tilde{F})$ commuting with the action of G. We state without proof the simple

Lemma 2.7 a) *Any G-invariant differential operator $\tilde{D} : C^\infty(\tilde{E}) \longrightarrow C^\infty(\tilde{F})$ is of the form*

$$\tilde{D} = \sum_i R(X_i) \otimes A_i \qquad (2.8)$$

above.
b) *The formal adjoint $\tilde{D}^*$ corresponds to*

$$d^* = \sum_i X_i^* \otimes A_i^* \in (\mathfrak{U}(\mathfrak{G}) \otimes \mathrm{Hom}\,(E, F))^K,$$

where $x \longrightarrow x^*$ denotes the conjugate-linear anti-automorphisms of $\mathfrak{U}(\mathfrak{G})$ such that $x^* = -\bar{x}$, $x \in \mathfrak{G}_c$. □

For a unitary representation $\pi : G \longrightarrow U(\mathcal{H}(\pi))$ and $d = \sum_i X_i \otimes A_i \in (\mathfrak{U}(\mathfrak{G}) \otimes \mathrm{Hom}\,(V_E, V_F))^K$ define $\pi(d) : \mathcal{H}(\pi)_\infty \otimes V_E \longrightarrow \mathcal{H}(\pi)_\infty \otimes V_F$ by

$$\pi(d) := \sum_i \pi(X_i) \otimes A_i.$$

Here $\mathcal{H}(\pi)_\infty$ denotes the space of C^∞-vectors of π. $\pi(d)$ induces an operator $d_\pi : (\mathcal{H}(\pi) \otimes V_E)^K \longrightarrow (\mathcal{H}(\pi) \otimes V_F)^K$.

Proposition 2.8 *Suppose that d is elliptic. Then*

$$\ker d_\pi = \{u \in (\mathrm{Hom}\,(\pi)_\infty \otimes V_E)^K \mid d_\pi u = 0\}$$

coincides with the orthogonal complement of

$$\mathrm{im}\, d_\pi^* = \{d_\pi^* v \mid v \in (\mathcal{H}(\pi)_\infty \otimes V_F)^K\}$$

in $(\mathcal{H}(\pi) \otimes V_E)^K$. □

Corollary 2.9 a) $\ker d_\pi$ is closed in $(\mathcal{H}(\pi) \otimes E)^K$.
b) *The closure of d_π^* coincides with the Hilbert space adjoint of d_π.* □

Corollary 2.10 *Suppose that d is elliptic and*

$$\pi = \int_\Lambda^\oplus \pi_\lambda \, d\lambda, \quad \mathcal{H}(\pi) = \int_\Lambda^\oplus \mathcal{H}(\pi_\lambda) \, d\lambda$$

is an integral decomposition of π. Then

$$\ker d_\pi = \int_\Lambda^\oplus \ker d_{\pi_\lambda} \, d\lambda. \tag{2.9}$$

□

Now we come to the main part of our present discussions, the locally symmetric case. Identifying $L_2(E)$ with $(L_2(\Gamma\backslash G) \otimes V_E)^K$, and taking into consideration the decompositions

$$R^\Gamma = R_d^\Gamma \oplus R_c^\Gamma, \quad L_2(\Gamma\backslash G) = L_{2,d}(\Gamma\backslash G) \oplus L_{2,c}(\Gamma\backslash G)$$

of the right quasi–regular representation R^Γ of G on $L_2(\Gamma\backslash G)$, we obtain the decomposition

$$\begin{aligned} L_2(E) &= L_{2,d}(E) \oplus L_{2,c}(E), \\ L_{2,d}(E) &= (L_{2,d}(\Gamma\backslash G) \otimes V_E)^K, \\ L_{2,c}(E) &= (L_{2,c}(\Gamma\backslash G) \otimes V_E)^K, \end{aligned}$$

similarly for $F = \Gamma\backslash \tilde{F}$.
Consider now the operators $D = d_{R^\Gamma}$ and $D_d = d_{R_d^\Gamma} : C_c^\infty(E) \longrightarrow C_c^\infty(F)$.

Theorem 2.11 *Under the assumptions above (on G, K, Γ),*

$$\ker D = \ker D_d \tag{2.10}$$

and

$$\dim \ker D < \infty. \tag{2.11}$$

Denote by $\tilde{G}_d^\Gamma$ the set of all equivalence classes of irreducible unitary representation π of G whose multiplicity $m_\Gamma(\pi)$ in R_d^Γ is nonzero. In particular $L_{2,d}(\Gamma\backslash G) = \sum_{\pi \in \tilde{G}_d^\Gamma} m_\Gamma(\pi)\mathcal{H}(\pi)$.

Theorem 2.12 *Let $K \subset G$ be maximal compact, $\Gamma \in G$ discrete and torsion free, $\tau_E : K \longrightarrow V_E$, $\tau_F : K \longrightarrow V_F$ unitary representations, $\tilde{E} = G/K \times_K V_E$, $\tilde{F} = G/K \times_K V_F$, $E = \Gamma\backslash\tilde{E}$, $F = \Gamma\backslash\tilde{F}$ and $D = d_{R^\Gamma}$ a corresponding locally invariant elliptic differential operator acting between $L_2(E)$ and $L_2(F)$. Then*

$$\mathrm{ind}_a D = \dim \ker D - \dim \ker D^*$$

is well defined and

$$\mathrm{ind}_a D = \sum_{\pi \in \tilde{G}_d^\Gamma} m_\Gamma(\pi)(\dim(\mathcal{H}(\pi) \otimes E)^K - \dim(\mathcal{H}(\pi) \otimes F)^K).$$

(2.12)

□

Corollary 2.13 *Let $X = \Gamma\backslash G/K$ be a locally symmetric space of negative curvature with finite volume and $L_2(E) \supset \mathcal{D}_D \xrightarrow{D} L_2(F)$ a locally symmetric elliptic differential operator then $\mathrm{ind}\, D$ is defined and depends only on the K-modules $K \longrightarrow U(V_E), U(V_F)$ which define $\tilde{E}, \tilde{F}$, $E = \Gamma\backslash\tilde{E}$, $F = \Gamma\backslash\tilde{F}$.* □

The value of the formula in theorem 2.12 is very limited since in general the $m_\Gamma(\pi)$ are not known. Hence there arises the task to find a meaningful expression for it. This has been done with great success e. g. in [22] and [51], [52] where they essentially restrict to generalized Dirac operators. To be more precise, we must briefly recall what is a manifold with cusps. Here we densely follow [50]. Let G be a semisimple Lie group with finite center, $K \subset G$ a maximal compact subgroup. P_a split rank one parabolic subgroup of G with split component A, $P = UAM$ the corresponding Langlands decomposition, where U is the unipotent radical of P, A a $\mathbb{R}$–split torus of dimension one and M centralizes A. Set $S = UM$ and let Γ be a discrete uniform torsion free subgroup of S. Then $Y = \Gamma\backslash\tilde{Y} = \Gamma\backslash G/K$ is called a complete cusp of rank one. Put $K_M = M \cap K$, K_M is a maximal compact subgroup of M. If $X_M = M/K_M$ there is a canonical diffeomorphism $\tilde{\xi}: \mathbb{R}_+ \times U \times X_M \longrightarrow \tilde{Y}$. Set for $t \geq 0$ $\tilde{Y}_t = \tilde{\xi}([t,\infty[\times U \times X_M)$ and call $Y_t = \Gamma\backslash\tilde{Y}_t$ a cusp of rank one. Another, even more explicit description is given as follows. Let $\Gamma_M = M \cap (U\Gamma)$, $Z = S/S \cap K$. Then there is a canonical fibration $P : \Gamma\backslash Z \longrightarrow \Gamma_M\backslash X_M$ with fibre $\Gamma \cap U\backslash U$ a compact nilmanifold and a canonical diffeomorphism $\xi : [t,\infty[\times\Gamma\backslash Z \xrightarrow{\cong} Y_t$. The induced metric on $[t,\infty[\times\Gamma_2\backslash Z$ looks locally as $ds^2 = dr^2 + dx^2 + e^{-br}du_\lambda^2(x) + e^{-4br}du_{2\lambda}^2(x)$, where $|b| = \lambda$, dx^2 is the invariant metric on X_M induced by restriction of the Killing form.

Now a complete Riemannian manifold is called a manifold with cusps of rank one if X has a decomposition $X = X_0 \cup X_1 \cup \cdots \cup X_s$ such that X_0 is a compact manifold with boundary, for $i,j \geq 1$, $i \neq j$ holds $X_i \cap X_j = \emptyset$ and each X_j, $j \geq 1$, is a cusp of rank one. The first general statement for generalized Dirac operators on rank one cusps manifold is

Theorem 2.14 *Let X be a rank one cusp manifold, $(E, h, \nabla, \cdot) \longrightarrow (X, g_X)$ a Clifford bundle and D its corresponding generalized Dirac operator. Then D is essentially self-adjoint and*

$$\dim(\ker \overline{D}) < \infty. \tag{2.13}$$

The spectrum of $H = \overline{D}^2$ consists of a point spectrum and an absolutely continuous spectrum. If $L_2(E) = L_{2,d}(E) \oplus L_{2,c}(E)$ is the corresponding decomposition of $L_2(E)$ and $H_d = H|_{L_{2,d}(E)}$ then for $t > 0$

$$e^{-zH_d} \text{ is of trace class}. \tag{2.14}$$

□

As we mentioned after corollary 2.13, the main task, main objective consists in the case of a $\mathbb{Z}_2$-grading to get an expression for $\text{ind}_a \overline{D}$. For the sake of simlicity we restrict to spaces

$X = X_0 \cup Y_1$ as above with one cusp Y_1, $Y_0 \cup Y_1 = Y = \Gamma \backslash G/K$. Let $(E = E^+ \oplus E^-, h, \nabla, \cdot) \longrightarrow (Y, g)$ be a Z_2-graded Clifford bundle such that $E^\pm|_{Y_1} = \Gamma \backslash \tilde{E}^\pm$, where $\tilde{E}^\pm$ are homogeneous vector bundles over G/K and let $D^+ : C^\infty(Y, E^+_+) \longrightarrow C^\infty(Y, E^-)$ the corresponding generalized Dirac operator. We recall $K_M = M \cap K$, $X_M = M/K_M$. D^+ induces an elliptic differential operator $D^+_0 : C^\infty(\mathbb{R}_+ \times \Gamma_M \backslash X_M, E^+_M) \longrightarrow C^\infty(\mathbb{R}_+ \times \Gamma_M \backslash X_M, E^-_M)$, where $E^\pm_M$ are locally homogenous vector bundles over $\Gamma_M \backslash X_M$. From this come a self–adjoint differential operator $D_M : C^\infty(\Gamma_M \backslash X_M, E^+_M) \longrightarrow C^\infty(\Gamma_M \backslash X_M, E^-_M)$ and a bundle isomorphism $\beta : E^+_M \longrightarrow E^-_M$ such that $D^+_0 = \beta \left(r \frac{\partial}{\partial r} + D_M \right)$. We set $\tilde{D}_M = D_M + \frac{m}{2} \mathrm{id}$, $m = \dim u_\lambda |\lambda| + 2 \dim u_{2\lambda} |\lambda|$, λ the unique simple root of the pair (P, A).

W. Müller then established in [50] the following general index theorem for a locally symmetric graded Dirac operator.

Theorem 2.15 *Assume* $\ker \tilde{D}_M = \{0\}$, *let* $\eta(0)$ *be the eta invariant of* $\tilde{D}_M$ *and* ω_{D^+} *the index form of* D^+. *Then*

$$\mathrm{ind}_a D^+ = \int_X \omega_{D^+} + \mathcal{U} + \frac{1}{2} \eta(0), \qquad (2.15)$$

where the term $\mathcal{U}$ *is essentially given by the value of an L-series at zero and an expression in the scattering matrix at zero.* □

Finally, application of an elaborated version of theorem 2.15 allows to prove the famous Hirzebruch conjecture for Hilbert modular varieties. This has been done by W. Müller in [51].

There is another approach to Fredholmness by Gilles Carron, which relies on an inequality quite similar to 2.2 d).

Let $(E, h, \nabla, \cdot) \longrightarrow (M^n, g)$ be a Clifford bundle over the complete Riemannian manifold (M^n, g) and $D : C^\infty(E) \longrightarrow C^\infty(E)$ the associated generalized Dirac operator. D is called non-parabolic at infinity if there exists a compact set $K \subset M$ such that for any open and relative compact $U \subset M \setminus K$ there exists a constant $C(U) > 0$ such that

$$C(U) |\varphi|_W \leq |D\varphi|_{L_2(E|_{M \setminus K})} \text{ for all } \varphi \in C^\infty_c(E|_{M \setminus K}). \qquad (2.16)$$

To exhibit the consequences of this inequality, we establish another characterization of it.

Proposition 2.16 *Let* $(E, h, \nabla, \cdot) \longrightarrow (M^n, g)$ *and* D *as above and let* $W(E)$ *be a Hilbert space of sections such that*
a) $C_c^\infty(E)$ *is dense in* $W(E)$ *and*
b) *the injection* $C_c^\infty(E) \hookrightarrow \Omega_{\text{loc}}^{2,1}(E, D)$ *extends continuously to* $W(E) \longrightarrow \Omega_{\text{loc}}^{2,1}(E, D)$.
Then $D : W(E) \longrightarrow L_2(E)$ *is Fredholm if and only if there exist a compact* $K \subset M$ *and a constant* $C(K) > 0$ *such that*

$$C(K) \cdot |\varphi|_W \leq |D\varphi|_{L_2(E|_{M\setminus K})} \text{ for all } \varphi \in C_c^\infty(E|_{M\setminus K}). \quad (2.17)$$

$\square$

Remark 2.17 The norm $\varphi \longrightarrow |D\varphi|_{L_2} = \mathcal{N}(\varphi)$ above is equivalent to the norm

$$\mathcal{N}_{\overline{U}(K)}(\cdot), \mathcal{N}_{\overline{U}(K)}(\varphi)^2 = |\varphi|^2_{L_2(E|_{\overline{U}(K)})} + |D\varphi|^2_{L_2(E)}. \quad (2.18)$$

Corollary 2.18 $D : C^\infty(E) \longrightarrow C^\infty(E)$ *is non–parabolic at infinity if and only if there exists a compact* $K \subset M$ *such that the completion of* $C_c^\infty(E)$ *w. r. t.* $\mathcal{N}_K(\cdot)$,

$$\mathcal{N}_K(\varphi)^2 = |\varphi|^2_{L_2(E|_K)} + |D\varphi|^2_{L_2} \quad (2.19)$$

yields a space $W(E)$ *such that the injection* $C_c^\infty(E) \longrightarrow \Omega_{\text{loc}}^{2,1}(E, D)$ *continuously extends to* $W(E)$. $\square$

The point now is that we know if D is non–parabolic at infinity then $D : W(E) \longrightarrow L_2(E)$ is Fredholm. We emphasize, this does not mean $L_2(E) \supset \mathcal{D}_D \longrightarrow L_2(E)$ is Fredholm. We get a weaker Fredholmness, not the desired one. But in certain cases this can be helpful too.

Suppose again a $\mathbb{Z}_2$-grading of E and D, $D = \begin{pmatrix} 0 & D^- \\ D^+ & 0 \end{pmatrix}$, $L_2(E) = L_2(E^+) \oplus L_2(E^-)$, $W(E) = W(E^+) \oplus W(E^-)$. Following Gilles Carron, we now define the extended index $\text{ind}_e D^+$

as

$$\begin{aligned}\operatorname{ind}_e D^+ &:= \dim \ker {}_W D^+ - \dim \ker {}_{L_2} D^- \\ &= \dim\{\varphi \in W(E^+) \mid D^+ \varphi = 0\} - \\ &\quad - \dim\{\varphi \in L_2(E^-) \mid D^- \varphi = 0\}. \quad (2.20)\end{aligned}$$

If we denote $h_\infty(D^+) := \dim(\ker_W D^+ / \ker_{L_2} D^+)$ then we can (2.20) rewrite as

$$\begin{aligned}\operatorname{ind}_e D^+ &= h_\infty(D^+) + \operatorname{ind}_{L_2} D^+ \\ &= h_\infty(D^+) + \dim \ker {}_{L_2} D^+ - \dim \ker {}_{L_2} D^-. \quad (2.21)\end{aligned}$$

The most interesting question now are applications and examples. For $D =$ Gauß–Bonnet operator, there are in fact good examples (cf. [12]). For the general case it is not definitely clear, is non–parabolicity really a practical sufficient criterion for Fredholmness since in concrete cases it will be very difficult it to establish. In some well known standard cases which have been presented by Carron and which we will discuss now it is of great use.

Proposition 2.19 *Let $D : C^\infty(E) \longrightarrow C^\infty(E)$ be a generalized Dirac operator and assume that outside a compact $K \subset M$ the smallest eigenvalue $\lambda_{\min}(x)$ of $\mathcal{R}_x$ in $D^2 = \nabla^* \nabla + \mathcal{R}$ is ≥ 0. Then D is non–parabolic at infinity.* □

We obtain from proposition 2.18

Corollary 2.20 *Assume the hypothesis of 2.18. Then $D : W_0(E) \longrightarrow L_2(E)$ is Fredholm.* □

Under certain additional assumptions the pointwise condition on $\lambda_{\min}(x)$ of $\mathcal{R}_x$ can be replaced by a (weaker) integral condition. Denote $\mathcal{R}_-(x) = \max\{0, -\lambda_{\min}(x)\}$, where $\lambda_{\min}(x)$ is the smallest eigenvalue of $\mathcal{R}_x$.

Theorem 2.21 *Suppose that for a $p > 2$ (M^n, g) satisfies the Sobolev inequality*

$$c_P(M) \left(\int_M |u|^{\frac{2p}{p-2}}(x) \, \mathrm{dvol}_x(g) \right)^{\frac{p-2}{2}}$$
$$\leq \int_M |du|^2(x) \, \mathrm{dvol}_x(g) \text{ for all } u \in C_c^\infty(M) \quad (2.22)$$

and

$$\int_M |\mathcal{R}_-|^{\frac{p}{2}}(x) \, \mathrm{dvol}_x(g) < \infty.$$

Then $D: W_0(E) \longrightarrow L_2(E)$ is Fredholm. □

Another important example are manifolds with a cylindriccal end which we already mentioned. In this case, there is a compact submanifold with boundary $K \subset M$ such that $(M \setminus K, g)$ is isometric to $(]0, \infty[\times \partial K, dr^2 + g_{\partial K})$. One assumes that $(E, h)|_{]0,\infty[\times \partial K}$ also has product structure and $D|_{M \setminus K} = \nu \cdot (\frac{\partial}{\partial r} + A)$, where $\nu \cdot$ is the Clifford multiplication with the exterior normal at $\{\gamma\} \times \partial K$ and A is first order elliptic and self–adjoint on $E|_{\partial K}$.

Proposition 2.22 *D is non–parabolic at infinity.*

Proof. There are two proofs. The first one refers to [5]. According to proposition 2.5 of [5], there exists on $M \setminus K$ a parametrix $Q: L_2(E|_{M \setminus K}) \longrightarrow \Omega_{\text{loc}}^{2,1} E|_{M \setminus K}, D)$ such that $QD\varphi = \varphi$ for all $\varphi \in C_c^\infty(E|_{M \setminus K})$. Hence for $C_c^\infty(E|_{M \setminus K})$, $U \supset M \setminus K$ bounded,

$$|\varphi|_{L_2(E|_U)} = |QD\varphi|_{L_2(E|_U)} \leq |Q|_{L_2 \to \Omega^{2,1}} \cdot |D\varphi|_{L_2}.$$

The other proof is really elementary calculus. For $\varphi \in C_c^\infty(E|_{M \setminus K})$, $|\varphi(r, y)| = \left| \int_0^r \frac{\partial \varphi}{\partial r} dr \right| \leq \sqrt{r} \cdot \left| \frac{\partial \varphi}{\partial r} \right|_{L_2}$. Hence

$$|\varphi|^2_{L_2(E|_{]0,T[\times\partial K})} \leq \tfrac{T^2}{2}\left|\tfrac{\partial\varphi}{\partial r}\right|^2_{L_2} \leq \tfrac{T^2}{2}\left(\left|\tfrac{\partial\varphi}{\partial r}\right|^2_{L_2} + |A\varphi|^2_{L_2}\right) = \tfrac{T^2}{2}|D\varphi|^2_{L_2}.$$
□

The authors of [5] define extended L_2-sections of $E|_{]0,\infty[\times\partial K}$ as sections $\varphi \in L_{2,\text{loc}}$, $\varphi(r,y) = \varphi_0(r,y) + \varphi_\infty(y)$, $\varphi_0 \in L_2$, $\varphi_\infty \in \ker A$.

Proposition 2.23 *The extended solutions of $D\varphi = 0$ are exactly the solutions of $D\varphi = 0$ in W.*

Proof. Let $\{\varphi_\lambda\}_{\lambda\in\sigma(A)}$ be a complete orthonormal system in $L_2(E|_{\partial K})$ consisting of the eigensections of A. Then we can a solution φ of $D\varphi = 0$ on $]0,\infty[\times\partial K$ decompose as

$$\varphi(r,y) = \sum_{\lambda\in\sigma(A)} c_\lambda e^{-\lambda r}\varphi_\lambda(y) \tag{2.23}$$

and $\varphi \in W$ if and only if $c_\lambda = 0$ for $\lambda < 0$. In this case

$$\varphi_0(r,y) = \sum_{\substack{\lambda\in\sigma(A)\\\lambda>0}} c_\lambda e^{-\lambda r}\varphi_\lambda(y), \quad \varphi_\infty(y) = \sum_{\lambda\in\sigma(A)} c_{0,i}\varphi_{0,i}(y).$$

□

This proposition can also be reformulated as

Proposition 2.24 *Denote by $P_{\leq 0}$ or $P_{<0}$ the spectral projection of A onto the sum of eigenspaces belonging to eigenvalues ≤ 0 or < 0, respectively. Then*
a) *φ is a solution in W of $D\varphi = 0$ if and only if*

$$D\varphi = 0 \text{ on } K$$

and

$$P_{<0}\varphi = 0 \text{ on } \partial K.$$

b) *φ is an L_2-solution of $D\varphi = 0$ if and only if*

$$D\varphi = 0 \text{ on } K$$

and
$$P_{\leq 0}\varphi = 0 \text{ on } \partial K.$$

□

There is a very general approach to index theory as established by Connes, Roe and others. The initial data are as follows: D an elliptic differential operator as above, $\mathfrak{B}$ an operator algebra, the K–theory $K_i(\mathfrak{B})$ of $\mathfrak{B}$, the cyclic cohomology $HC^*(\mathfrak{B})$ of $\mathfrak{B}$. Then one constructs the diagram

$$\begin{array}{ccc} D & \longrightarrow & \operatorname{Ind} D \in K_i(\mathfrak{B}) \\ \downarrow & & \downarrow \\ I_D & \longrightarrow \langle I_D, m \rangle = \operatorname{ind}_t D \stackrel{?}{=} \operatorname{ind}_a D = \langle \operatorname{Ind} D, \zeta \rangle \end{array}$$

Here I_D is of cohomological nature, m a fundamental class, $\langle I_D, m \rangle$ a pairing, $\operatorname{Ind} D$ comes from ellipticity and the 6 term exact sequence of K–theory, $\zeta \in HC^*(\mathfrak{B})$ and $\langle \operatorname{Ind} D, \zeta \rangle$ is the Connes' pairing.

Choice of $\mathfrak{B}, i, \zeta, m, \operatorname{Ind}_D$ yields a concrete index theory. We refer to [60], [61], [62], [74] for details. The classical index theory on closed manifolds is given by the choice $i = 0$, $\mathfrak{B} =$ ideal K of compact operators, $\operatorname{Ind} D \in K_0(K) =$ projectors - projectors, $HC^0 \ni \zeta =$ trace, $\operatorname{trInd} D = \operatorname{ind}_a D$, $I_D =$ classical index from, $m = [M]$. The lack of all these (absolute) index theories for open manifolds is that they either refer to very special cases or there are not enough serious applications. This was one of the motivations for us to establish a general relative index theory as in chapter V.

II Non–linear Sobolev structures

Relative invariants $i(P_1, P_2)$ are invariants defined for pairs (P_1, P_2), where P_2 has to be considered as a perturbation of P_1, where P e.g. stands for a triple (manifold, bundle, differential operator).

Our approach consists in defining a metrizable Sobolev uniform structure for the set of Ps and defining generalized components $\operatorname{gen comp}(P) = \{P' | \operatorname{distance}(P, P') < \infty\}$. The admitted perturbations of P are just the $P' \in \operatorname{gen comp}(P)$. In special cases, the generalized components coincide with path components which was the motivation for the notation "generalized component".

To perform the indicated program, we have to collect results on Sobolev spaces for open manifolds, to introduce uniform spaces and to connect both concepts in the theory of non–linear Sobolev structures of manifolds and Clifford bundles. Since Sobolev maps enter into the definitions, we enclose an outline of manifolds of maps between open manifolds. The notion of generalized components of manifolds offers a certain approach to the classification problem. In a first step one "counts or classifies" the generalized compoenents $\operatorname{gen comp}(M^n, g)$. In a second step one "counts or classifies" the manifolds (M'^n, g') inside a generalized component $\operatorname{gen comp}(M^n, g)$. Here we understand under "count or classify" not a complete classification but steps toward this goal. A "complete classification" is nowaday even for closed manifolds not available.

This chapter is organized as follows. We start in section 1 with Sobolev spaces, embedding theorems, module structure theorems for open manifolds and establish some theorems which will be essential in chapter IV. The general approach to uniform structures of proper metric spaces will be presented in section 2. Completed manifolds of maps, in particular completed diffeomorphism groups, are the content of the third section. The heart of chapter II is section 4, where we introduce uniform structures of manifolds and Clifford bundles. As a first application, we give an outline of the classification approach for open manifolds. In particular, we define new (co–)homologies, define relative characteristic numbers and apply this to bordism theory for open manifolds.

1 Clifford bundles, generalized Dirac operators and associated Sobolev spaces

In the first part of this section, we consider the Sobolev spaces associated to Clifford bundles, the $W-$ and H–spaces. In the second part we introduce Sobolev spaces for arbitrary Riemannian vector bundles, the Ω–spaces which we essentially need in section 3.

We recall for completeness very briefly the basic properties of generalized Dirac operators on open manifolds. Let (M^n, g) be a Riemannian manifold, $m \in M$, $Cl(T_m M, g_m)$ the corresponding Clifford algebra at m. $Cl(T_m, g_m)$ shall be complexified or not, depending on the other bundles and structure under consideration. A Hermitean vector bundle $E \to M$ is called a bundle of Clifford modules if each fibre E_m is a Clifford module over $Cl(T_m, g_m)$ with skew symmetric Clifford multiplication. We assume E to be endowed with a compatible connection ∇^E, i.e. ∇^E is metric and

$$\nabla^E_X (Y \cdot \Phi) = (\nabla^g_X Y) \cdot \Phi + Y \cdot (\nabla^E_X \Phi)$$

$X, Y \in \Gamma(TM), \Phi \in \Gamma(E)$. Then we call the pair (E, ∇^E) a Clifford bundle. The composition

$$\Gamma(E) \xrightarrow{\nabla} \Gamma(T^*M \otimes E) \xrightarrow{g} \Gamma(TM \otimes E) \longrightarrow \Gamma(E)$$

shall be called the generalized Dirac operator D. We have $D = D(g, E, \nabla)$. If $X_1, \ldots X_n$ is an orthonormal basis in $T_m M$ then

$$D = \sum_{i=1}^n X_i \cdot \nabla^E_{X_i}.$$

D is of first order elliptic, formally self-adjoint and
$$D^2 = \Delta^E + R,$$
where $\Delta^E = (\nabla^E)^*\nabla^E$ and $R \in \Gamma(End(E))$ is the bundle endomorphism
$$R\Phi = \frac{1}{2}\sum_{i,j=1}^n X_i X_j R^E(X_i, X_j)\Phi.$$

Next we recall some associated functional spaces and their properties if we assume bounded geometry. These facts are contained in [8], [36] and partially in [27].

Let $E \to M$ be a Clifford bundle, $\nabla = \nabla^E$, D the generalized Dirac operator. Then we define for $\Phi \in \Gamma(E), p \geq 1, r \in \mathbf{Z}, r \geq 0$,

$$|\Phi|_{W^{p,r}} := \left(\int \sum_{i=0}^r |\nabla^i\Phi|_x^p \, dvol_x(g)\right)^{\frac{1}{p}},$$

$$|\Phi|_{H^{p,r}} := \left(\int \sum_{i=0}^r |D^i\Phi|_x^p \, dvol_x(g)\right)^{\frac{1}{p}},$$

$$W_r^p(E) := \{\Phi \in \Gamma(E) | |\Phi|_{W^{p,r}} < \infty\},$$
$$W^{p,r}(E) := \text{completion of } W_r^p \text{ w. r. t. } |\ |_{W^{p,r}},$$
$$H_r^p(E) := \{\Phi \in \Gamma(E) | |\Phi|_{H^{p,r}} < \infty\},$$
$$H^{p,r}(E) := \text{completion of } H_r^p \text{ w. r. t. } |\ |_{H^{p,r}}.$$

In a big part of our consideration we restrict to $p = 1, 2$. In the case $p = 2$ we write $W^{2,r} \equiv W^r, H^{2,r} \equiv H^r$ etc.. If $r < 0$ then we set

$$W^r(E) := \left(W^{-r}(E)\right)^*,$$
$$H^r(E) := \left(H^{-r}(E)\right)^*.$$

Assume (M^n, g) complete. Then $C_c^\infty(E)$ is a dense subspace of $W^{p,1}(E)$ and $H^{p,1}(E)$. If we use this density and the fact
$$|D\Phi(m)| \leq C \cdot |\nabla\Phi(m)|,$$

we obtain $|\Phi|_{H^{p,1}} \leq C' \cdot |\Phi|_{W^{p,1}}$ and a continious embedding

$$W^{p,1}(E) \hookrightarrow H^{p,1}(E).$$

For $r > 1$ this cannot be established, and we need further assumptions. Consider as in the introduction the following conditions

(I) $r_{inj}(M,g) = \inf_{x \in M} r_{inj}(x) > 0,$
$(B_k(M,g))$ $|(\nabla^g)^i R^g| \leq C_i, \ 0 \leq i \leq k,$
$(B_k(E, \nabla^E))$ $|(\nabla^g)^i R^E| \leq C_i, \ 0 \leq i \leq k.$

It is a well known fact that for any open manifold and given $k, 0 \leq k \leq \infty$, there exists a metric g satisfying (I) and $(B_k(M,g))$. Moreover, (I) implies completeness of g.

Lemma 1.1 *Assume (M^n, g) with (I) and (B_k). Then $C_c^\infty(E)$ is a dense subset of $W^{p,r}(E)$ and $H^{p,r}(E)$ for $0 \leq r \leq k+2$.*

See [27], proposition 1.6 for a proof. $\square$

Lemma 1.2 *Assume (M^n, g) with (I) and (B_k). Then there exists a continuous embedding*

$$W^{p,r}(E) \longrightarrow H^{p,r}(E), \quad 0 \leq r \leq k+1.$$

Proof. According to 1.1, we are done if we can prove

$$|\Phi|_{H^{p,r}} \leq C \cdot |\Phi|_{W^{p,r}}$$

for $0 \leq r \leq k+1$ and $\Phi \in C_c^\infty(E)$. Perform induction. For $r = 0$ $|\Phi|_{H^{p,0}} = |\Phi|_{W^{p,0}}$. Assume $|\Phi|_{H^{p,r}} \leq C \cdot |\Phi|_{W^{p,r}}$. Then

$$\begin{aligned}|\Phi|_{H^{p,r+1}} &\leq C \cdot (|\Phi|_{H^{p,r}} + |D^r D\Phi|_{H^{p,r}}) \\ &\leq C \cdot (|\Phi|_{W^{p,r}} + |D\Phi|_{W^{p,r}}).\end{aligned}$$

Let $\frac{\partial}{\partial x^i}, i = 1, \ldots, n$ coordinate vectors fields which are orthonormal in $m \in M$. Then with $\nabla_i = \nabla_{\frac{\partial}{\partial x^i}}$

$$|\nabla^s D\Phi|_m^p \leq C \cdot \sum_{i_1,\ldots,i_s,j} |\nabla_{i_1} \ldots \nabla_{i_s} \frac{\partial}{\partial x^j} \cdot \nabla_j \Phi|^p.$$

Now we apply the Leibniz rule and use the fact that in an atlas of normal charts the Christoffel symbols have bounded euclidean derivatives up to order $k - 1$. This yields

$$|\nabla^r D\Phi|_m^p \leq C \cdot \sum_{i_1,\ldots,i_{r+1}} |\nabla_{i_1}\ldots\nabla_{i_{r+1}}\Phi|_m^p \text{ for } r \leq k,$$

i. e.
$$|D\Phi|_{W^{p,r}} \leq C \cdot |\Phi|_{W^{p,r+1}}$$

altogether
$$|\Phi|_{H^{p,r+1}} \leq C \cdot |\Phi|_{W^{p,r+1}}.$$

□

Remark 1.3 For $p = 2$ this proof is contained in [8]. □

Theorem 1.4 *Assume (M^n, g) with (I) and (B_k) and (E, ∇) with (B_k) and $p = 2$. Then for $r \leq k$*

$$H^{2,r}(E) \equiv H^r(E) \cong W^r(E) \equiv W^{2,r}(E)$$

as equivalent Hilbert spaces.

Proof. According to 1.2, $W^r(E) \subseteq H^r(E)$ continuously. Hence we have to show $H^r(E) \subseteq W^r(E)$ continuously. The latter follows from the local elliptic inequality, a uniformly locally finite cover by normal charts of fixed radius, uniform trivializations and the existence of uniform elliptic constants. The proof is performed in [8]. Another proof is contained in [66]. □

Remark 1.5 1.4 holds for $1 \leq p < \infty$ (cf. [27]). □

As it is clear from the definition that the spaces $W^{p,k}(E)$ can be defined for any Riemannian vector bundle (E, h_E, ∇^E). We assume this more general case and define additionally

$$^{b,s}W(E) := \left\{ \varrho \in C^s(E) \ \Big| \ ^{b,s}|\varrho| := \sum_{i=0}^{s} \sup_{x \in M} |\nabla^i \varrho|_x < \infty \right\}$$

and in the case of a Clifford bundle

$$^{b,s}H(E) := \left\{ \varrho \in C^s(E) \ \Big| \ ^{b,s,D}|\varrho| := \sum_{i=0}^{s} \sup_{x \in M} |D^i \varrho|_x < \infty \right\}.$$

$^{b,s}W(E)$ is a Banach space and coincides with the completion of the space of all $\varrho \in \Gamma(E)$ with $^{b,s}|\varrho| < \infty$ with respect to $^{b,s}\|$.

Theorem 1.6 Let (E, h, ∇^E) be a Riemannian vector bundle satisfying (I), $(B_k(M^n, g))$, $B_k(E; \nabla))$.
a) Assume $k \geq r, k \geq 1, r - \frac{n}{p} \geq s - \frac{n}{q}, r \geq s, q \geq p$, then

$$W^{p,r}(E) \hookrightarrow W^{q,s}(E) \tag{1.1}$$

continuously.
b) If $k \geq r > \frac{n}{p} + s$ then

$$W^{p,r}(E) \hookrightarrow {}^{b,s}W(E) \tag{1.2}$$

continuously.

We refer to [36] for the proof. $\square$

Corollary 1.7 Let $E \to M$ be a Clifford bundle satisfying (I), $(B_k(M))$, $(B_k(E))$, $k \geq r$, $r > \frac{n}{2} + s$. Then

$$H^r(E) \hookrightarrow {}^{b,s}H(E) \tag{1.3}$$

continuously.

Proof. We apply 1.4, (1.1) and obtain

$$H^r(E) \hookrightarrow {}^{b,s}W(E). \tag{1.4}$$

Quite similar as in the proof of 1.2,

$$H^r(E) \hookrightarrow {}^{b,s}W(E). \tag{1.5}$$

continuously. $\square$

A key role for everything in the sequel plays the module structure theorem for Sobolev spaces.

Theorem 1.8 Let $(E_i, h_i, \nabla_i) \to (M^n, g)$ be Clifford bundles with (I), $(B_k(M^n, g))$, $(B_k(E_i, \nabla_i))$, $i = 1, 2$. Assume $0 \leq r \leq r_1, r_2 \leq k$. If $r = 0$ assume

$$\left\{\begin{array}{rcl} r - \frac{n}{p} & < & r_1 - \frac{n}{p_1} \\ r - \frac{n}{p} & < & r_2 - \frac{n}{p_2} \\ r - \frac{n}{p} & \leq & r_1 - \frac{n}{p_1} + r_2 - \frac{n}{p_2} \\ \frac{1}{p} & \leq & \frac{1}{p_1} + \frac{1}{p_2} \end{array}\right\} \quad or$$

$$\left\{\begin{array}{rcl} r - \frac{n}{p} & \leq & r_1 - \frac{n}{p_1} \\ 0 & < & r_2 - \frac{n}{p_2} \\ \frac{1}{p} & \leq & \frac{1}{p_1} \end{array}\right\} \quad or \quad \left\{\begin{array}{rcl} 0 & < & r_1 - \frac{n}{p_1} \\ r - \frac{n}{p} & \leq & r_2 - \frac{n}{p_2} \\ \frac{1}{p} & \leq & \frac{1}{p_2} \end{array}\right\}. \quad (1.6)$$

If $r > 0$ assume $\frac{1}{p} \leq \frac{1}{p_1} + \frac{1}{p_2}$ and

$$\left\{\begin{array}{rcl} r - \frac{n}{p} & < & r_1 - \frac{n}{p_1} \\ r - \frac{n}{p} & < & r_2 - \frac{n}{p_2} \\ r - \frac{n}{p} & \leq & r_1 - \frac{n}{p_1} + r_2 - \frac{n}{p_2} \end{array}\right\}$$

$$or \quad \left\{\begin{array}{rcl} r - \frac{n}{p} & \leq & r_1 - \frac{n}{p_1} \\ r - \frac{n}{p} & \leq & r_2 - \frac{n}{p_2} \\ r - \frac{n}{p} & < & r_1 - \frac{n}{p_1} + r_2 - \frac{n}{p_2} \end{array}\right\}. \quad (1.7)$$

Then the tensor product of sections defines a continuous bilinear map

$$W^{p_1, r_1}(E_1, \nabla_1) \times W^{p_2, r_2}(E_2, \nabla_2) \longrightarrow W^{p, r}(E_1 \otimes E_2, \nabla_1 \otimes \nabla_2).$$

We refer to [36] for the proof. □

Define for $u \in C^0(M), c > 0$

$$\overline{u}_c(x) := \frac{1}{vol B_c(x)} \int_{B_c(x)} u(y) \, dvol_y(g).$$

Proposition 1.9 Let (M^n, g) be complete, $Ric(g) \geq k$, $k \in \mathbb{R}$. Then there exists a positive constant $C = C(n, k, R)$, depending only on k, R such that for any $c \in]0, R[$ and any $u \in W^{1,1}(M) \cap C^1(M)$

$$\int_M |u - \bar{u}_c| dvol_x(g) \leq C \cdot c \cdot \int_M |\nabla u| \, dvol_x(g). \tag{1.8}$$

Proof. For $u \in C_c^\infty(M)$ the proof is performed in [40], p. 31–33. But what is only needed in the proof is $\int |u| dx, \int |\nabla u| \, dx < \infty$ (even only $\int |\nabla u| \, dx < \infty$).
The key is a the lemma of Buser (cf. [10]),

$$\int_{B_c(x)} |u - \bar{u}_c(x)| \, dy \leq C \cdot c \cdot \int_{B_c(x)} |\nabla u| \, dy \tag{1.9}$$

for all $x \in M$, $x \in]0, 2R[$ and $u \in C^1(\overline{B_c(x)})$. □

But for completeness, we will give a complete proof of the proposition since it plays in chapter IV a crucial role. For this, we recall two well known lemmas which are contained e.g. in [40], [41].

Lemma 1.10 *let (M^n, g) be open, complete $\mathrm{Ric}\,(g) \geq k$, $\varrho > 0$. Then there exists a sequence $(x_i)_i$ of points $x_i \in M$, such that for any $c > \varrho$ there holds*
a) *The family $(B_c(x_i))_i$ is a uniformly locally finite cover of M and an upper bound N for the number of non–empty local intersections is given by $N = N(n, \varrho, c, k)$.*
b) *For any $i \neq j$, $B_{\frac{\varrho}{2}}(x_i) \cap B_{\frac{\varrho}{2}}(x_j) = \emptyset$.* □

Lemma 1.11 *Let (M^n, g) be open, complete, $\mathrm{Ric}\,(g) \geq k$. Then, for any $0 < r < R$ and any $x \in M$*

$$\mathrm{vol}(B_R(x)) \leq \frac{\mathrm{vol}_k(R)}{\mathrm{vol}_k(r)} \mathrm{vol}(B_r(x)), \tag{1.10}$$

where $\mathrm{vol}_k(t)$ denotes the volume of a ball of radius t in the complete simply connected Riemannian n–manifold of constant curvature k. In particular, for any $r > 0$ and any $x \in M$

$$\mathrm{vol}(B_r(x)) \leq \mathrm{vol}_k(r). \tag{1.11}$$

Corollary 1.12 *For any $x \in M$, $0 < r < R$,*

$$\operatorname{vol}(B_r(x)) \geq e^{-\sqrt{(n-1)|k|}R} \cdot \left(\frac{r}{R}\right)^n \operatorname{vol}(B_R(x)). \qquad (1.12)$$

Proof. This follows from (1.10), (1.11) and the standard estimate for $\operatorname{vol}_k(t)$. □

Now we proceed with the proof of proposition 1.9. Let now, according to lemma 1.10, $c \in]0, R[$, $(x_i)_{i \in I}$ such that $M = \bigcup_i B_{2c}(x_i)$, $B_c(x_i) \cap B_c(x_j) = \emptyset$ for $i \neq j$ and

$$\#\{i \in I \mid x \in B_{2c}(x_i)\} \leq N = N(n, k, R) \leq 16^n e^{8\sqrt{(n-1)|k|} \cdot R}.$$

Then, for $u \in W^{1,1}(M) \cap C^1(M)$, $dx = \operatorname{dvol}_x(g)$, $dy = \operatorname{dvol}_y(g)$,

$$\int_M |u - \overline{u}_c|\,dy \leq \sum_i \int_{B_c(x_i)} |u - \overline{u}_c|\,dy$$

$$\leq \sum_i \int_{B_c(x_i)} |u - \overline{u}_c(x_i)|\,dy$$

$$+ \sum_i \int_{B_c(x_i)} |\overline{u}_c(x_i) - \overline{u}_{2c}(x_i)|\,dy$$

$$+ \sum_i \int_{B_c(x_i)} |\overline{u}_c - \overline{u}_{2c}(x_i)|\,dy. \qquad (1.13)$$

According to (1.9)

$$\sum_i \int_{B_c(x_i)} |u - \overline{u}_c(x_i)|\,dy \leq C \cdot c \sum_i \int_{B_c(x_i)} |\nabla u|\,dy \leq N \cdot C \cdot c \int_M |\nabla u|\,dy.$$

$$(1.14)$$

Moreover, we obtain

$$\sum_i \int_{B_c(x_i)} |\overline{u}_c(x_i) - \overline{u}_{2c}(x_i)| dy$$

$$= \sum_i \text{vol}(B_c(x_i)) |\overline{u}_c(x_i) - \overline{u}_{2c}(x_i)|$$

$$\leq \sum_i \int_{B_c(x_i)} |u - \overline{u}_{2c}(x_i)| dy$$

$$\leq \sum_i \int_{B_{2c}(x_i)} |u - \overline{u}_{2c}(x_i)| dy$$

$$\leq 2N \cdot C \cdot c \int_M |\nabla u| dy \qquad (1.15)$$

and

$$\sum_i \int_{B_c(x_i)} |\overline{u}_c - \overline{u}_{2c}(x_i)| dy$$

$$\leq \sum_i \int_{B_c(x_i)} \left\{ \frac{1}{\text{vol}(B_c(x))} \int_{y \in B_c(x_i)} |u(y) - \overline{u}_{2c}(x_i)| dy \right\} dx$$

$$\leq \sum_i \int_{x \in B_c(x_i)} \left\{ \frac{1}{\text{vol}(B_c(x))} \int_{y \in B_{2c}(x_i)} |u(y) - \overline{u}_{2c}(x_i)| dy \right\} dx$$

$$= \sum_i \int_{B_{2c}(x_i)} |u(y) - u_{2c}(x_i)| dy \cdot \int_{B_c(x_i)} \frac{1}{\text{vol} B_c(x)} dx. \qquad (1.16)$$

Using (1.19), we estimate

$$\int_{B_{2c}(x_i)} |u(y) - \overline{u}_{2c}(x_i)| dy \leq 2C \cdot c \int_{B_{2c}(x_i)} |\nabla u| dy. \qquad (1.17)$$

We infer from lemma 1.11

$$\frac{1}{\operatorname{vol}B_c(x))} \le \frac{K}{\operatorname{vol}B_{2c}(x))}, \quad K = K(n,k,R) = 2^n e^{2\sqrt{(n-1)|k|} \cdot R}. \tag{1.18}$$

Moreover, for $x \in B_c(x_i)$, $B_c(x_i) \subset B_{2c}(x)$, hence

$$\int_{B_c(x_i)} \frac{1}{\operatorname{vol}B_c(x)} dx \le K. \tag{1.19}$$

(1.16), (1.17), (1.19) yield

$$\sum_i \int_{B_c(x_i)} |\bar{u}_c - \bar{u}_{2c}(x_i)| dy \le K \cdot C \cdot N \cdot c \int_M |\nabla u| dy, \tag{1.20}$$

and (1.13) – (1.15), (1.20) imply

$$\int_M |u - u_c| \, d\mathrm{vol}(g) \le 3(1+K) N \cdot C \cdot c \int_M |\nabla u| \, d\mathrm{vol}(g).$$

□

Corollary 1.13 *Let* $(E, h, \nabla) \longrightarrow (M^n, g)$ *be a Riemannian vector bundle,* (M^n, g) *with* (I), (B_0), $r > n+1$, $0 < c < r_{inj}$ *and* $\eta \in W^{1,r}(E)$. *Then* $\overline{|\eta|}_c \in W^{1,0}(M) \equiv L_1(M)$, *where*

$$\overline{|\eta|}_c(x) := \frac{1}{\operatorname{vol}B_c(x)} \int_{B_c(x)} |\eta(y)| dy.$$

Proof. Set $u(x) = |\eta(x)|$. Then $\overline{u(x)} = \overline{|\eta|}_c(x)$ and, according to Kato's inequality,

$$\int |\nabla u| dx = \int |\nabla |\eta|| dx \le \int |\nabla \eta| dx < \infty.$$

Hence we obtain from $|u| = |\eta| \in L_1$ and 1.9, $|\eta| - \overline{|\eta|}_c \in L_1$,

$$\overline{|\eta|}_c \in L_1. \tag{1.21}$$

□

In sections 3, 4 and 5, we consider not only Clifford bundles, Clifford connections but arbitrary Riemannian vector bundles. For this reason, we introduce very briefly general Sobolev spaces and the corresponding main theorems which are completely parallel to the preceding ones for Clifford bundles. Details and proofs can be found e.g. in [27], chapter I 3.

Let $(E, h, \nabla^h) \longrightarrow (M^n, g)$ be a Riemannian vector bundle. Then the Levi-Civita connection ∇^g and ∇^h define metric connections ∇ in all tensor bundles $T_v^u \otimes E$. Denote smooth sections as above by $C^\infty(T_v^u \otimes E)$, by $C_c^\infty(T_v^u \otimes E)$ those with compact support. In the sequel we shall write E instead of $T_v^u \otimes E$, keeping in mind that E can be an arbitrary vector bundle. Now we define for $p \in \mathbb{R}$, $1 \leq p < \infty$ and r a non–negative integer

$$|\varphi|_{p,r} := \left(\int \sum_{i=0}^r |\nabla^i \varphi|_x^p \, \mathrm{dvol}_x(g) \right)^{1/p},$$

$$\begin{aligned}
\Omega_r^{0,p}(E) \equiv \Omega_r^p(E) &= \{\varphi \in C^\infty(E) | |\varphi|_{p,r} < \infty\}, \\
\bar{\Omega}^{0,p,r}(E) \equiv \bar{\Omega}^{p,r}(E) &= \text{completion of } \Omega_r^p(E) \\
&\quad \text{with respect to } |\cdot|_{p,r}, \\
\mathring{\Omega}^{0,p,r}(E) \equiv \mathring{\Omega}^{p,r}(E) &= \text{completion of } C_c^\infty(E) \\
&\quad \text{with respect to } |\cdot|_{p,r} \text{ and} \\
\Omega^{0,p,r}(E) \equiv \Omega^{p,r}(E) &= \{\varphi \mid \varphi \text{ measurable distributional} \\
&\quad \text{section with } |\varphi|_{p,r} < \infty\}.
\end{aligned}$$

Here we use the standard identification of sections of a vector bundle E with E–valued zero–forms. $\Omega^{q,p,r}(E)$ stands for a Sobolev space of q–forms with values in E. For $p = 2$, we often use the notations $||_{2,0} = ||_{L_2} = ||_2$. Furthermore, we define

$$^{b,m}|\varphi| := \sum_{i=0}^{m} |\nabla^i \varphi|_x,$$

$$^{b,m}\Omega(E) = \{\varphi \mid \varphi \; C^m\text{-section and } {}^{b,m}|\varphi| < \infty\}, \text{ and}$$

$$^{b,m}\overset{\circ}{\Omega}(E) = \text{completion of } C_c^\infty(E) \text{ with respect to } {}^{b,m}|\cdot|.$$

$^{b,m}\Omega(E)$ equals the completion of

$$^{b}_{m}\Omega(E) = \{\varphi \in C^\infty(E) \mid {}^{b,m}|\varphi| < \infty\}$$

with respect to $^{b,m}|\cdot|$.

Denote by $^{b,\infty}\Omega(E)$ the locally convex space of smooth sections φ such that $\nabla^s \varphi$ is bounded for $s = 0, 1, 2, \ldots$.

Proposition 1.14 *The spaces $\overset{\circ}{\Omega}{}^{p,r}(E)$, $\bar{\Omega}^{p,r}(E)$, $\Omega^{p,r}(E)$, $^{b,m}\overset{\circ}{\Omega}(E)$, $^{b,m}\Omega(E)$ are Banach spaces and there are inclusions*

$$\overset{\circ}{\Omega}{}^{p,r}(E) \subseteq \bar{\Omega}^{p,r}(E) \subseteq \Omega^{p,r}(E),$$
$$^{b,m}\overset{\circ}{\Omega}(E) \subseteq {}^{b,m}\Omega(E).$$

If $p = 2$, then $\overset{\circ}{\Omega}{}^{2,r}(E)$, $\bar{\Omega}^{2,r}(E)$, $\Omega^{2,r}(E)$ are Hilbert spaces. □

$\overset{\circ}{\Omega}{}^{p,r}(E)$, $\bar{\Omega}^{p,r}(E)$, $\Omega^{p,r}(E)$ are different from one another in general.

Proposition 1.15 *If (M^n, g) satisfies (I) and (B_k), then*

$$\overset{\circ}{\Omega}{}^{p,r}(E) = \bar{\Omega}^{p,r}(E) = \Omega^{p,r}(E), \quad 0 \le r \le k+2.$$

□

Embedding theorems are of great importance in non–linear global analysis and even more the module structure theorem which we present now.

Theorem 1.16 Let $(E, h, \nabla^E) \longrightarrow (M^n, g)$ be a Riemannian vector bundle satisfying (I), $(B_k(M^n, g))$, $(B_k(E, \nabla))$, $k \geq 1$.
a) Assume $k \geq r$, $r - \frac{n}{p} \geq s - \frac{n}{q}$, $r \geq s$, $q \geq p$. Then

$$\Omega^{p,r}(E) \hookrightarrow \Omega^{q,s}(E) \tag{1.22}$$

continuously.
b) If $r - \frac{n}{p} > s$, then

$$\Omega^{p,r}(E) \hookrightarrow {}^{b,s}\Omega(E) \tag{1.23}$$

continuously.

Now we come to the general module structure theorem.

Theorem 1.17 Let $(E_i, h_i, \nabla_i) \longrightarrow (M^n, g)$ be vector bundles with (I), $(B_k(M^n, g))$, $(B_k(E_i, \nabla_i))$, $i = 1, 2$. Assume $0 \leq r \leq r_1, r_2 \leq k$. If $r = 0$ assume

$$\left\{ \begin{array}{l} r - \frac{n}{p} < r_1 - \frac{n}{p_1} \\ r - \frac{n}{p} < r_2 - \frac{n}{p_2} \\ r - \frac{n}{p} \leq r_1 - \frac{n}{p_1} + r_2 - \frac{n}{p_2} \\ \frac{1}{p} \leq \frac{1}{p_1} + \frac{1}{p_2} \end{array} \right\} \text{ or }$$

$$\left\{ \begin{array}{l} r - \frac{n}{p} \leq r_1 - \frac{n}{p_1} \\ 0 < r_2 - \frac{n}{p_2} \\ \frac{1}{p} \leq \frac{1}{p_1} \end{array} \right\} \text{ or } \left\{ \begin{array}{l} 0 < r_1 - \frac{n}{p_1} \\ r - \frac{n}{p} \leq r_2 - \frac{n}{p_2} \\ \frac{1}{p} \leq \frac{1}{p_2} \end{array} \right\}.$$

If $r > 0$, assume $\frac{1}{p} \leq \frac{1}{p_1} + \frac{1}{p_2}$ and

$$\left\{ \begin{array}{l} r - \frac{n}{p} < r_1 - \frac{n}{p_1} \\ r - \frac{n}{p} < r_2 - \frac{n}{p_2} \\ r - \frac{n}{p} \leq r_1 - \frac{n}{p_1} + r_2 - \frac{n}{p_2} \end{array} \right\}$$

$$\text{or } \left\{ \begin{array}{l} r - \frac{n}{p} \leq r_1 - \frac{n}{p_1} \\ r - \frac{n}{p} \leq r_2 - \frac{n}{p_2} \\ r - \frac{n}{p} < r_1 - \frac{n}{p_1} + r_2 - \frac{n}{p_2} \end{array} \right\}.$$

Then the tensor product of sections defines a continuous bilinear map

$$\Omega^{p_1,r_1}(E_1, \nabla_1) \times \Omega^{p_2,r_2}(E_2, \nabla_2) \longrightarrow \Omega^{p,r}(E_1 \otimes E_2, \nabla_1 \otimes \nabla_2). \tag{1.24}$$

$\square$

Corollary 1.18 *Assume $r = r_1 = r_2$, $p = p_1 = p_2$, $r > \frac{n}{p}$.*
(a) If $E_1 = M \times \mathbb{R}$, $E_2 = E$, then $\Omega^{p,r}(E)$ is a $\Omega^{p,r}(M \times \mathbb{R})$-module.
(b) If $E_1 = M \times \mathbb{R} = E_2$, then $\Omega^{p,r}(M \times \mathbb{R})$ is a commutative, associative Banach algebra.
(c) If $E_1 = E = E_2$, then the tensor product of sections defines a continuous map

$$\Omega^{p,r}(E) \times \Omega^{p,r}(E) \longrightarrow \Omega^{p,r}(E \otimes E).$$

$\square$

The invariance properties of Sobolev spaces will be the topic of our next considerations.

Given $(E, h, \nabla^E) \longrightarrow (M^n, g)$, for fixed $E \longrightarrow M$, $r \geq 0$, $p \geq 1$, the Sobolev space $\Omega^{p,r}(E) = \Omega^{p,r}(E, h, \nabla^E, g, \nabla^g, \mathrm{dvol}_x(g))$ depends on h, $\nabla = \nabla^E$ and g. Moreover, if we choose another sequence of differential operators with injective symbol, e. g. $D, D^2, \ldots$ in case of a Clifford bundle, we should get other Sobolev spaces. Hence two questions arise, namely

1) the dependence on the choice of h, ∇^E, g,

2) the dependence on the sequence of differential operators.

For later applications we pose a third question, that is

3) must h, ∇^E, g be smooth?

We start with the first issue and investigate the dependence upon the metric connection $\nabla = \nabla^E$ of (E, h). If $\nabla' = \nabla'^E$ is another metric connection then $\eta = \nabla' - \nabla$ is a 1–form with values in $\mathfrak{G}_E$, $\nabla' - \nabla \in \Omega^1(\mathfrak{G}) = \Omega(T^*M \otimes \mathfrak{G}_E)$. Here $\mathfrak{G}$ is the bundle of the skew–symmetric endomorphisms. $\nabla = \nabla^E$ induces a connection $\nabla = \nabla^{\mathfrak{G}_E}$ in $\mathfrak{G}_E$ and hence a Sobolev norm $|\nabla' - \nabla|_{\nabla, p, r} = |\nabla' - \nabla|_{h, \nabla, g, \nabla^g, p, r}$.

Theorem 1.19 *Assume* $(E, h, \nabla^E) \longrightarrow (M^n, g)$ *with* (I), $(B_k(M))$, $(B_k(E, \nabla^E))$, $k \geq r > \frac{n}{p} + 1$. *Let* $\nabla' = \nabla'^E$ *be a second metric connection with* $(B_k(E, \nabla'^E))$ *and suppose*

$$|\nabla' - \nabla|_{\nabla, p, r-1} < \infty. \tag{1.25}$$

Then

$$\Omega^{p,\varrho}(E, h, \nabla, g) = \Omega^{p,\varrho}(E, h, \nabla', g), \quad 0 \leq \varrho \leq r \tag{1.26}$$

as Sobolev spaces.

Proof. First we remark that our assumptions imply $(B_k(\mathfrak{G}_E, \nabla^{\mathfrak{G}_E}))$, $(B_k(\mathfrak{G}_E, \nabla'^{\mathfrak{G}_E}))$. We start with the inclusion

$$\Omega^{p,\varrho}(E, \nabla) \subset \Omega^{p,\varrho}(E, \nabla') \tag{1.27}$$

and show its continuity. We are done if we prove (1.27) for smooth elements $\varphi \in \Omega^p_\varrho(E, \nabla)$. For $\varrho = 0$, there is nothing to prove,

$$|\varphi|_{\nabla, p, 0} = |\varphi|_{\nabla', p, 0} = |\varphi|_p = |\varphi|_{L_p}.$$

Start with $\varrho = 1$, $\varphi \in \Omega^p_1(E, \nabla)$. Then

$$\nabla' \varphi = (\nabla' - \nabla)\varphi + \nabla \varphi.$$

We apply the module structure theorem to $(\nabla' - \nabla)\varphi$: $\nabla' - \nabla \in \Omega^{p, r-1}(\mathfrak{G}_E, \nabla)$, $\varphi \in \Omega^p_1(E, \nabla)$ and $(\nabla' - \nabla)\varphi$ should be in $\Omega^{0,p}(T^* \otimes E)$,

$$r - 1 - \frac{n}{p} > 0 - \frac{n}{p},$$

$$1 - \frac{n}{p} > 0 - \frac{n}{p},$$

$$r - 1 - \frac{n}{p} + 1 - \frac{n}{p} = \left(r - \frac{n}{p}\right) - \frac{n}{p} > 0 - \frac{n}{p} \text{ since } r > \frac{n}{p},$$

$$\frac{1}{p} \leq \frac{1}{p} + \frac{1}{p}.$$

Hence

$$|(\nabla' - \nabla)\varphi|_p \le C_1 \cdot |\nabla' - \nabla|_{\nabla,p,r-1} \cdot |\varphi|_{\nabla,p,1},$$
$$|\nabla'\varphi|_p \le C_1 \cdot |\nabla' - \nabla|_{\nabla,p,r-1} \cdot |\varphi|_{\nabla,p,1} + |\nabla\varphi|_p,$$
$$|\nabla'\varphi|_p \le C_2 \cdot |\varphi|_{\nabla,p,1}.$$
$$|\varphi|_{\nabla',p,1} = |\varphi|_p + |\nabla'\varphi|_p \le C_3 \cdot |\varphi|_{\nabla,p,1}, \quad (1.28)$$
$$C_3 = C_3(|\nabla' - \nabla|_{\nabla,p,r-1}),$$
$$\Omega^{p,1}(E, \nabla) \subseteq \Omega^{p,1}(E, \nabla') \text{ continuously.}$$

We conclude similarly for $\varrho = 2$,

$$\nabla'^2\varphi = (\nabla' - \nabla)(\nabla' - \nabla)\varphi + (\nabla' - \nabla)\nabla\varphi + \nabla(\nabla' - \nabla)\varphi + \nabla^2\varphi \quad (1.29)$$

and apply the module structure theorem to the single terms of (1.29).

In $(\nabla' - \nabla)(\nabla' - \nabla)$ the left hand term $\nabla' - \nabla \in \Omega(T^{*2} \otimes \mathfrak{G}_E)$ is not the same as the right hand one $\nabla' - \nabla \in \Omega(T^{*2} \otimes \mathfrak{G}_E)$, but it is canonically is an element of $\Omega^{p,r-1}(T^{*2} \otimes \mathfrak{G}_E)$ since

$$(\nabla^g \otimes \nabla' - \nabla^g \otimes \nabla)(t^* \otimes \psi) = t^* \otimes (\nabla' - \nabla)\psi = (1 \otimes (\nabla' - \nabla))(t \otimes \psi). \quad (1.30)$$

The covariant differentiation rule for the composition of bundle morphisms $\psi_2 \circ \psi_1$, $\nabla(\psi_2 \circ \psi_1) = \nabla\psi_2 \circ \psi_1 + \psi_2 \circ \nabla\psi_1$, the pointwise norm estimate $|\psi_2 \circ \psi_1| \le |\psi_2||\psi_1|$ and the module structure theorem yield $(\nabla' - \nabla)(\nabla' - \nabla) \in \Omega^{p,r-1}(T^* \otimes \mathfrak{G}_E)$. Similarly for higher–order compositions.

Considering $(\nabla' - \nabla)(\nabla' - \nabla)$ we have $r - 1 - \frac{n}{p} > 0 - \frac{n}{p}$, $2 - \frac{n}{p} > 0 - \frac{n}{p}$, $r - 1 - \frac{n}{p} + 2 - \frac{n}{p} > 0 - \frac{n}{p}$

$$|(\nabla' - \nabla)(\nabla' - \nabla)\varphi|_p \le C_4 \cdot |\nabla' - \nabla|^2_{\nabla,p,r-1} \cdot |\varphi|_{\nabla,p,2}. \quad (1.31)$$

For $(\nabla'-\nabla)\nabla\varphi$: $r-1-\frac{n}{p} > 0-\frac{n}{p}$, $1-\frac{n}{p} > 0-\frac{n}{p}$, $r-1-\frac{n}{p}+1-\frac{n}{p} > 0-\frac{n}{p}$

$$\begin{aligned}|(\nabla' - \nabla)(\nabla' - \nabla)\nabla\varphi|_p &\le C_5 \cdot |\nabla' - \nabla|_{\nabla,p,r-1} \cdot |\nabla\varphi|_{\nabla,p,1} \\ &\le C_5 \cdot |\nabla' - \nabla|_{\nabla,p,r-1} \cdot |\varphi|_{\nabla,p,2}.\end{aligned} \quad (1.32)$$

For $(\nabla(\nabla' - \nabla))\varphi$: $r - 2 - \frac{n}{p} > 0 - \frac{n}{p}$, $2 - \frac{n}{p} > 0 - \frac{n}{p}$, $r - 2 - \frac{n}{p} + 2 - \frac{n}{p} > 0 - \frac{n}{p}$

$$\begin{aligned}|(\nabla(\nabla' - \nabla))\varphi|_p &\leq C_6 \cdot |\nabla(\nabla' - \nabla)|_{\nabla,p,r-2} \cdot |\varphi|_{\nabla,p,2} \\ &\leq C_6 \cdot |\nabla' - \nabla|_{\nabla,p,r-1} \cdot |\varphi|_{\nabla,p,2}. \quad (1.33)\end{aligned}$$

So $(\nabla' - \nabla)\nabla\varphi$ is done.

We obtain from (1.29)–(1.33)

$$|\nabla'^2 \varphi|_p \leq C_7 \cdot |\varphi|_{\nabla,p,2},$$
$$|\varphi|_{\nabla',p,2} \leq C_8 \cdot |\varphi|_{\nabla,p,2}, \quad C_8 = C_8(|\nabla' - \nabla|_{\nabla,p,r-1}).$$

Now we turn to the general case. Assume the result holds for $\nabla'\varphi, \ldots, \nabla'^{\varrho-1}\varphi$, i. e.

$$|\nabla'^j \varphi|_p \leq C_j \cdot |\varphi|_{\nabla,p,j}, \quad 0 \leq j \leq \varrho - 1, \quad C_j = C_j(|\nabla' - \nabla|_{\nabla,p,r-1}).$$

Then

$$\nabla'^\varrho \varphi = \sum_{i=1}^{\varrho} \nabla^{i-1}(\nabla' - \nabla)\nabla'^{\varrho-i}\varphi + \nabla^\varrho \varphi. \quad (1.34)$$

By assumption, $|\nabla^\varrho \varphi|_p < \infty$. There remains to consider the terms

$$\nabla^{i-1}(\nabla' - \nabla)\nabla'^{\varrho-i}\varphi.$$

Iterating the procedure, i. e. applying it to $\nabla'^{\varrho-i}$ and so on, we have to estimate expressions of the kind

$$\nabla^{i_1}(\nabla' - \nabla)^{i_2} \cdots \nabla^{i_{\varrho-2}}(\nabla' - \nabla)^{i_{\varrho-1}} \nabla^{i_\varrho}\varphi$$

with $i_1 + \cdots + i_\varrho = \varrho$, $i_\varrho < \varrho$. Applying the corresponding version of the Leibniz rule, we finally have to estimate

$$(\nabla^{n_1}(\nabla' - \nabla))(\nabla^{n_2}(\nabla' - \nabla)) \cdots (\nabla^{n_{s-1}}(\nabla' - \nabla))\nabla^{n_s}\varphi$$

such that $n_1 + 1 + n_2 + 2 + \cdots + n_{s-1} + 1 + n_s = \varrho < r$, $n_s < \varrho$, $n_1, \ldots, n_{s-1} > 0$. The module structure theorem can now be applied in virtue of the sequence of inequalities

$$r-1-n_1-\frac{n}{p} > r-1-(n_1+n_2)-\frac{n}{p},$$

$$r-1-n_2-\frac{n}{p} > r-1-(n_1+n_2)-\frac{n}{p},$$

$$r-1-n_2-\frac{n}{p}+r-1-n_2-\frac{n}{p}$$

$$=r-1-\frac{n}{p}+r-1-(n_1+n_2)-\frac{n}{p}$$

$$> r-1-(n_1+n_2)-\frac{n}{p};$$

$$r-1-(n_1+n_2)-\frac{n}{p} > r-1-(n_1+n_2+n_3)-\frac{n}{p},$$

$$r-1-n_3-\frac{n}{p} > r-1-(n_1+n_2+n_3)-\frac{n}{p},$$

$$r-1-(n_1+n_2)-\frac{n}{p}+r-1-n_3-\frac{n}{p}$$

$$=r-1-\frac{n}{p}+r-1-(n_1+n_2+n_3)-\frac{n}{p}$$

$$> r-1-(n_1+n_2+n_3)-\frac{n}{p};$$

$$\vdots$$

$$r-1-(n_1+\cdots+n_{s-2})-\frac{n}{p} > r-1-(n_1+\cdots+n_{s-1})-\frac{n}{p},$$

$$r-1-n_{s-1}-\frac{n}{p} > r-1-(n_1+\cdots+n_{s-1})-\frac{n}{p},$$

$$r-1-(n_1+\cdots+n_{s-2})-\frac{n}{p}+r-1-n_{s-1}-\frac{n}{p}$$

$$=r-1-\frac{n}{p}+r-1-(n_1+\cdots+n_{s-1})-\frac{n}{p}$$

$$> r-1-(n_1+\cdots+n_{s-1})-\frac{n}{p};$$

$$r-1-(n_1+\cdots+n_{s-1})-\frac{n}{p} > 0-\frac{n}{p},$$

$$\varrho-n_s-\frac{n}{p} > 0-\frac{n}{p},$$

$$r - 1 - (n_1 + \cdots + n_{s-1}) - \frac{n}{p} + \varrho - n_s - \frac{n}{p}$$
$$= r - 1 - \frac{n}{p} + \varrho - (n_1 + \cdots + n_s) - \frac{n}{p}$$
$$> 0 - \frac{n}{p}.$$

This yields

$$|(\nabla^{n_1}(\nabla' - \nabla)) \cdots (\nabla^{n_{s-1}}(\nabla' - \nabla))(\nabla^{n_s}\varphi)|_p$$
$$\leq C_9 \cdot |\nabla' - \nabla|_{\nabla,p,r-1}^{(s-1)(r-1)-(n_1+\cdots+n_{s-1})} \cdot |\nabla^{n_s}\varphi|_{\nabla,p,\varrho-n_s}$$
$$\leq C_9 \cdot |\nabla' - \nabla|_{\nabla,p,r-1}^{(s-1)(r-1)-(n_1+\cdots+n_{s-1})} \cdot |\varphi|_{\nabla,p,\varrho},$$
$$|\nabla'^{\varrho}\varphi|_p \leq C_{10} \cdot |\varphi|_{\nabla,p,\varrho}$$
$$|\varphi|_{\nabla',p,\varrho} \leq C_{11}|\varphi|_{\nabla,p,\varrho}, \quad C_{11} = C_{11}(|\nabla' - \nabla|_{\nabla,p,r-1}),$$
$$\Omega^{p,\varrho}(E, \nabla) \subseteq \Omega^{p,\varrho}(E, \nabla') \text{ continuously}, 0 \leq \varrho \leq r.$$

The other continuous inclusion would follow from

Lemma 1.20 *Under the hypotheses of 1.19,*

$$|\nabla' - \nabla|_{\nabla,p,r-1} < \infty \text{ implies } |\nabla' - \nabla|_{\nabla',p,r-1} < \infty.$$

Proof. This is only a special case of what we just have proven. Set $\eta = \nabla' - \nabla$, assume $|\nabla' - \nabla|_{\nabla,p,r-1} = |\eta|_{\nabla,p,r-1} < \infty$, then $|\nabla' - \nabla|_{\nabla',p,r-1} = |\eta|_{\nabla',p,r-1} \leq P(|\nabla' - \nabla|_{\nabla,p,r-1}) \cdot |\eta|_{\nabla,p,r-1} < \infty$, where P is a polynomial in the indicated variable. □

Now let $|\nabla' - \nabla|_{\nabla',p,r-1} < \infty$ and repeat the first part of the proof exchanging ∇ and ∇' in all formulas and arguments, to obtain

$$\Omega^{p,\varrho}(E, \nabla') \subseteq \Omega^{p,\varrho}(E, \nabla) \text{ continuously}, 0 \leq \varrho \leq r.$$

This finishes the proof of 1.19. □

Corollary 1.21 *The conditions* $|\nabla' - \nabla|_{\nabla,p,r-1} < \infty$ *and* $|\nabla' - \nabla|_{\nabla',p,r-1} < \infty$ *are equivalent.*

□

The next step is to understand changes of the metric g.

Theorem 1.22 *Assume* $(E, h, \nabla) \longrightarrow (M^n, g)$ *with* (I), $(B_k(M^n, g))$, $(B_k(E, \nabla))$, $k \geq r > \frac{n}{p} + 1$, *and* g' *with* $C \cdot g \leq g' \leq D \cdot g$, $(B_k(M^n, g))$ *and* $|\nabla^{g'} - \nabla^g|_{g,p,r-1} = (\int |g - g'|_{g,x}^p \, \mathrm{dvol}_x(g))^{\frac{1}{p}} + \sum_{i=0}^{r-1} |(\nabla^{g'})^i (\nabla^{g'} - \nabla^g)|_p < \infty$. *Then*

$$\Omega^{p,\varrho}(E, h, \nabla, g) = \Omega^{p,\varrho}(E, h, \nabla, g'), \quad 0 \leq \varrho \leq r,$$

as euivalent Sobolev spaces.

Proof. The assumption $C \cdot g \leq g' \leq D \cdot g$ implies $C_1 \, \mathrm{dvol}_x(g) \leq \mathrm{dvol}_x(g') \leq D_1 \, \mathrm{dvol}_x(g)$ and $C_2 |t|_{g,x} \leq |t|_{g',x} \leq D_2 |t|_{g,x}$ for all tensors t of degree $\leq r$. Hence

$$\Omega^{p,\varrho}(E, h, \nabla, g, \nabla^g, \mathrm{dvol}_x(g)) = \Omega^{p,\varrho}(E, h, \nabla, g', \nabla^{g'}, \mathrm{dvol}_x(g')),$$

$0 \leq \varrho \leq r$, and we must check only the result when replacing ∇^g by $\nabla^{g'}$. $\nabla^g, \nabla^{g'}$ come first into the calculations with the second vector bundle derivatives:

$$(\nabla^{g'} \otimes \nabla)\nabla\varphi = ((\nabla^{g'} - \nabla^g) \otimes 1)\nabla\varphi + (\nabla^g \otimes \nabla)\nabla\varphi.$$

Now use the estimates of the proof of 1.19, but applied to $\nabla^{g'} - \nabla^g$. $\nabla^{g'}$ is no longer a metric connection with respect to g but for $\nabla^g - \nabla^{g'}$ as an element of $\Omega^1(\mathrm{End}\, T)$ all estimates remain valid. □

Since the estimate of $(\nabla^{g'} \otimes \nabla)^i$ involves the j–th derivatives of $\nabla^{g'} - \nabla^g$, $0 \leq j \leq i-1$, only $|\nabla^{g'} - \nabla^g|_{\nabla^g, p, r-2}$ seemingly enters the estimates. But we need in fact $|\nabla^{g'} - \nabla^g|_{\nabla^g, p, r-1}$ to apply the module structure theorem. One sees this e. g. that $r > \frac{n}{p} + 1$,

$$r - 2 - n_1 - \frac{n}{p} > r - 2 - (n_1 + n_2) - \frac{n}{p},$$
$$r - 2 - n_2 - \frac{n}{p} > r - 2 - (n_1 + n_2) - \frac{n}{p}$$

do not imply

$$r - 2 - \frac{n}{p} + r - 2 - (n_1 + n_2) - \frac{n}{p} > r - 2 - (n_1 + n_2) - \frac{n}{p}$$

and so on, since not necessary $r - 2 - \frac{n}{p} > 0$. Sufficient for this and the whole module structure procedure would be $r > \frac{n}{p} + 2$. Thus

Theorem 1.23 *Assume* $(E, h, \nabla) \longrightarrow (M^n, g)$ *with* (I), $(B_k(M^n, g))$, $(B_k(E, \nabla))$, $k \geq r > \frac{n}{p} + 2$ *and* g' *with* $C \cdot g \leq g' \leq D \cdot g$, $(B_k(M^n, g'))$, (I) *and* $|\nabla^{g'} - \nabla^g|_{g, p, r-1} < \infty$. *Then*

$$\Omega^{p, \varrho}(E, h, \nabla, g) = \Omega^{p, \varrho}(E, h, \nabla, g), \quad 0 \leq \varrho \leq r,$$

as equivalent Sobolev spaces. □

Finally we study what happens by replacing $(h, \nabla^h) \longrightarrow (h', \nabla^{h'})$.

Theorem 1.24 *Let* $(E, h, \nabla^h) \longrightarrow (M^n, g)$ *be a Riemannian vector bundle with* (I), $(B_k(M^n, g))$, $(B_k(E, \nabla^h))$, $k \geq r > \frac{n}{p} + 1$, h' *a second fibre metric with metric connection* $\nabla^{h'}$ *and with* $(B_k(E, \nabla^{h'}))$, $C \cdot h \leq h' \leq D \cdot h$ *and* $|\nabla^{h'} - \nabla^h|_{h, \nabla^h, g, p, r-1} < \infty$. *Then*

$$\Omega^{p, \varrho}(E, h, \nabla^h, g) = \Omega^{p, \varrho}(E, h', \nabla^{h'}, g), \quad 0 \leq \varrho \leq r,$$

as equivalent Sobolev spaces.

Proof. $\nabla^{h'}$ is not necessarily metric with respect to h but the Sobolev space $\Omega^{p, \varrho}(E, h, \nabla', g)$ is nevertheless well defined. Then $C \cdot h \leq h' \leq D \cdot h$ implies

$$\Omega^{p, \varrho}(E, h, \nabla^h, g) = \Omega^{p, \varrho}(E, h', \nabla^h, g)$$
$$\text{and } \Omega^{p, \varrho}(E, h', \nabla^{h'}, g) = \Omega^{p, \varrho}(E, h, \nabla^{h'}, g).$$

$\nabla^{h'} - \nabla^h$ is no longer a section of $T^* \otimes \mathfrak{G}_E$ but still a section of $T^* \otimes \operatorname{End} E$, by our assumption $\nabla^{h'} - \nabla^h \in$

$\Omega^{p,r-1}(\operatorname{End} E, h, \nabla^h, g)$. We decompose $\nabla^{h'}\varphi = (\nabla^{h'} - \nabla^h)\varphi + \nabla^h\varphi$ etc. and conclude as in the proof of 1.19

$$\begin{aligned}\Omega^{p,\varrho}(E, h, \nabla^h, g) &\subseteq \Omega^{p,\varrho}(E, h, \nabla^{h'}, g) \\ &= \Omega^{p,\varrho}(E, h', \nabla^{h'}, g), \quad 0 \le \varrho \le r,\end{aligned}$$

continuously. The same inclusion then holds for the Sobolev spaces of E^* and $\operatorname{End} E = E^* \otimes E$, all endowed with the induced metrics and connections. In particular

$$\begin{aligned}\Omega^{p,r-1}(\operatorname{End} E, h, \nabla^h, g) &\subseteq \Omega^{p,r-1}(\operatorname{End} E, h, \nabla^{h'}) \\ &= \Omega^{p,r-1}(\operatorname{End} E, h', \nabla^{h'}, g),\end{aligned}$$

which implies $|\nabla^{h'} - \nabla^h|_{h', \nabla^{h'}, p, r-1} < \infty$. From this we get

$$\Omega^{p,\varrho}(E, h', \nabla^{h'}, g) \subseteq \Omega^{p,\varrho}(E, h, \nabla^h, g), \quad 0 \le \varrho \le r,$$

continuously. $\square$

Combining theorems 1.19 – 1.24, we obtain as main result:

Theorem 1.25 *Let* $(E, h, \nabla) \longrightarrow (M^n, g)$ *be a Riemannian vector bundle with* (I), $(B_k(M^n, g))$, $(B_k(E, \nabla))$, $k \ge r > \frac{n}{p} + 1$. *Suppose* h' *is a fibre metric on* E *with metric connection* ∇' *and* g' *a metric on* M^n *with* (I), $(B_k(M^n, g'))$, $(B_k(E, \nabla'))$ *satisfying* $C \cdot h \le h' \le D \cdot h$, $C_1 \cdot g \le g' \le C_2 \cdot g$, $|\nabla' - \nabla|_{h, \nabla, g, p, r-1} < \infty$, $|\nabla^{g'} - \nabla^g|_{g, p, r-1} < \infty$. *Then*

$$\Omega^{p,\varrho}(E, h, \nabla, g) = \Omega^{p,\varrho}(E, h', \nabla', g'), \quad 0 \le \varrho \le r,$$

as equivalent Sobolev spaces. $\square$

We are left with the dependence on the sequence of differential operators.

The most important and frequently used cases are sequences based on ∇^i, $(\nabla^* \nabla)^i$, Δ^i, D^i respectively. We will present far reaching results. First we compare the Sobolev spaces based on ∇ (up to now the main point of interest) and on the Bochner

Laplacian $\Delta = \Delta_B = \nabla^*\nabla$. Concerning this issue, there is a remarkable contribution from Gorm Salomonsen [66] which we fit into our discussions.

Let again $(E, h, \nabla) \longrightarrow (M^n, g)$ be a Riemannian or Hermitean vector bundle, and ∇^* the formal adjoint to ∇ with respect to the canonical Hilbert scalar product, $\langle \nabla^*\varphi, \psi \rangle = \langle \varphi, \nabla\psi \rangle$, $\varphi \in C_c^\infty(T^*M \otimes E)$, $\psi \in C_c^\infty(E)$, $\langle \nabla^*\varphi, \psi \rangle = \int h(\nabla^*\varphi, \psi)_x \, \mathrm{dvol}_x(g)$. Locally this can be written as $\nabla^* = -g^{ij}\nabla_i$. We do not assume (I) or completeness for (M^n, g). Until otherwise said, we denote by $\Delta = \nabla^*\nabla$ the Bochner Laplacian, i.e. the Friedrichs' extension of Δ.

For our purpose we must indicate the differential operators defining the Sobolev spaces. Therefore we write $\Omega^{p,r}(E, h, \nabla, g) = \Omega^{p,r}(E, \nabla)$ or $\Omega^{p,2s}(E, \Delta)$, respectively. We restrict to the case $p = 2$ because here and there we apply Hilbert space methods. Since we do not assume (I), (B_k) for (M^n, g), the Sobolev spaces $\overset{\circ}{\Omega}{}^{2,r} \subseteq \bar{\Omega}^{2,r} \subseteq \Omega^{2,r}$ do not automatically coincide. Explicitly

$$\Omega^{2,r}(E, \nabla) = \{\varphi | \varphi \text{ distributional section of } E$$
$$\text{s. t. } \nabla^i \varphi \in L_2(T^{*i} \otimes E), 0 \leq i \leq r\},$$

$$|\varphi|_{\nabla,2,r} = \left(\sum_{i=0}^r |\nabla^i\varphi|^2_{L_2}\right)^{\frac{1}{2}} \sim \sum_{i=0}^r |\nabla^i\varphi|_{L_2},$$

$$\Omega^{2,2s}(E, \Delta) = \{\varphi | \varphi \text{ distributional section of } E,$$
$$\Delta^i \varphi \in L_2(E), 0 \leq i \leq s\},$$

$$|\varphi|_{\Delta,2,2s} = \left(\sum_{i=0}^s |\Delta^i\varphi|^2_{L_2}\right)^{\frac{1}{2}} \sim \sum_{i=0}^s |\Delta^i\varphi|_{L_2},$$

$$\Omega^2_{2s}(E, \Delta) = \{\varphi \in C^\infty(E) | \; |\varphi|_{\Delta,2,2s} < \infty\},$$
$$\bar{\Omega}^{2,2s}(E, \Delta) = \bar{\Omega}^2_{2s}(E, \Delta),$$
$$\overset{\circ}{\Omega}{}^{2,2s}(E, \Delta) = \overline{C_c^\infty(E)}^{|\,|_{\Delta,2,2s}}.$$

We consider Δ as a self-adjoint operator on the domain given by Friedrichs' extension.

Theorem 1.26 *Assume for* $(E, h, \nabla) \longrightarrow (M^n, g)$ $(B_\infty(M,g))$ *and* $(B_\infty(E, \nabla^E))$. *Then*

$$\overset{\circ}{\Omega}{}^{2,2k}(E, \nabla) = \overset{\circ}{\Omega}{}^{2,2k}(E, \Delta), \quad k = 0, 1, 2, \ldots$$

as equivalent Sobolev spaces.

We refer to [66] for the proof. $\square$

Since $\Delta = \Delta_F$ is self–adjoint, $\Delta^{\frac{j}{2}}$ and

$$|\varphi|_{\Delta,2,k} = \sum_{j=0}^{k} |\Delta^{\frac{j}{2}} \varphi|_{L_2}, \quad \varphi \in C_c^\infty(E)$$

and

$$\overset{\circ}{\Omega}{}^{2,k}(E, \Delta) = \overline{C_c^\infty(E)}^{||\,|_{\Delta,2,k}}$$

are well defined.

One can now easily prove an analogous result to 1.26.

Theorem 1.27 *Assume* $(E, h, \nabla) \longrightarrow (M^n, g)$ *with* $(B_\infty(M, g))$ *and* $(B_\infty(E, \nabla))$. *Then*

$$\Omega^{2,k}(E, \Delta) = \Omega^{2,k}(E, \nabla)$$

as equivalence of Sobolev spaces, also for k odd. $\square$

By means of spectral calculus we can define $\overset{\circ}{\Omega}{}^{2,s}(E, \Delta)$ as $\overline{C_c^\infty(E)}^{||\,|_{2,(1+\Delta)^{\frac{s}{2}}}}$ even for $s \in \mathbb{R}_+$, where $|\varphi|^2_{(1+\Delta)^{\frac{s}{2}}} = \langle (1+\Delta)^{\frac{s}{2}} \varphi, (1+\Delta)^{\frac{s}{2}} \varphi \rangle = \langle (1+\Delta)^s \varphi, \varphi \rangle$. $\Omega^{2,-s}(E, \Delta)$ will be defined as the dual of $\overset{\circ}{\Omega}{}^{2,s}(E, \Delta)$. Refering to [66], we state without proof

Theorem 1.28 *With the assumptions of 1.27 and* $s \in \mathbb{R}$, ∇ *maps* $\Omega^{2,s}(E, \Delta)$ *into* $\Omega^{2,s-1}(T^* \otimes E, \Delta)$. *If* $\Psi \in {}^{b,\infty}\Omega(\mathrm{Hom}\,(E, F))$, $(E, h_F, \nabla^F) \longrightarrow (M^n, g)$, *then* Ψ *maps* $\Omega^{2,s}(E)$ *into* $\Omega^{2,s}(F)$.

$\square$

Other canonical differentiable operators in geometric analysis used for the definition of Sobolev spaces are the generalized Dirac operators D and the full Laplacian Δ. As (the graded) Δ is a special case of D^2, we concentrate on D. Let $(E, h, \nabla, \cdot) \longrightarrow (M^n, g)$ be a Clifford bundle, D its generalized Dirac operator. Then we define as above

$$\Omega^{2,r}(E, D) := \{\varphi \,|\, \varphi \text{ a distributional section} \\ \text{and } D^i \varphi \in L_2(D),\ 0 \leq i \leq r\},$$

$$|\varphi|_{D,2,r} = \sqrt{\sum_{i=0}^{r} |D^i \varphi|_{L_2}^2} \sim \sum_{i=0}^{r} |D^i \varphi|_{L_2},$$

$$\Omega_r^2(E, D) := \{\varphi \in C^\infty(E) \,|\, |\varphi|_{D,2,r} < \infty\} \equiv H_r(E),$$
$$\overline{\Omega}^{2,r}(E, D) := \overline{\Omega_r^2(E, D)}^{|\,|_{D,2,r}} \equiv H^{2,r}(E),$$
$$\overset{\circ}{\Omega}{}^{2,r}(E, D) := \overline{C_c^\infty(E)}^{|\,|_{D,2,r}} \equiv \overset{\circ}{H}{}^{2,r}(E).$$

There is a sequence of closed subspaces

$$\overset{\circ}{\Omega}{}^{2,r}(E, D) \subseteq \overline{\Omega}^{2,r}(E, D) \subseteq \Omega^{2,r}(E, D).$$

For $p = 2$, lemma 1.2 can be sharpened as

Lemma 1.29 *Let $(E, h, \nabla, \cdot) \longrightarrow (M^n, g)$ be a Clifford bundle. There are continuous inclusions*

$$\Omega^{2,r}(E, \nabla) \hookrightarrow \Omega^{2,r}(E, D), \tag{1.35}$$

$$\overset{\circ}{\Omega}{}^{2,r}(E, \nabla) \hookrightarrow \overset{\circ}{\Omega}{}^{2,r}(E, D), \quad r = 0, 1, 2, \ldots. \tag{1.36}$$

If (M^n, g) is complete then

$$\overset{\circ}{\Omega}{}^{2,r}(E, D) = \Omega^{2,r}(E, D).$$

We refer to [66] for the proof. □

Theorem 1.4 can be modified and sharpened as follows.

Theorem 1.30 *Assume* $(E, h, \nabla, \cdot) \longrightarrow (M^n, g)$ *with* $(B_\infty(M^n, g))$ *and* $(B_\infty(E, \nabla))$. *Then*

$$\mathring{\Omega}^{2,r}(E, \nabla) = \mathring{\Omega}^{2,r}(E, D), \quad r = 0, 1, 2, \ldots$$

as equivalence of Sobolev spaces.

We refer to [66] for the proof. □

The (graded) Laplace operator $\Delta = (\Delta_0, \ldots, \Delta_n)$ of (M^n, g) (with Weitzenboeck terms) is a special case of D^2 a generalized Dirac operator D. This yields

Theorem 1.31 *Let* (M^n, g) *be an open Riemannian manifold satisfying* $(B_\infty(M^n, g))$. *Then*

$$\mathring{\Omega}^{q,2,2s}(M, \nabla) = \mathring{\Omega}^{q,2,2s}(M, \Delta), \quad 0 \leq q \leq n, \quad s = 0, 1, 2, \ldots.$$

as equivalence of Sobolev spaces. □

Here the Ω's are Sobolev spaces of forms.

Theorem 1.32 *Let* $(E, h, \nabla, \cdot) \longrightarrow (M^n, g)$ *be a Clifford bundle satisfying* $(B_\infty(M^n, g))$ *and* $(B_\infty(E, \nabla))$. *If* (M^n, g) *is complete then*

$$\Omega^{2,r}(E, \nabla) = \Omega^{2,r}(E, D), \quad r = 0, 1, 2, \ldots$$

as equivalent Sobolev spaces.

Proof. We know from 1.29 that $\Omega^{2,r}(E, \nabla) \subseteq \Omega^{2,r}(E, D)$, $\mathring{\Omega}^{2,r}(E, \nabla) \subseteq \mathring{\Omega}^{2,r}(E, D)$. For complete (M^n, g) we have $\mathring{\Omega}^{2,r}(E, D) = \Omega^{2,r}(E, D)$, hence $\mathring{\Omega}^{2,r}(E, \nabla) = \mathring{\Omega}^{2,r}(E, D) = \Omega^{2,r}(E, D) \supseteq \Omega^{2,r}(E, \nabla) \supseteq \mathring{\Omega}^{2,r}(E, \nabla)$, so all these spaces must coincide. □

Corollary 1.33 Let (M^n, g) be complete and satisfy $(B_\infty(M^n, g))$. Then

$$\Omega^{q,2,2s}(M, \nabla) = \Omega^{q,2,2s}(M, \Delta), \quad q = 0, \ldots, n, \quad s = 0, 1, \ldots$$

as equivalent Sobolev spaces. □

1.33 has still a slight generalization. If $(E, h, \nabla, \cdot) \longrightarrow (M^n, g)$ is a Clifford bundle and $(F, h_F, \nabla^F) \longrightarrow (M^n, g)$ is another (Riemannian or Hermitean) bundle, then $E \otimes F$ has a canonical Clifford bundle structure. Applying 1.33 to this bundle with $E =$ Clifford bundle of graded forms, we obtain

Corollary 1.34 Let (M^n, g) be complete, $(F, h_F, \nabla^F) \longrightarrow (M^n, g)$ a Riemannian or Hermitean bundle, both satisfying (B_∞). Then

$$\Omega^{q,2,2s}(F, \nabla) = \Omega^{q,2,2s}(F, \Delta), \quad q = 0, \ldots, n, \quad s = 0, 1, \ldots$$

as equivalent Sobolev spaces. □

We finish at this point our short review of Sobolev spaces since we established what is needed in the sequel.

2 Uniform structures of metric spaces

As we already indicated in the preface to this chapter, the key of our whole approach to define relative number valued invariants are Sobolev uniform structures. They allow to introduce natural intrinsic topologies in the considered set of geometric/analytic objects and to define admitted perturbations which are the elements of a generalized component.

We start with a brief outline of the fundamental notions connected with uniform structures. Thereafter we present some basic examples as uniform structures of metric spaces, of sections of vector bundles, of conformal factors and of Riemannian metrics.

Let X be a set. A filter F on X is a system of subsets which satisfies

(F_1) $\quad M \in F, M_1 \supseteq M$ implies $M_1 \in F$.
(F_2) $\quad M_1, \ldots, M_n \in F$ implies $M_1 \cap \cdots \cap M_n \in F$.
(F_3) $\quad \emptyset \notin F$.

A system $\mathfrak{U}$ of subsets of $X \times X$ is called a uniform structure on X if it satisfies (F_1), (F_2) and

(U_1) $\quad$ Every $U \in \mathfrak{U}$ contains the diagonal $\Delta \subset X \times X$.
(U_2) $\quad V \in \mathfrak{U}$ implies $V^{-1} \in \mathfrak{U}$.
(U_3) $\quad$ If $V \in \mathfrak{U}$ then there exists $W \in \mathfrak{U}$ s.t. $W \circ W \subset V$.

The sets of $\mathfrak{U}$ are called neighbourhoods of the uniform structure and $(X, \mathfrak{U})$ is called the uniform space.

$\mathfrak{B} \subset \mathfrak{P}(X \times X)$ ($=$ sets of all subsets of $X \times X$) is a basis for a uniquely determined uniform structure if and only if it satisfies the following conditions:

(B_1) $\quad$ If $V_1, V_2 \in \mathfrak{B}$ then $V_1 \cap V_2$ contains an element of $\mathfrak{B}$.
(U_1') $\quad$ Each $V \in \mathfrak{B}$ contains the diagonal $\Delta \subset X \times X$.
(U_2') $\quad$ For each $V \in \mathfrak{B}$ there exists $V' \in \mathfrak{B}$ s.t. $V' \subseteq V^{-1}$.
(U_3') $\quad$ For each $V \in \mathfrak{B}$ there exists $W \in \mathfrak{B}$ s.t. $W \circ W \subset V$.

Every uniform structure $\mathfrak{U}$ induces a topology on X. Let $(X, \mathfrak{U})$ be a uniform space. Then for every $x \in X$, $\mathfrak{U}(x) = \{V(x)\}_{V \in \mathfrak{U}}$ is the neighbourhood filter for a uniquely determined topology on X. This topology is called the uniform topology generated by the uniform structure $\mathfrak{U}$. We refer to [68] for the proofs and further informations on uniform structures. We ask under which conditions $\mathfrak{U}$ is metrizable. A uniform space $(X, \mathfrak{U})$ is called Hausdorff if $\mathfrak{U}$ satisfies the condition

$(U_1 H)$ $\quad$ The intersection of all sets $\in \mathfrak{U}$
$\qquad\qquad$ is the diagonal $\Delta \subset X \times X$.

Then the uniform space $(X, \mathfrak{U})$ is Hausdorff if and only if the corresponding topology on X is Hausdorff. The following criterion answers the question above.

Proposition 2.1 *A uniform space $(X, \mathfrak{U})$ is metrizable if and only if $(X, \mathfrak{U})$ is Hausdorff and $\mathfrak{U}$ has a countable basis $\mathfrak{B}$.* □

Next we have to consider completions. Let $(X, \mathfrak{U})$ be a uniform space, V a neighbourhood. A subset $A \subset V$ is called small of order V if $A \times A \subset V$. A system $\mathfrak{G} \subset \mathfrak{P}(X)$ has arbitrary small sets if for every $V \in \mathfrak{U}$ there exists $M \in \mathfrak{G}$ such that M is small of order V, i.e. $M \times M \subset V$. A filter on X is called a Cauchy filter if it has arbitrary small sets. A sequence $(x_\nu)_\nu$ is called a Cauchy sequence if the associated elementary filter $(= \{x_\nu | \nu \geq \nu_0\}_{\nu_0})$ is a Cauchy filter. Every convergent filter on X is a Cauchy filter. A uniform space is called complete if every Cauchy filter converges, i.e. is finer than the neighbourhood filter of a point.

Proposition 2.2 *Let $(X, \mathfrak{U})$ be a uniform space. Then there exists a complete uniform space $(\overline{X}^\mathfrak{U}, \overline{\mathfrak{U}})$ such that X is isomorphic to a dense subset of $\overline{X}$. If $(X, \mathfrak{U})$ is also Hausdorff then there exists a complete Hausdorff uniform space $(\overline{X}^\mathfrak{U}, \overline{\mathfrak{U}})$, uniquely determined up to an isomorphism, such that X is isomorphic to a dense subset of $\overline{X}$. $(\overline{X}^\mathfrak{U}, \overline{\mathfrak{U}})$ is called the completion of $(X, \mathfrak{U})$.*

For the proof we refer to [68], p. 126/127. □

We define an ε_0–locally metrized set. Let X be a set, $\varepsilon_0 > 0$. X is called ε_0–locally metrized if for $x \in X$ there exists $d_x(\cdot, \cdot)$, $X \times X \supset D_{d_x} \longrightarrow [0, \varepsilon_0[$ such that $(U_\varepsilon(x) = \{y | d_x(x,y) < \varepsilon\}, d_x)$ is a metric space, $0 < \varepsilon < \varepsilon_0$ and

For given $0 < \varepsilon < \varepsilon_1$ there exists $\varepsilon' = \varepsilon'(\varepsilon) > 0$
s.t. $d_y(y, x) < \varepsilon'$ implies $y \in U_\varepsilon(x)$, (2.1)
for given $0 < \varepsilon < \varepsilon_1$ there exists $\delta = \delta(\varepsilon) > 0$
s.t. $y \in U_{\delta(x)}, z \in U_{\delta(y)}$ implies $z \in U_\varepsilon(x)$,
i. e. $d_x(x, y) < \delta, d_y(y, z) < \delta$ implies $d_x(x, z) < \varepsilon$. (2.2)

We admit the case $\varepsilon_0 = \infty$.

Proposition 2.3 *Let X be ε_0-locally metrized, $0 < \delta < \varepsilon_0$ and set*
$$V_\delta = \{(x,y) \in X^2 | d_x(x,y) < \delta\}.$$
Then $\mathfrak{B} = \{V_\delta\}_{0<\delta<\varepsilon_0}$ is a basis for a metrizable uniform structure $\mathfrak{U}(X)$.

Proof. (2.1) assures (U_2'), (1.2) assures (U_3'). The other conditions are trivially satisfied. □

For the metric of $(X, \mathfrak{U})$ there is a standard procedure. First one defines an almost metric (distance can be $= \infty$) d^a by

$$d^a(x,y) = \inf \sum_{i=1}^{p} d_{z_{i-1}}(z_{i-1}, z_i) \qquad (2.3)$$

where the infimum is taken over all finite sequences $z_0, z_1, \ldots, z_p$ s.t. $z_0 = x$, $z_p = y$ and $p = 1, 2, 3, \ldots$.

Proposition 2.4 *If in (2.3) $d_x(x,y)$ is defined then*

$$\frac{1}{2}d_x(x,y) \le d^a(x,y) \le d_x(x,y). \qquad (2.4)$$

Proof. This is just (2.6) in [68], p. 117. □

d^a can be transformed into a metric d, e.g. by $d = \frac{d^a}{1+d^a}$. Let $(Y, \mathfrak{U}_Y)$ be a Hausdorff uniform space, $X \subset Y$ a dense subspace. If X is metrizable by a metric ϱ then ϱ may be extended to a metric ϱ on Y which metrizes the uniform space $(Y, \mathfrak{U}_Y)$. In conclusion, if $(X, \mathfrak{U})$ is a metrizable uniform space and $(\overline{X}^\varrho, \overline{\mathfrak{U}}_\varrho)$ or $(\overline{X}^\mathfrak{U}, \overline{\mathfrak{U}})$ are uniform or metric completions, respectively, then

$$\overline{X}^\mathfrak{U} = \overline{X}^\varrho \qquad (2.5)$$

as metrizable topological spaces.

We present now some important examples. Let $(E, h, \nabla) \longrightarrow (M^n, g)$ be a Riemannian vector bundle, $1 \leq p < \infty$, $r > 0$, $\delta > 0$. Set

$$V_\delta = \{(\varphi, \varphi') \in C^\infty(E)^2 \mid |\varphi' - \varphi|_{p,r}$$
$$= \left(\int_M \sum_{i=0}^r |\nabla^i(\varphi' - \varphi)|_\alpha^p \, \mathrm{dvol}_x(g) \right)^{\frac{1}{2}} < \delta \}.$$

Then it is evident that $\mathfrak{B} = \{V_\delta\}_{\delta>0}$ is a basis for a metrizable uniform structure $\mathfrak{U}^{p,r}(C^\infty(E))$, $(C^\infty(E), \mathfrak{U}^{p,r}(C^\infty(E)))$ is a uniform space. Let $(\overline{C^\infty(E)}^{\mathfrak{U}^{p,r}}, \overline{\mathfrak{U}^{p,r}})$ be the completion. We started with the local metric $d_\varphi(\varphi, \varphi') = d(\varphi, \varphi') = |\varphi - \varphi'|_{p,r}$ in the dense subspace $C^\infty(E)$ which can be extended to a complete local metric $|\cdot - \cdot|_{p,r}$ what is by (2.4) locally equivalent to the existence of a global one.

Lemma 2.5 $\overline{C^\infty(E)}$ *is locally arcwise connected.*

Proof. Let $\varphi_0 \in \overline{C^\infty(E)}$. There exists a neighbourhood $U(\varphi_0)$ such that $U(\varphi_0)$ is metrized by the extended $|\cdot - \cdot|_{p,r} = |\cdot - \cdot|_{p,r}^{ext}$. Let $\varphi_1 \in U$. Then $\{\varphi_t\}_{0 \leq t \leq 1}$, $\varphi(t) = (1-t)\varphi_0 + t\varphi_1$ is an arc between φ_0 and φ_1: $|\varphi_{t_1} - \varphi_{t_2}|_{p,r}^{ext} = |(1-t_1)\varphi_0 - t_1\varphi_1 - (1-t_2)\varphi_0 + t_2\varphi_1|_{p,r}^{ext} = |(t_2 - t_1)\varphi_0 - (t_2 - t_1)\varphi_1|_{p,r}^{ext} = |(t_2 - t_1)(\varphi_0 - \varphi_1)|_{p,r}^{ext}$. We would be done if $|(t_2 - t_1)(\varphi_0 - \varphi_1)|_{p,r}^{ext} \leq |t_2 - t_1||\varphi_1 - \varphi_0|_{p,r}^{ext}$. But this holds. We approximate φ_0 by $(\varphi_{0,\nu})_\nu$, φ_1 by $(\varphi_{1,\nu})_\nu$, $\varphi_{0,\nu}, \varphi_{1,\nu} \in C^\infty(E)$. Then $|\varphi_{t_1,\nu} - \varphi_{t_2,\nu}|_{p,r} = |(t_2 - t_1)(\varphi_{0,\nu} - \varphi_{1,\nu})| \leq |(t_2 - t_1)||(\varphi_{0,\nu} - \varphi_{1,\nu})|_{p,r}$ and this extends to the limit.
□

Corollary 2.6 a) *In* $\overline{C^\infty(E)}^{\mathfrak{U}^{p,r}}$ *coincide components and arc components.*

b) $\overline{C^\infty(E)}^{\mathfrak{U}^{p,r}}$ *has a representation as a topological sum*

$$\overline{C^\infty(E)}^{\mathfrak{U}^{p,r}} = \sum_{i \in I} \mathrm{comp}^{p,r}(\varphi_i)$$

c)
$$\begin{aligned}\operatorname{comp}^{p,r}(\varphi) &= \{\varphi' \in \overline{C^\infty(E)}^{\mathfrak{U}^{p,r}} \mid |\varphi - \varphi'|_{p,r}^{ext} < \infty\} \\ &= \varphi + \overline{\Omega}^{p,r}(E,\nabla),\end{aligned} \quad (2.6)$$

i.e. each component is an affine Sobolev space.

Proof. a) and b) immediately follow from lemma 2.5. Consider $\operatorname{comp}^{p,r}(\varphi_0)$, $\varphi_1 \in \operatorname{comp}^{p,r}(\varphi_0)$ and let $\{\varphi_t\}_{0 \le t \le 1}$ be an arc between φ_0 and φ_1. $\{\varphi_t\}_{0 \le t \le 1}$ is compact and can be covered by a finite number of ε–balls $U_\varepsilon(\varphi_{t_0}) = U_\varepsilon(\varphi_0), U_\varepsilon(\varphi_{t_1}), \ldots, U_\varepsilon(\varphi_{t_m}) = U_\varepsilon(\varphi_1)$, where ε is chosen sufficiently small that (2.4) is valid. Let $\varphi_{t_{i,i+1}} \in U_\varepsilon(\varphi_{t_i}) \cap U_\varepsilon(\varphi_{t_{i+1}})$. We obtain $|\varphi_{t_i} - \varphi_{t_{i+1}}|_{p,r} = |\varphi_{t_i} - \varphi_{t_{i,i+1}} + \varphi_{t_{i,i+1}} - \varphi_{t_{i+1}}|_{p,r} \le |\varphi_{t_i} - \varphi_{t_{i,i+1}}| + |\varphi_{t_{i,i+1}} - \varphi_{t_{i+1}}| < \varepsilon + \varepsilon = 2\varepsilon$, altogether we obtain that $|\varphi_0 - \varphi_1|_{p,r}$ is defined and $< m\varepsilon < \infty$. The last equation (2.6) follows from $\varphi_1 = \varphi_0 + (\varphi_1 - \varphi_0)$. $\square$

Corollary 2.7 *On a compact manifold there is only one (arc) component, namely*

$$\operatorname{comp}^{p,r}(0) = \overline{\Omega}^{p,r}(E,\nabla).$$

Proof. On a compact manifold always holds $|\varphi_0 - \varphi_1|_{p,r} < \infty$ for $\varphi_0, \varphi_1 \in C^\infty(E)$. This inequality extends to the difference of Cauchy sequences (since they are bounded) and to the limit by continuity. In particular this holds for $\varphi_0 = 0$. $\square$

We write $\Omega^{p,r}(C^\infty(E))$ for $\overline{C^\infty(E)}^{\mathfrak{U}^{p,r}(C^\infty(E))}$.

Remarks 2.8 a) We see from (2.6) for $\varphi = 0$ that the zero component $\operatorname{comp}^{p,r}(0)$ coincides with the Sobolev space $\overline{\Omega}^{p,r}(E,\nabla)$. Insofar our approach yields a generalization of Sobolev spaces.

b) It is very easy to see that the index set I is uncountable if (M^n, g) is open. "Each growth generalies its component." On compact manifolds, there is only one growth, namely no growth, hence there is only one component.

c) Let $\{X_i\}_{i\in I}$ be a family of disjoint metric spaces, d_i the metric on X_i. Then there exists a metric d on the topological sum $X = \sum_{i\in I} X_i$ s.t. d induces the uniform structure on X_i which belongs to d_i (cf. [68], p. 120). This is the situation in corollary 2.6 and c). □

Let us consider other choices of V_δ. Set $V_\delta = \{(\varphi, \varphi') \in L_{1,\mathrm{loc}}(E)^2 \mid |\varphi - \varphi'|_{p,r} < \delta\}$. In $|\varphi - \varphi'|_{p,r}$ we take distributional derivatives. Then $\mathfrak{B} = \{V_\delta\}_{\delta > 0}$ is a basis for a metrizable uniform structure $\mathfrak{U}^{p,r}(L_{1,\mathrm{loc}}(E))$ which is already complete, $(\overline{L_{1,\mathrm{loc}}(E)}, \overline{\mathfrak{U}}^{p,r}(L_{1,\mathrm{loc}})) = (L_{1,\mathrm{loc}}, \mathfrak{U}^{p,r}(L_{1,\mathrm{loc}}(E)))$. The background for completeness is the fact that the Sobolev spaces of distributions are already complete. We write $\Omega^{p,r}(L_{1,\mathrm{loc}}(E))$ for $(L_{1,\mathrm{loc}}(E), \mathfrak{U}^{p,r}(L_{1,\mathrm{loc}}))$ which is a complete uniform space.

Proposition 2.9 a) $\Omega^{p,r}(L_{1,\mathrm{loc}}(E))$ *is locally arcwise connected.*
b) *In* $\Omega^{p,r}(L_{1,\mathrm{loc}}(E))$ *coincide components and arc components.*
c) $\Omega^{p,r}(L_{1,\mathrm{loc}}(E))$ *has a representation as a topological sum*

$$\Omega^{p,r}(L_{1,\mathrm{loc}}(E)) = \sum_{j \in J} \mathrm{comp}^{p,r}(\varphi_j).$$

d)

$$\begin{aligned}\mathrm{comp}^{p,r}(\varphi) &= \{\varphi' \in \Omega^{p,r}(L_{1,\mathrm{loc}}(E)) \mid |\varphi - \varphi'|_{p,r} < \infty\} \\ &= \varphi + \Omega^{p,r}(E, \nabla).\end{aligned} \quad (2.7)$$

The proof is the same as that of corollary 2.6. □

Consider finally

$$V_\delta = \{(\varphi, \varphi') \in C_c^\infty(E)^2 \mid |\varphi - \varphi'|_{p,r} < \delta\}.$$

We obtain $\mathfrak{U}^{p,r}(C_c^\infty(E))$, $(\overline{C_c^\infty(E)}^{\mathfrak{U}^{p,r}}, \overline{\mathfrak{U}}^{p,r}(C_c^\infty(E))) = \Omega^{p,r}(C_c^\infty(E))$, locally arcwise connectedness, a topological sum representation and

$$\operatorname{comp}^{p,r}(\varphi) = \varphi + \overset{\circ}{\Omega}{}^{p,r}(E, \nabla). \tag{2.8}$$

But $\Omega^{p,r}(C_c^\infty(E))$ consists of one component as the following remark shows.

Remark 2.10 If $\varphi \in C_c^\infty(E)$ then in $\Omega^{p,r}(C_c^\infty(E))$

$$\operatorname{comp}(\varphi) = \operatorname{comp}^{p,r}(0) = \overset{\circ}{\Omega}{}^{p,r}(E, \nabla).$$

□

Corollary 2.11 If $(E, h, \nabla) \longrightarrow (M^n, g)$ satisfies (I), $(B_k(M^n, g))$, $(B_k(E, \nabla))$ then

$$\Omega^{p,r}(L_{1,\text{loc}}(E)) = \Omega^{p,r}(C^\infty(E)) \tag{2.9}$$

for $0 \leq r \leq k + 2$. □

Remark 2.12 A long further series of equivalences for uniform spaces can be stated if we apply all of our invariance properties of Sobolev spaces. □

The advantage of this approach is that we can develop e.g. a Sobolev theory of PDEs without decay conditions for the sections. The classical theory is a theory between the zero components comp(0). Our framework allows a quite parallel theory as maps between other components. Clearly ∇ maps $\operatorname{comp}^{p,r}(\varphi)$ into $\operatorname{comp}^{p,r-1}(\varphi)$. If A is a differential operator which maps $\Omega^{p,r}(E)$ into $\Omega^{p,r-m}(F)$ then A maps $\operatorname{comp}^{p,r}(\varphi) \subset \Omega^{p,r}(L_{1,\text{loc}}(E))$ into $\operatorname{comp}^{p,r-m}(A\varphi) \subset \Omega^{p,r-m}(L_{1,\text{loc}}(F))$. A necessary condition for the solvability of $A\varphi = \psi$, $\varphi \in \operatorname{comp}^{p,r}(\varphi_0)$ to find, $\psi \in \Omega^{p,r-m}(L_{1,\text{loc}}(F))$ or $\in \Omega^{p,r-m}(C^\infty(F))$ given, is that $A\varphi_0 \in \operatorname{comp}^{p,r-m}(\psi)$ etc. We will here not establish the complete PDE–theory for this setting. It should appear elsewhere.

A similar setting can be established for the Banach–Hölder theory. Set

$$V_\delta = \{(\varphi,\varphi') \in C^\infty(E)^2 \mid {}^{b,m}|\varphi-\varphi'| = \sum_{i=0}^{m} \sup_x |\nabla^i(\varphi-\varphi')|_x < \delta\}$$

and $\mathfrak{B} = \{V_\delta\}_{\delta>0}$. $\mathfrak{B}$ is a basis for a metrizable uniform structure ${}^{b,m}\mathfrak{U}(C^\infty(E))$. Let $(\overline{C^\infty(E)}^{b,m\mathfrak{U}}, {}^{b,m}\overline{\mathfrak{U}})$ be the completion. Then we get properties absolutely parallel to the assertions 2.5 – 2.9,

$$\overline{C^\infty(E)}^{b,m\mathfrak{U}} \equiv {}^{b,m}\Omega(C^\infty(E)) = \sum_{j\in J} {}^{b,m}\mathrm{comp}(\varphi_j),$$

$$\begin{aligned}{}^{b,m}\mathrm{comp}(\varphi) &= \{\varphi' \in {}^{b,m}\Omega(C^\infty(E)) \mid |\varphi-\varphi'| < \infty\} \\ &= \varphi + {}^{b,m}\Omega(E,\nabla).\end{aligned}$$

Similarly

$$V_\delta = \{(\varphi,\varphi') \in C^\infty(E)^2 \mid {}^{b,m,\alpha}|\varphi-\varphi'| < \delta\}$$

and $\mathfrak{B} = \{V_\delta\}_{\delta>0}$ define ${}^{b,m,\alpha}\mathfrak{U}(C^\infty(E))$, the completion ${}^{b,m,\alpha}\Omega(C^\infty(E)) = \sum_{j\in J} {}^{b,m,\alpha}\mathrm{comp}(\varphi_j)$ and

$$^{b,m,\alpha}\mathrm{comp}(\varphi) = \varphi + {}^{b,m,\alpha}\Omega(E,\nabla). \tag{2.10}$$

Here ${}^{b,m,\alpha}|\varphi|$ is defined as

$$^{b,m}|\varphi| + \sup_{x,y\in M} \sup_{c\in G(x,y)} \frac{|\tau(c)\nabla^m\varphi(x) - \nabla^m\varphi(y)|}{d(x,y)^\alpha},$$

where $G(x,y) = \{$ length minimizing geodesics joining x and $y\}$, $\tau(c)$ is parallel translation along c from $\pi^{-1}(x)$ to $\pi^{-1}(y)$ and $d(x,y)$ is the distance from x to y.

Remark 2.13 Our Sobolev embedding theorems from section 1 induce embedding theorems for components of corresponding uniform spaces, e.g. if we have (I), (B_k), $r > \frac{n}{p} + m$ then

$$\Omega^{p,r}(C^\infty(E)) \supset \mathrm{comp}^{p,r}(\varphi) \hookrightarrow {}^{b,m}\mathrm{comp}(\varphi) \subset {}^{b,m}\Omega(C^\infty(E)).$$

$\square$

We discuss another example, which is important in Teichmüller theory for open surfaces. That is the space of bounded conformal factors, adapted to a Riemannian metric g.
Let

$$\mathcal{P}_m(g) = \{\varphi \in C^\infty(M) \mid \inf_{x \in M} \varphi(x) > 0, \sup_{x \in M} \varphi(x) < \infty,$$
$$|\nabla^i \varphi|_{g,x} \leq C_i, \ 0 \leq i \leq m\}.$$

Set for $1 \leq p < \infty$, $r \in \mathbb{Z}_+$, $\delta > 0$,

$$V_\delta = \{(\varphi, \varphi') \in \mathcal{P}_m(g)^2 \mid ||\varphi - \varphi'|_{g,p,r}$$
$$:= \left(\int \sum_{i=0}^{r} |(\nabla^g)^i (\varphi - \varphi')|_{g,x}^p \, \mathrm{dvol}_x(g) \right)^{\frac{1}{2}} < \delta\}.$$

Then $\mathfrak{B} = \{V_\delta\}_{\delta > 0}$ is a basis for a metrizable uniform structure. Let $\overline{\mathcal{P}}^p_{m,r}(g)$ the completion,

$$C^1\mathcal{P} = \{\varphi \in C^1(M) \mid \inf_{x \in M} \varphi(x) > 0, \sup_{x \in M} \varphi(x) < \infty\}$$

and set

$$\mathcal{P}^{p,r}_m(g) = \overline{\mathcal{P}}^p_{m,r}(g) \cap C^1\mathcal{P}.$$

$\mathcal{P}^{p,r}_m(g)$ is locally contractible, hence locally arcwise connected and hence components coincide with arc components. Let

$$U^{p,r}_m(\varphi) = \{\varphi' \in \mathcal{P}^{p,r}_m(g) \mid ||\varphi - \varphi'|_{g,p,r} < \infty\} \qquad (2.11)$$

and denote by $\mathrm{comp}(\varphi) = \mathrm{comp}^{p,r}_m(\varphi)$ the component of φ in $\mathcal{P}^{p,r}_m(g)$. $||_{g,p,r}$ in (2.10) means the local extended metric, i.e. it is defined by taking distributional derivatives.

Theorem 2.14 *$\mathcal{P}^{p,r}_m(g)$ has a representation as topological sum*

$$\mathcal{P}^{p,r}_m(g) = \sum_{i \in I} \mathrm{comp}(\varphi_i)$$

and

$$\mathrm{comp}(\varphi) = U^{p,r}_m(\varphi).$$

$\square$

Remark 2.15 On a compact manifold there is only one component, the component comp(1).

□

We introduce now uniform structures of Riemannian metrics. Let M^n be an open smooth manifold, $\mathcal{M} = \mathcal{M}(M)$ be the space of all Riemannian metrics. We want to endow $\mathcal{M}$ with a canonical intrinsic topology either in the C^m- or Sobolev setting, depending on the subsequent investigation.
Let $g \in \mathcal{M}$. We define

$${}^bU(g) = \{g' \in \mathcal{M} \mid {}^b|g-g'| := \sup_{x \in M} |g-g'|_{g,x} < \infty, {}^b|g-g'|_{g'} < \infty\}.$$

Then, it is easy to see that ${}^bU(g)$ coincides with the quasi isometry class of g, i.e., $g' \in {}^bU(g)$ if and only if there exist $C, C' > 0$ such that

$$C \cdot g' \leq g \leq C' \cdot g' \qquad (2.12)$$

holds in the sense of positive definite forms. In particular $g' \in {}^bU(g)$ if and only if $g \in {}^bU(g')$. We introduced ${}^bU(g)$ in chapter I after remark 1.17 as $\mathcal{US}(g)$. For a tensor t we define ${}^b|t|_g := \sup_{x \in M} |t|_{g,x}$. $g \in {}^bU(g')$ implies the existence of bounds $A_k(g,g'), B_k(g,g') > 0$ such that, for every (r,s)–tensor field t with $r + s = k$,

$$A_k \cdot |t|_{g,x} \leq |t|_{g',x} \leq B_k \cdot |t|_{g,x}, \quad A_k \cdot {}^b|t|_g \leq {}^b|t|_{g'} \leq B_k \cdot {}^b|t|_g.$$
$$(2.13)$$

To endow $\mathcal{M}$ with canonical topologies we use the language of uniform structures.
Set for $m \geq 1$, $\delta > 0$, $C(n,\delta) = 1 + \delta + \delta\sqrt{2n(n-1)}$

$$V_\delta = \{(g,g') \in \mathcal{M} \mid g' \in {}^bU(g), C(n,\delta)^{-1} \cdot g \leq g' \leq C(n,\delta) \cdot g$$

and $\ {}^{b,m}|g - g'|_g := {}^b|g - g'|_g + \sum_{j=0}^{m-1} {}^b|(\nabla^g)^j(\nabla^g - \nabla^{g'})|_g < \delta\}.$

Proposition 2.16 *The set $\mathfrak{B} = \{V_\delta\}_{\delta > 0}$ is a basis for a metrizable uniform structure on $\mathcal{M}$.*

Proof. The proof is already modelled by the proof of chapter I, lemma 1.1. and chapter II, theorem 1.19. But we present for completeness a complete proof. (B_1) and (U'_1) are trivial. We start to prove (U'_2). This would be proved if we could show

$$^{b,m}|g - g'|_{g'} \leq P(^{b}|g - g'|_g, {}^{b}|(\nabla^g)^i(\nabla^g - \nabla^{g'}|_g) = P_{m,g}, \quad (2.14)$$

where $P_{m,g}$ is a polynomial in $^{b}|g - g'|_g$, $^{b}|(\nabla^g)^i(\nabla^g - \nabla^{g'}|_g)$, $0 \leq i \leq m - 1$, without constant term. In fact, (2.14) implies (U'_2) as follows. Given $\delta > 0$, there exists $\delta' = \delta'(\delta)$ such that for $^{b,m}|g - g'|_g < \delta'$, $P_{m,g} < \delta$. According to (2.14), exchanging the role of g and g', $^{b,m}|g - g'|_{g'} < \delta'$ implies $^{b,m}|g - g'|_g < \delta$, $(g, g') \in V_\delta$, $(g', g) \in V_\delta^{-1}$. Therefore we have to prove (2.14). We set $\nabla^g = \nabla$, $\nabla^{g'} = \nabla'$. From (2.2) follows

$$^{b}|g - g'|_{g'} \leq C_0 \cdot {}^{b}|g - g'|_g \text{ and } {}^{b}|\nabla - \nabla'|_{g'} \leq C_1 \cdot {}^{b}|\nabla - \nabla'|_g,$$

which gives the assertion for $m = 0, 1$. Set $m = 2$. Then, using (2.14),

$$|\nabla'(\nabla - \nabla')|_{g',x} \leq |(\nabla - \nabla')(\nabla - \nabla')|_{g',x} + |\nabla(\nabla - \nabla')|_{g',x},$$
$$^{b}|\nabla'(\nabla - \nabla')|_{g'} \leq C_2({}^{b}|\nabla - \nabla'|_g^2 + |\nabla(\nabla - \nabla')|_g),$$
$$^{b,2}|g - g'|_{g'} \leq C_0 \cdot {}^{b}|g - g'|_g + C_1 \cdot {}^{b}|\nabla - \nabla'|_g$$
$$+ C_2 \cdot ({}^{b}|\nabla - \nabla'|_g^2 + {}^{b}|\nabla(\nabla - \nabla')|_g) = P_{2,g}.$$

Assume now for $\mu \leq m - 1$

$$^{b,\mu}|g - g'|_{g'} \leq P_{\mu,g},$$

set $\nabla' - \nabla = \eta$ and consider $\nabla'^\lambda \eta$, $\lambda \leq m - 1$. Observe that $\nabla'^{\mu-1}\eta$ enters as highest derivative into $^{b,\mu}|g - g'|_{g'}$.
Starting with

$$\nabla'^\lambda \eta = (\nabla'^\lambda - \nabla \nabla'^{\lambda-1})\eta + \nabla \nabla'^{\lambda-1}\eta = (\nabla' - \nabla)\nabla'^{\lambda-1}\eta + \nabla \nabla'^{\lambda-1}\eta, \quad (2.15)$$

one obtains by an easy induction

$$\nabla'^\lambda \eta = \sum_{i=1}^{\lambda} \nabla^{i-1}(\nabla' - \nabla)\nabla'^{\lambda-i}\eta + \nabla^\lambda \eta, \quad (2.16)$$

$$|\nabla'^\lambda \eta|_{g',x} \leq \sum_{i=1}^{\lambda} |\nabla^{i-1}(\nabla' - \nabla)\nabla'^{\lambda-i}\eta|_{g',x} + |\nabla^\lambda \eta|_{g',x}. \quad (2.17)$$

For $i = 1$ we are done in (2.6) according to our induction assumption

$$^b|(\nabla' - \nabla)\nabla'^{\lambda-1}\eta|_{g'} \leq C_{\lambda-1} \cdot {}^b|\nabla - \nabla'|_{g'} \cdot P_{\lambda,g}. \quad (2.18)$$

Therefore, we have to consider the terms

$$\nabla^{i-1}(\nabla' - \nabla)\nabla'^{\lambda-i}\eta, \quad i > 1. \quad (2.19)$$

Iterating the procedure (2.15), (2.16) to $\nabla'^{\lambda-1}$ in $\nabla^{i-1}(\nabla' - \nabla)\nabla'^{\lambda-1}\eta$, we have finally to estimate the expressions

$$\nabla^{i_1}(\nabla' - \nabla)^{i_2}\nabla^{i_3} \cdots (\nabla' - \nabla)^{i_{\lambda-1}}\nabla^{i_\lambda}\eta, \quad (2.20)$$

$i_1 + \cdots + i_\lambda = \lambda$, $i_1 \geq 1$, $i_2 \geq 1$, $0 \leq i_\lambda \leq \lambda - 1$. By the Leibniz rule $\nabla((\nabla' - \nabla)\eta) = (\nabla(\nabla' - \nabla))\eta + (\nabla' - \nabla)\nabla\eta$ each term of (2.20) splits into a sum of products (= compositions), where each factor can be estimated by

$$C_j \cdot |\nabla^j(\nabla' - \nabla)|_{g',x} \leq D_j|\nabla^j\eta|_{g,x}, \quad 0 \leq j \leq \lambda - 1. \quad (2.21)$$

Taking $\sup_{x \in M}$ we are done.

Next, we have to prove (U_3'). Given $\delta > 0$, we have to show that there exists $\delta' = \delta'(\delta) > 0$ such that $V_{\delta'} \circ V_{\delta'} \subset V_\delta$, i.e., if $(g_1, g_2) \in V_{\delta'} \circ V_{\delta'} = \{(g_1', g_2') \in \mathcal{M} \times \mathcal{M} \mid$ there exists g such that $(g_1', g) \in V_{\delta'}$ and $(g, g_2') \in V_{\delta'}\}$, then

$$^{b,m}|g_1 - g_2|_{g_1} < \delta.$$

(U_3') would be proved if we could show for g_1, g, g_2 with ${}^{b,m}|g_1 - g|_{g_1} < \infty$, ${}^{b,m}|g - g_2|_g < \infty$

$$^{b,m}|g_1 - g_2|_{g_1} \leq P_m({}^{b,i}|g_1 - g|_{g_1}, {}^{b,j}|g - g_2|_g), \quad (2.22)$$

where P_m is a polynomial in ${}^{b,i}|g_1-g|_{g_1}$, ${}^{b,j}|g-g_2|_g$, $i, j = 0, \ldots, m$ without constant term. We start with

$$|g_1 - g_2|_{g_1} \leq |g_1 - g|_{g_1} + |g - g_2|_{g_1},$$
$$^b|g_1 - g_2|_{g_1} \leq C_0 \cdot ({}^b|g_1 - g|_{g_1} + {}^b|g - g_2|_g)$$
$$= P_{00}({}^b|g_1 - g|_{g_1}, {}^b|g - g_2|_g), \quad (2.23)$$

where we used $g \in {}^b U(g_1)$. We set $\nabla^{g_1} = \nabla_1$, $\nabla^{g_2} = \nabla_2$, $\nabla^g = \nabla$. Since, according to our assumptions, for the pointwise norms
$$||_{g_1} \sim ||_{g_2} \sim ||_g,$$
we simply write $||$ and multiply by constants. This gives
$${}^b|\nabla_1 - \nabla_2| \leq C \cdot ({}^b|\nabla_1 - \nabla| + {}^b|\nabla - \nabla_2|) = P_{11},$$
together with (2.23),
$${}^{b,1}|g_1 - g_2| \leq P_1({}^{b,i}|g_1 - g|_{g_1}, {}^{b,j}|g - g_2|_g) = P_0 + P_{11}, i,j = 0,1. \tag{2.24}$$
Consider next ${}^{b,2}|g_1 - g_2|$ which amounts, using (2.23), (2.24), to the estimate of $\nabla_1(\nabla_1 - \nabla_2)$. We have
$$|\nabla_1(\nabla_1 - \nabla_2)| \leq C \cdot (|\nabla_1(\nabla - \nabla_2)| + |\nabla_1(\nabla_1 - \nabla)|). \tag{2.25}$$
The critical point in (2.14) is the estimate of $|\nabla_1(\nabla - \nabla_2)|$. But
$$|\nabla_1(\nabla - \nabla_2)| \leq C_0 \cdot (|(\nabla_1 - \nabla)(\nabla - \nabla_2)| + |\nabla(\nabla - \nabla_2)|),$$
$${}^b|\nabla_1(\nabla - \nabla_2)| \leq C_1 \cdot ({}^b|\nabla_1 - \nabla| \cdot {}^b|\nabla - \nabla_2|$$
$$+ {}^b|\nabla(\nabla - \nabla_2)|),$$
$${}^b|\nabla_1(\nabla_1 - \nabla_2)| \leq C_2 \cdot ({}^b|\nabla_1(\nabla_1 - \nabla)| + {}^b|\nabla_1 - \nabla|$$
$$\cdot {}^b|\nabla - \nabla_2| + {}^b|\nabla(\nabla - \nabla_2)|)$$
$$= P_{22}({}^b|\nabla_1 - \nabla|, {}^b|\nabla_1(\nabla_1 - \nabla)|, {}^b|\nabla - \nabla_2|, {}^b|\nabla(\nabla - \nabla_2)|),$$
together with (2.24)
$${}^{b,2}|g_1 - g_2|_{g_1} \leq P_2({}^{b,i}|g_1 - g|_{g_1}, {}^{b,j}|g - g_2|_g) = P_1 + P_{22}, i,j = 0,1,2. \tag{2.26}$$
In the general case
$$|\nabla_1^r(\nabla_1 - \nabla_2)| \leq C_0 \cdot (|\nabla_1^r(\nabla_1 - \nabla)| + |\nabla_1^r(\nabla - \nabla_2)|), \tag{2.27}$$
holds and we have to estimate $\nabla_1^r(\nabla - \nabla_2) \equiv \nabla_1^r \eta$. We start as above
$$\nabla_1^r \eta = (\nabla_1 - \nabla)\nabla_1^{r-1}\eta + \nabla\nabla_1^{r-1}\eta,$$
$$\nabla_1^r \eta = \sum_{i=1}^{r} \nabla^{i-1}(\nabla_1 - \nabla)\nabla_1^{r-i}\eta + \nabla^r \eta. \tag{2.28}$$

Interating the procedure, we have to estimate expressions of the kind

$$\nabla^{i_1}(\nabla_1 - \nabla)^{i_2} \cdots \nabla^{i_{r-2}}(\nabla_1 - \nabla)^{i_{r-1}} \nabla^{i_r} \eta \qquad (2.29)$$

with $i_1 + \cdots + i_r = r$, $i_r = r$. (2.18) splits into a sum of terms each of them can be estimated by

$$C_{n_1 \ldots n_s} \cdot |\nabla^{n_1}(\nabla_1 - \nabla)| \cdots |\nabla^{n_{s-1}}(\nabla_1 - \nabla)| |\nabla^{n_s}|, \qquad (2.30)$$

where $n_1 + 1 + \cdots + n_s + 1 = r + 1$, $n_s < r$, $n_1 + \cdots + n_{s-1} \leq r - 1$, $s \geq 2$. As an example, the term $i = r$ coming from (2.17) leads to

$$\nabla^{r-1}((\nabla_1 - \nabla)\eta) = (\nabla^{r-1}(\nabla_1 - \nabla))(\eta) + (\nabla_1 - \nabla)(\nabla^{r-1}\eta).$$

We have two terms, each with $s = 2$. $|\nabla^{r-1}(\nabla_1 - \nabla)| |\eta|$ corresponds to $n_1 = r - 1$, $n_2 = 0$, $|\nabla_1 - \nabla| |\nabla^{r-1}\eta|$ to $n_1 = 0$, $n_2 = r - 1$. According to the proof of (U_2')

$$^b|\nabla^{n_i}(\nabla_1 - \nabla)|_{g_1} \leq Q_{n-i}(^b|\nabla_1^j(\nabla_1 - \nabla)|), \quad 0 \leq j \leq n_i, \qquad (2.31)$$

where Q_{n_i} is a polynomial in the indicated variables without constant terms. (2.27) – (2.31) yield

$$^b|\nabla_1^r(\nabla_1 - \nabla_2)|_{g_1} \leq P_{r+1,r+1}(^b|\nabla_1^i(\nabla_1 - \nabla)|_{g_1},$$
$$^b|\nabla^j(\nabla - \nabla_2)|_g),$$
$$^{b,m}|g_1 - g_2| \leq C_0 \cdot (^b|g_1 - g|_{g_1} + {}^b|g - g_2|_g)$$
$$+ \sum_{r=0}^{m-1} P_{r+1,r+1}$$
$$= P_m(^{b,i}|g_1 - g|_{g_1}, {}^{b,j}|g - g_2|_g).$$

Therefore, we established (U_3'). Denote by $^{b,m}\mathcal{U}(\mathcal{M})$ the corresponding uniform structure. It is metrizable since it is trivially Hausdorff and $\{V_{1/n}\}_{n \geq n_0}$ is a countable basis. □

We see, the proof of 2.16 is parallel to that of 1.19, replacing $^{b,m}||$ by $||_{p,r}$.

Denote $^b_m\mathcal{M} = (\mathcal{M}, {}^{b,m}\mathfrak{U}(\mathcal{M}))$ and by $^{b,m}\overline{\mathcal{M}}$ the completion. It has been proven in [202] that $^{b,m}\overline{\mathcal{M}}$ still consists of positive definite elements, which are of class C^m.

Remark 2.17 We endowed by our procedure $\mathcal{M}$ with a canonical intrinsic C^m-topology without choice of a cover or a special g to define the C^m-distance. According to the definition of the uniform topology, for $g \in {}^{b,m}\mathcal{M}$

$$\{{}^{b,m}U_\varepsilon\}_{\varepsilon>0} = \{g' \in {}^{b,m}\mathcal{M} \mid {}^b|g - g'|_{g'} < \infty, {}^{b,m}|g - g'|_g < \infty\}$$

is a neighbourhood basis in this topology. □

Proposition 2.18 *The space $^{b,m}\mathcal{M}$ is locally contractible.*

For a proof we refer to [27], [35] and to proposition 1.9 of chapter VII. □

Corollary 2.19 *In $^{b,m}\mathcal{M}$ components and arc components coincide.* □

Set

$$^{b,m}U(g) = \{g' \in {}^{b,m}\mathcal{M} \mid {}^b|g - g'|_{g'} < \infty, {}^{b,m}|g - g'|_g < \infty\}.$$

Lemma 2.20 $g' \in {}^{b,m}U(g)$ *if and only if* $g \in {}^{b,m}U(g')$.

Proof. Assume $g' \in {}^{b,m}U(g)$. Then $^b|g - g'|_{g'} < \infty$, $^{b,m}|g - g'|_g < \infty$. We have to show $^{b,m}|g - g'|_{g'} < \infty$, i. e. to estimate

$$\nabla'^r(\nabla' - \nabla), \quad 0 \leq r \leq m - 1.$$

This has been done in the proof of (U'_2). We conclude

$$^{b,m-1}|\nabla' - \nabla|_{g'} < \infty, {}^{b,m}|g' - g|_{g'} < \infty.$$

All the arguments are symmetric in g and g' which proves the assertion. □

Lemma 2.21 *If $g' \in {}^{b,m}U(g)$ then ${}^{b,m}U(g') = {}^{b,m}U(g)$.*

Proof. Assume $h \in {}^{b,m}U(g')$. Then ${}^{b}|g'-h|_h < \infty$, ${}^{b,m}|g'-h|_{g'} < \infty$. By assumption ${}^{b}|g-g'|_{g'} < \infty$, ${}^{b,m}|g-g'|_g < \infty$. This implies ${}^{b}|g-h|_h < \infty$ and, according to the proof of (U_3'), ${}^{b,m}|g-h|_g < \infty$, $h \in {}^{b,m}U(g)$, ${}^{b,m}U(g') \subseteq {}^{b,m}U(g)$. The other inclusion follows in the same manner. □

Corollary 2.22 *If ${}^{b,m}U(g') \cap {}^{b,m}U(g) \neq \emptyset$ then ${}^{b,m}U(g') = {}^{b,m}U(g)$.* □

Proposition 2.23 *Denote by $\mathrm{comp}(g)$ the component of $g \in {}^{b,m}\mathcal{M}$. Then*

$$\mathrm{comp}(g) \equiv {}^{b,m}\mathrm{comp}(g) = {}^{b,m}U(g).$$

□

Proof. We start with $\mathrm{comp}(g) = {}^{b,m}U(g)$. Let $g' \in \mathrm{comp}(g)$ and $\{g_t\}_{0 \leq t \leq 1}$ be an arc between g and g'. For any $\varepsilon > 0$ this can be covered by a finite number of ε-neighbourhoods with respect to the local metric ${}^{b,m}|\cdot - \cdot|_g$, ${}^{b,m}U_\varepsilon(g_{t_\sigma})$, $\sigma = 0, \ldots, s$, $g_{t_0} = g$, $g_{t_s} = g'$. We set $g_{t_\sigma} = g_\sigma$. Then ${}^{b,m}U_\varepsilon(g_0) \cap {}^{b,m}U_\varepsilon(g_1) \neq \emptyset$. According to corollary 2.22, $g_1 \in {}^{b,m}U(g_0)$. Performing a simple induction, $g_s = g' \in {}^{b,m}U(g_0) \equiv {}^{b,m}U(g_0)$, $\mathrm{comp}(g) \subseteq {}^{b,m}U(g)$. Let $g' \in {}^{b,m}U(g)$. Then $\{tg' + (1-t)g\}_{0 \leq t \leq 1}$ is an arc in ${}^{b,m}U(g)$ connecting g and g'. This follows from 3.2, 3.10 in [35]. See also chapter VII section 1. □

By construction, ${}^{b,m}U(g)$ is open in ${}^{b,m}\mathcal{M}$ and, according to proposition 2.23, closed. Therefore we have

Theorem 2.24 *The space ${}^{b,m}\mathcal{M}$ has a representation as a topological sum*

$$^{b,m}\mathcal{M} = \sum_{i \in I} {}^{b,m}U(g_i). \qquad (2.32)$$

□

Remark 2.25 On a compact manifold $\mathcal{M}$, the index set I consists of one element. One has only one component. On compact manifolds the notion of growth at infinity does not exist. □

For later use we restrict ourselves additionally to metrics with bounded geometry. Let (M^n, g) be open. Consider the conditions (I) and (B_k) and

$$\begin{aligned} \mathcal{M}(I) &= \{g \in \mathcal{M} \,|\, g \text{ satisfies}(I)\}, \\ \mathcal{M}(B_k) &= \{g \in \mathcal{M} \,|\, g \text{ satisfies}(B_k)\}, \\ \mathcal{M}(I, B_k) &= \mathcal{M}(I) \cap \mathcal{M}(B_k). \end{aligned}$$

Lemma 2.26 *If $g \in \mathcal{M}(I)$, then (M^n, g) is complete.*

Proof. We have to show that every Cauchy sequence converges. Let $r = r_{\text{inj}}(M) > 0$ and $(x_i)_i$ be a Cauchy sequence. There exists i_0 such that $\varrho(x_i, x_j) < \frac{r}{4}$ for all $i, j \geq i_0$. Let $U_{\frac{r}{2}}(x_{i_0}) = \exp(B_{\frac{r}{2}}(O_{x_{i_0}}))$ be the geodesic ball of radius $\frac{r}{2}$ centered at x_{i_0}. Then, $x_i \in \overline{U_{\frac{r}{4}}(x_{i_0})}$ for all $i \geq i_0$. But $\overline{U_{\frac{r}{4}}(x_{i_0})} \subset U_{\frac{r}{2}}(x_{i_0})$ is complete and $(x_i)_{i \geq i_0} \longrightarrow x \in \overline{U_{\frac{r}{4}}(x_{i_0})}$. □

Remark 2.27 There are several other proofs of this simple fact. □

It is trivially clear that (B_k) does not imply completeness. Nevertheless we have

Corollary 2.28 *If $g \in \mathcal{M}(I, B_k)$, then g is complete.* □

Proposition 2.29 *Let $g \in \mathcal{M}(B_k)$, $g' \in \mathcal{M}$ and $^{b,k+2}|g - g'|_g < \infty$. Then $g' \in \mathcal{M}(B_k)$.*

We refer to [32]. □

Corollary 2.30 *The space $\mathcal{M}(B_k)$ is open in $^b_m\mathcal{M}$, $m \geq k+2$.*
□

Now we can restrict our uniform structure $^{b,m}\mathcal{U}(\mathcal{M})$ to $\mathcal{M}(B_k) \times \mathcal{M}(B_k)$ and obtain a completed metrizable space. But for $m < k+2$ there doesn't exist a good description of components. Therefore it would be better to consider the case $m \geq k+2$.

Lemma 2.31 *Let $g \in {}^{b,m}\mathcal{M}$, comp$(g)$ the component of g in $^{b,m}\mathcal{M}$ and $m \geq k+2$. If comp(g) contains a C^m-metric satisfying (B_k), then all metrics of comp(g) satisfy (B_k).*

This follows from propositions 2.34. □

Corollary 2.32 *Let $m \geq k+2$. There exists a representation as topological sum*

$$^{b,m}\mathcal{M}(B_k) = \sum_{j \in J} {}^{b,m}U(g_j),$$

$J \subset I$ an index set. Each component of $^{b,m}\mathcal{M}(B_k)$ is a Banach manifold.

We refer to [32] for the proof that each component is a Banach manifold.

Proposition 2.33 *Assume g satisfies (I) and (B_k), $k \geq 0$. Then there exists $\varepsilon > 0$ such that g' satifies (I) for all $g' \in {}^{b,k}U_\varepsilon(g)$.*

We refer to [32]. □

Proposition 2.34 *Assume g satisfies (I) and (B_k), $g' \in$ comp$(g) \subset {}^{b,m}\mathcal{M}$, $m \geq k+2$. Then g' satifies (I) and (B_k).*

We refer to [32] for the proof. □

Now we want to introduce Sobolev uniform structures into the space of metrics. Let now $k \geq r > \frac{n}{p} + 1$, $\delta > 0$, $C(n,\delta) = 1 + \delta + \delta\sqrt{2n(n-1)}$,

$$V_\delta = \big\{(g,g') \in \mathcal{M}(I,B_k) \times \mathcal{M}(I,B_k) \,|\,$$
$$C(n,\delta)^{-1} g \leq g' \leq C(n,\delta) g \text{ and}$$
$$|g - g'|_{g,p,r} = \Big(\int (|g - g'|_{g,x}^p$$
$$+ \sum_{i=0}^{r-1} |(\nabla^g)^i(\nabla^{g'} - \nabla^g)|_{g,x}^p) \,\mathrm{dvol}_x(g)\Big)^{\frac{1}{p}} < \delta\big\}.$$

Proposition 2.35 *The set $\{V_\delta\}_{\delta > 0}$ is a basis for a metrizable uniform structure on $\mathcal{M}(I, B_k)$.*

The proof is quite analogous to that of 2.16. We replace $^b|\,|$ by $|\,|_{p,r}$ and apply in (2.17) – (2.31) the module structure theorem. □

Denote $\mathcal{M}_r^p(I, B_k)$ as $(\mathcal{M}(I, B_k), \mathfrak{U}^{p,r}(\mathcal{M}(I, B_k)))$ and by $\mathcal{M}^{p,r}(I, B_k)$ the completion. It was proven [67] that the completion yields only positive definite elements, i.e. we still remain in the space of C^1 Riemannian metrics.

For $g \in \mathcal{M}^{p,r}(I, B_k)$

$$\{U_\varepsilon^{p,r}(g)\}_{\varepsilon > 0} = \big\{\{g' \in \mathcal{M}^{p,r}(I, B_k) \,|\, ^b|g - g'|_g < \infty,$$
$$^b|g - g'|_{g'} < \infty, |g - g'|_{g,p,r} < \varepsilon\}\big\}_{\varepsilon > 0}$$

is a neighbourhood basis in the uniform topology. There arises a small difficulty. $g \in \mathcal{M}^{p,r}(I, B_k)$ must not be smooth and hence $|g - g'|_{g,p,r}$ must not be defined immediately. But in this case we use the density of $\mathcal{M}(I, B_k) \subset \mathcal{M}^{p,r}(I, B_k)$ and refer to [27] for further details.

Proposition 2.36 *The space $\mathcal{M}^{p,r}(I, B_k)$ is locally contractible.*

For the proof we refer to [35], lemma 3.8 or the chapter VII. □

Proposition 2.37 *In $\mathcal{M}^{p,r}(I, B_k)$ components and arc components coincide.* □

Set for $g \in \mathcal{M}^{p,r}(I, B_k)$

$$U_\varepsilon^{p,r}(g) = \{g' \in \mathcal{M}^{p,r}(I, B_k) \mid {}^b|g - g'|_g < \infty, {}^b|g - g'|_{g'} < \infty,$$
$$|g - g'|_{g,p,r} < \infty\}.$$

Lemma 2.38 *We have: $g' \in U^{p,r}(g)$ if and only if $g \in U^{p,r}(g')$.*

Proof. Assume $g' \in U^{p,r}(g)$. We have to show $|g - g'|_{g',p,r} < \infty$. Since $|g - g'|_{g,p} < \infty$ and g and g' are quasi isometric, we conclude $|g - g'|_{g',p} < \infty$. There remains to show $|\nabla'^i(\nabla' - \nabla)|_{g',p} < \infty$, $0 \le i \le r - 1$. But this can literally be done as in the proof of (U_2'). All arguments are symmetric in g and g' which proves the assertion. □

Lemma 2.39 *If $g' \in U^{p,r}(g)$ then $U^{p,r}(g') = U^{p,r}(g)$.*

Proof. Assume $h \in U^{p,r}(g')$. Then ${}^b|g' - h|_{g'} < \infty$, $|g' - h|_{g',p,r} < \infty$. By assumption ${}^b|g - g'|_{g'} < \infty$, $|g - g'|_{g,p,r} < \infty$. This implies ${}^b|g - h|_h < \infty$ and, according to the proof of (U_3'), $|g - h|_{g,p,r} < \infty$, $h \in U^{p,r}(g)$, $U^{p,r}(g') \subseteq U^{p,r}(g)$. The other inclusion follows quite similar.
□

Corollary 2.40 *If $U^{p,r}(g) \cap U^{p,r}(g') \ne \emptyset$, then $U^{p,r}(g') = U^{p,r}(g)$.*

Proposition 2.41 *Denote by $\text{comp}(g)$ the component of $g \in \mathcal{M}^{p,r}(I, B_k)$. Then,*

$$\text{comp}(g) \equiv \text{comp}^{p,r}(g) = U^{p,r}(g).$$

Proof. We start with $\text{comp}(g) \subseteq U^{p,r}(g)$. Let $g' \in \text{comp}(g)$ and $\{g_t\}_{0 \le t \le 1}$ be an arc between g and g'. For any $\varepsilon > 0$ this arc can be covered by a finite number of ε-balls $U_\varepsilon^{p,r}(g_{t_\sigma})$, $\sigma = 0, \ldots, s$, $g_{t_0} = g$, $g_{t_s} = g'$. We set $g_{t_\sigma} = g_\sigma$ and assume without loss of generality $U_\varepsilon^{p,r}(g_\sigma) \cap U_\varepsilon^{p,r}(g_{\sigma+1}) \ne \emptyset$. If this would not be the case, then we would choose a corresponding ordering. Then $U_\varepsilon^{p,r}(g_0) \cap U_\varepsilon^{p,r}(g_1) \ne \emptyset$. According to corollary 2.39, $g_1 \in U_\varepsilon^{p,r}(g_0)$. An easy induction gives $g_s = g' \in U^{p,r}(g)$, $\text{comp}(g) \subseteq U^{p,r}(g)$. Then $\{tg' + (1-t)g\}_{0 \le t \le 1}$ is an arc in $U^{p,r}(g)$ connecting g and g'. This follows from lemma 3.8 of [35] or chapter VII. □

By construction, $U^{p,r}(g)$ is open and, according to proposition 2.36, closed. Therefore, we have

Theorem 2.42 *Let M^n be open, $k \ge r > \frac{n}{p} + 1$. Then $\mathcal{M}^{p,r}(I, B_k)$ has a representation as a topological sum*

$$\mathcal{M}^{p,r}(I, B_k) = \sum_{i \in I} U^{p,r}(g_i).$$

□

We finish at this point the example of uniform structures of Riemannian metrics and turn to our last class of examples, uniform structures of metric spaces. We start with the Gromov–Hausdorff uniform structure. Let $Z = (Z, d_Z)$ be a metric space, $X, Y \subset Z$ subsets, $\varepsilon > 0$, define $U_\varepsilon(X) := \{z \in Z \mid \text{dist}(z, X) < \varepsilon\}$, analogously $U_\varepsilon(Y)$. Then the Hausdorff distance $d_H(X, Y) = d_H^Z(X, Y)$ is defined as

$$d_H^Z(X, Y) := \inf\{\varepsilon \mid X \subset U_\varepsilon(Y), Y \subset U_\varepsilon(X)\}.$$

If there is no such ε then we set $d_H^Z(X, Y) := \infty$. Then d_H^Z factorized by $d(\cdot, \cdot) = 0$ is an almost metric on all closed subsets, i.e. it has values in $[0, \infty]$ but satisfies all other conditions of a metric. If Z is compact then d_H^Z is a metric on the the set of all closed subsets. A metric space (X, d) is called *proper* if the closed balls $\overline{B_\varepsilon(x)}$ are compact for all $x \in X$, $\varepsilon > 0$. This implies that X is separable, complete and locally compact. In the sequel we restrict our attention to proper metric spaces. Let $X = (X, d_X)$, $Y = (Y, d_Y)$ be metric spaces, $X \sqcup Y$ their disjoint union. A metric d on $X \sqcup Y$ is called *admissible* if d restricts to d_X and d_Y, respectively. The *Gromov–Hausdorff distance* $d_{GH}(X, Y)$ is defined as

$$d_{GH}(X, Y) := \inf\{d_H(X, Y) \mid d \text{ is admissible on } X \sqcup Y\}.$$

Note that the Gromov–Hausdorff distance can be infinity. Originally Gromov defined d_{GH} as

$$d_{GH}(X, Y) := \inf\{d_H^Z(i(X), j(Y)) i : X \to Z, j : Y \to Z \\ \text{isometric embeddings into a metric space } Z\}.$$

It is a well known fact and can simply be proved that both definitions coincide.

Lemma 2.43 *If X and Y are compact metric spaces and $d_{GH}(X, Y) = 0$ then X and Y are isometric.*

This follows from the definition and an Arzela–Ascoli argument. □

Remark 2.44 The class of all metric spaces, even only up to isometry, is not a set. But the class of isometry classes of proper metric spaces is, so we only consider proper metric spaces in this chapter. A proper metric space X can be covered by a countable number of compact metric balls of fixed radius. Each such ball is isometric to a subset of $L^\infty([0, 1])$ and we obtain X after identification of the overlappings, i.e. we can understand X as a subset of $(\prod_{i=1}^\infty L^\infty([0, 1]))^2$. □

Denote by $\mathfrak{M}$ the set of all isometry classes $[X]$ of proper metric spaces X and $\mathfrak{M}_{GH} = \mathfrak{M}/\sim$ where $[X] \sim [Y]$ if $d_{GH}([X], [Y]) = 0$.

Proposition 2.45 d_{GH} defines an almost metric on $\mathfrak{M}$, i.e. it is a metric with values in $[0, \infty]$. □

We write in the sequel $X = [X]_{GH}$ if there cannot arise any confusion. Now we define the uniform structure. Let $\delta > 0$ and set
$$V_\delta := \{(X, Y) \in \mathfrak{M}_{GH}^2 \mid d_{GH}(X, Y) < \delta\}.$$

Lemma 2.46 $\mathfrak{B} = \{V_\delta\}_{\delta > 0}$ is a basis for a metrizable uniform structure $\mathfrak{U}_{GH}(\mathfrak{M})$.

Proof. $\mathfrak{B}$ is defined by a local metric. Hence it satisfies all desired conditions. □

Let $\overline{\mathfrak{M}}_{GH}$ be the completion of $\mathfrak{M}_{GH}$ with respect to $\mathfrak{U}_{GH}$ and denote the metric in $\overline{\mathfrak{M}}$ by $\overline{d}_{GH}$.

Lemma 2.47 $\mathfrak{M}_{GH} = \overline{\mathfrak{M}}_{GH}$ as. sets and d_{GH} and $\overline{d}_{GH}$ are locally equivalent, i.e. for any $X \in \mathfrak{M}_{GH} = \overline{\mathfrak{M}}_{GH}$ there exist equivalent neighbourhood bases by metric balls.

Proof. Cauchy sequences with respect to d_{GH} and $\mathfrak{U}_{GH}(\mathfrak{M}_{GH})$ coincide. But according to [56], proposition 10.1.7, p. 277, any Cauchy sequence in $\mathfrak{M}$ with respect to d_{GH} converges in $\mathfrak{M}_{GH}$. This proposition is formulated there for compact metric spaces, but the proof does not use compactness at any stage. The assertion concerning the local equivalence of d_{GH} and $\overline{d}_{GH}$ follows immediately from the formulae (2.4)–(2.6) in [68], II.2.7, p. 117. We refer also to 2.3, 2.4. □

Let $X \in \mathfrak{M}_{GH} = \overline{\mathfrak{M}}_{GH}$ and denote by $\text{comp}(X)$ and $\text{arccomp}(X)$ the component and arc component of X in $\mathfrak{M}_{GH}$, respectively.

A key role for all what follows is played by

Proposition 2.48 $\mathfrak{M}_{GH} = \overline{\mathfrak{M}}_{GH}$ is locally arcwise connected.

We refer to [33] for the proof. □

Corollary 2.49 *In $\mathfrak{M}_{GH}$ components and arc components coincide. Moreover, each component is open and $\overline{\mathfrak{M}}_{GH} = \mathfrak{M}_{GH}$ is the topological sum of its components,*

$$\mathfrak{M} = \sum_{i \in I} \operatorname{comp}(X_i).$$

□

Proposition 2.50 *Let $X \in \overline{\mathfrak{M}}_{GH} = \mathfrak{M}_{GH}$. Then $\operatorname{comp}(X)$ is given by*

$$\operatorname{comp}(X) = \{Y \in \mathfrak{M}_{GH} \mid d_{GH}(X, Y) < \infty\}.$$

Proof. Let $Y \in \operatorname{comp}(X) = \operatorname{arccomp}(X)$. Then there exists an arc between X and Y. For given $\varepsilon > 0$ this can be covered by a finite number, say r, of ε-balls. Hence $d_{GH}(X,Y) \le 2r\varepsilon < \infty$. If $d_{GH}(X,Y) = \varepsilon < \infty$ then we can construct an arc from X to Y as given in the proof of proposition 2.48. □

Hence we have a quite natural splitting of $\overline{\mathfrak{M}}_{GH} = \mathfrak{M}_{GH}$ into its components = arc components and a canonical topology and convergence inside each component. This is not – as usual until now – the pointed convergence of all metric balls but uniform convergence. We discuss this later. Any complete Riemannian manifold determines a unique component. First we give another characterization of components.

We call a map $\Phi : X \to Y$ *metrically semilinear* if it satisfies the following two conditions.

1. It is uniformly metrically proper, i.e. for each $R > 0$ there is an $S > 0$ such that the inverse image under Φ of a set of diameter R is a set of diameter at most S.

2. There exists a constant $C_\Phi \ge 0$ such that for all $x_1, x_2 \in X$

$$d(\Phi(x_1), \Phi(x_2)) \le d(x_1, x_2) + C_\Phi.$$

Two metric spaces X and Y are called *metrically semilinearly equivalent* if there exist metrically semilinear maps $\Phi : X \to Y$, $\Psi : Y \to X$ and constants D_X and D_Y, such that for all $x \in X$ and $y \in Y$

$$d(x, \Psi\Phi x) \leq D_X, \qquad d(\Phi\Psi y, y) \leq D_Y.$$

Proposition 2.51 $Y \in \text{comp}(X)$, *i.e.* $d_{GH}(X,Y) < \infty$ *if and only if X and Y are metrically semilinearly equivalent.*

We refer to [33] for the proof. □

The other class of uniform structures of particular meaning are Lipschitz uniform structures. A map $\Phi : X \longrightarrow X$ is called Lipschitz if there is a constant $C > 0$ such that

$$d(\Phi(x_1), \Phi(x_2)) \leq C \cdot d(x_1, x_2)$$

for all $x_1, x_2 \in X$.

Restricting now to Lipschitz maps, we create a local metric which takes into account the measure of expansivity.

Define for a Lipschitz map $\Phi : X \to Y$

$$\text{dil } \Phi := \sup_{\substack{x_1, x_2 \in X \\ x_1 \neq x_2}} \frac{d(\Phi x_1, \Phi x_2)}{d(x_1, x_2)}.$$

Set

$$\begin{aligned} d_L(X,Y) := \inf\{&\max\{0, \log \text{dil } \Phi\} + \max\{0, \log \text{dil } \Psi\} \\ &+ \sup_{x \in X} d(\Psi\Phi x, x) + \sup_{y \in Y} d(\Phi\Psi y, y) \\ &\mid \Phi : X \to Y, \Psi : Y \to X \text{ Lipschitz maps}\}, \end{aligned}$$

if $\{\ldots\} \neq \emptyset$ and $\inf\{\ldots\}$ is $< \infty$ and set $d_L(X,Y) = \infty$ in the other case. Then $d_L \geq 0$, symmetric and $d_L(X,Y) = 0$ if X and Y are isometric. Set $\mathcal{M}_L = \mathcal{M}/\sim$, where $X \sim Y$ if $d_L(X,Y) = 0$. That $\sim$ is in fact an equivalence relation follows from 2.52 below.

Let $\delta > 0$ and define

$$V_\delta = \{(X,Y) \in \mathcal{M}_L^2 \mid d_L(X,Y) < \infty\}.$$

Proposition 2.52 $\mathfrak{B} = \{V_\delta\}_{\delta>0}$ *is a basis for a metrizable uniform structure* $\mathfrak{U}_L(\mathcal{M}_L)$.

We refer to [33] for the proof. □

Denote by $\overline{\mathfrak{M}}_L^{\mathfrak{U}_L}$ the completion of $\mathfrak{M}_L$ with respect to $\mathfrak{U}_L$. We come back to the topological properties of $\overline{\mathfrak{M}}_L$ below. Before doing this we introduce still another important Lipschitz uniform structure. Define for $X, Y \in \mathfrak{M}$

$$d_{L,top}(X,Y) := \inf\{\max\{0, \log\operatorname{dil}\Phi\} + \max\{0, \log\operatorname{dil}\Phi^{-1}\} \mid \Phi: X \to Y \text{ is a bi-Lipschitz homeomorphism}\}$$

if $\{\ldots\} \neq \emptyset$ and $= \infty$ in the other case.
Then $d_{L,top}(X,Y) \geq 0$, symmetric and $d_{L,top}(X,Y) = 0$ if X and Y are isometric. Set $\mathfrak{M}_{L,top} := \mathfrak{M}/\sim$ where $X \sim Y$ if $d_{L,top}(X,Y) = 0$. Then $d_{L,top}$ is an almost metric on $\mathfrak{M}_{L,top}$.

Remark 2.53 Gromov defined in [39]

$$d_{L,top}(X,Y) := \inf\{|\log\operatorname{dil}\Phi| + |\log\operatorname{dil}\Phi^{-1}| \mid \Phi: X \to Y \text{ is a bi-Lipschitz homeomorphism.}\}$$

But this does not work since this $d_{L,top}$ does not satisfy the triangle inequality. $\log(x \cdot y) = \log x + \log y$, log is monotone increasing, but $z_1 \leq z_2 + z_3$ does not imply $|z_1| \leq |z_2| + |z_3|$. There are explicit counterexamples to the triangle inequality of the local $|\log\operatorname{dil}(\cdot)|$–metric. □

Set for $\delta > 0$

$$V_\delta = \{(X,Y) \in \mathfrak{M}_{L,top}^2 \mid d_{L,top}(X,Y) < \delta\}.$$

Proposition 2.54 $\mathfrak{B} = \{V_\delta\}_{\delta>0}$ *is a basis for a metrizable uniform structure* $\mathfrak{U}_{L,top}$. □

Denote by $\overline{\mathfrak{M}}_{L,top}^{\mathfrak{U}_{L,top}}$ the completion of $\mathfrak{M}_{L,top}$ with respect to $\mathfrak{U}_{L,top}$.

Proposition 2.55 $\mathfrak{M}_{L,top}$ *is complete with respect to* $\mathfrak{U}_{L,top}$, *i.e.* $\overline{\mathfrak{M}}_{L,top} = \mathfrak{M}_{L,top}$.

We refer to [33] for the proof. □

Next we return to $\mathfrak{M}_L$. Above we discussed $\mathfrak{M}_{L,top}$ since the proofs for $\mathfrak{M}_L$ are modelled by the proofs for $\mathfrak{M}_{L,top}$. The next result is valid only for noncompact proper metric spaces (= classes of spaces, exactly spoken). Such spaces have infinite diameter. The restriction to noncompact proper spaces is no restriction for us since we have in mind them only.

Denote by $\mathfrak{M}_L(nc)$ the (classes of) noncompact proper metric spaces. We proved in [33]

Proposition 2.56 $\mathfrak{M}_L(nc)$ *is complete with respect to* $\mathfrak{U}_L(\mathfrak{M}_L)$ *restricted to* $\mathfrak{M}_L(nc)$. □

Remark 2.57 If we restrict to compact metric spaces then by Arzela–Ascoli arguments $d_{L,top}(X,Y) = 0$ or $d_L(X,Y) = 0$ always imply X isometric to Y. Hence the factorization by $\underset{L,top}{\sim}$ or $\underset{L}{\sim}$ would not be necessary. For noncompact proper metric spaces the corresponding question is open (at least for us). If a sequence $(\Phi)_i : X \to Y$ step by step realizes $d_{L,top}(X,Y) = 0$ then the composition with an unbounded sequence of isometries $\chi_i : Y \to Y$ yields the same constants in all estimates but it could destroy the existence of converging (on compact subsets) subsequences of maps. This does not mean that X is not isometric to Y but it only destroys the natural Arzela–Ascoli argument. □

The next step is to prove that $\mathfrak{M}_L(nc)$ and $\mathfrak{M}_{L,top}$ are locally arcwise connected. We start with $\mathfrak{M}_{L,top}$ since the proof is easier and the proof for $\mathfrak{M}_L(nc)$ is in certain sense modelled by that for $\mathfrak{M}_{L,top}$.

Proposition 2.58 $\mathfrak{M}_{L,top}$ *is locally arcwise connected.*

The proof is performed in [33]. □

Corollary 2.59 *In $\mathfrak{M}_{L,top}$ components and arc components coincide.* □

Theorem 2.60 a) $\mathfrak{M}_{L,top}$ *has a representation as a topological sum*

$$\mathfrak{M}_{L,top} = \sum_i \operatorname{comp} X_i.$$

b) $\operatorname{comp} X = \{Y \in \mathfrak{M}_{L,top} | d_{L,top}(X,Y) < \infty\}$.

Proof. a) follows immediately from 2.59. b) Assume $d_{L,top}(X,Y) < \varepsilon$. The proof of 2.58 provides an arc $\{X_t\}_t$ between X and Y, i. e. $Y \in \operatorname{comp} X$. Conversely, suppose X and Y can be connected by an arc $\{X_t\}_t$ in $\mathfrak{M}_{L,top}$. For any $\delta > 0$ this can be covered by r δ-balls (w. r. t. $d_{L,top}$). Then $d_{L,top}(X,Y) < \infty$. □

A crucial step are the same assertions for $\mathfrak{M}_L$ which we prove now. We started with $\mathfrak{M}_{L,top}$ since the proofs for $\mathfrak{M}_L$ are (slightly) modelled by them for $\mathfrak{M}_{L,top}$.

Proposition 2.61 $\overline{\mathfrak{M}}_L(nc) = \mathfrak{M}_L(nc)$ *is locally arcwise connected.*

We refer to [33] for the rather long proof. □

We immediately conclude from 2.61

Theorem 2.62 *In $\mathfrak{M}_L$ and $\mathfrak{M}_L(nc)$ components coincide with arc components. $\mathfrak{M}_L$ and $\mathfrak{M}_L(nc)$ have topological sum representations*

$$\mathfrak{M}_L = \operatorname{comp}_L(\text{point}) + \sum_{i \in I} \operatorname{comp}_L(X_i),$$

$$\mathfrak{M}_L(nc) = \sum_{i \in I} \operatorname{comp}_L(X_i)$$

and
$$\operatorname{comp}_L(X) = \{Y \in \mathfrak{M}_L | d_L(X,Y) < \infty\}.$$

In particular all compact spaces lie in the component of the 1-point-space. □

Finally we define still three further uniform structures which measure or express the homotopy neighbourhoods and, secondly, admit only compact deviations of the spaces inside one component.

Define

$$d_{L,h}(X,Y) := \inf \Big\{ \max\{0, \log \operatorname{dil} \Phi\} + \max\{0, \log \operatorname{dil} \Psi\} \\ + \sup_X d(\Psi\Phi x, x) + \sup_Y d(\Phi\Psi y, y) \\ | \Phi : X \longrightarrow Y, \Psi : Y \longrightarrow X \text{ are (uniformly} \\ \text{proper) Lipschitz homotopy equivalences,} \\ \text{inverse to each other} \Big\}$$

if there exist such homotopy equivalences and set $d_{L,h}(X,Y) = \infty$ in the other case. Here and in the sequel we require from the homotopies to id_X or id_Y, respectively, that they are uniformly proper and Lipschitz.

$d_{L,h} \geq 0$, $d_{L,h}$ is symmetric and $d_{L,h}(X,Y) = 0$ if X and Y are isometric. Define $\mathfrak{M}_{L,h} = \mathfrak{M}/\sim$, $X \sim Y$ if $d_{L,h}(X,Y) = 0$ and set

$$V_\delta = \{(X,Y) \in \mathfrak{M}_{L,h}^2 | d_{L,h}(X,Y) < \delta\}.$$

That this $\sim$ is in fact an equivalence relation follows from 2.63 below which we will only state here. The proof consists only of a generalized triangle inequality and is completely parallel to that of 2.52. We only restrict the class of admitted maps to uniformly proper Lipschitz homotopy equivalences.

Proposition 2.63 $\mathfrak{B} = \{V_\delta\}_{\delta > 0}$ *is a basis for a metrizable uniform structure* $\mathfrak{U}_{L,h}(\mathfrak{M}_{L,h})$. □

Proposition 2.64 $\overline{\mathfrak{M}}_{L,h}^{\mathfrak{U}_{L,h}}(nc) = \mathfrak{M}_{L,h}(nc)$.

We omit the proof which is again parallel to that of 2.56. □

Denote by $\mathrm{arccomp}_{L,h}(X)$ the arc component of X in $\mathfrak{M}_{L,h}$.

Proposition 2.65 *If $Y \in \mathrm{arccomp}_{L,h}(X)$ then X and Y are (uniformly proper) Lipschitz homotopy equivalent. In particular $d_{L,h}(X,Y) < \infty$.*

Proof. Cover the arc between X and Y by a finite number of $d_{L,h}$-balls and use transitivity of homotopy equivalence. □

Let $\Phi : X \longrightarrow Y$, $\Psi : Y \longrightarrow X$ be Lipschitz maps. We say that Φ and Ψ are stable Lipschitz homotopy equivalences at ∞ inverse to each other if there exists a compact set $K_X^0 \subset X$ s.t. for any $K_X^0 \subset K_X$ there exists $K_Y \subset Y$ s.t. $\Phi|_{X \setminus K_X} : X \setminus K_X \longrightarrow Y \setminus K_Y$ is a Lipschitz h.e. with homotopy inverse $\Psi|_{Y \setminus K_Y}$ and Ψ has the analogous property.
Set
$$d_{L,h,rel}(X,Y) := \inf \Big\{ \max\{0, \log \mathrm{dil}\, \Phi\} + \max\{0, \log \mathrm{dil}\, \Psi\} \\ + \sup_X d(\Psi\Phi x, x) + \sup_Y d(\Phi\Psi y, y) \\ \mid \Phi : X \longrightarrow Y, \Psi : Y \longrightarrow X \text{ are stable} \\ \text{Lipschitz homotopy equivalences at } \infty \\ \text{inverse to each other} \Big\}$$
if there exist such maps Φ, Ψ and set $d_{L,h,rel}(X,Y) = \infty$ in the other case. Then $d_{L,h,rel} \geq 0$, symmetric and $d_{L,h,rel} = 0$ if X and Y are isometric. Set $\mathfrak{M}_{L,h,rel} = \mathfrak{M}/_\sim$, $X \sim Y$ if $d_{L,h,rel}(X,Y) = 0$ and set
$$V_\delta = \{(X,Y) \in \mathfrak{M}^2_{L,h,rel} | d_{L,h,rel}(X,Y) < \delta\}.$$

Proposition 2.66 $\mathfrak{B} = \{V_\delta\}_{\delta>0}$ *is a basis for a metrizable uniform structure $\mathfrak{U}_{L,h,rel}$.* □

Proposition 2.67 *If $Y \in \mathrm{arccomp}_{L,h,rel}(X)$ then X and Y are stably Lipschitz homotopy equivalent at ∞. In particular $d_{L,h,rel}(X,Y) < \infty$.* □

The last uniform structure $\mathfrak{U}_{L,top,rel}$ is defined by $d_{L,top,rel}(X,Y)$, where we require that $\Phi : X \longrightarrow Y$, $\Psi : Y \longrightarrow X$ are outside compact sets bi–Lipschitz homeomorphisms, inverse to each other. We obtain $\mathfrak{M}_{L,top,rel}$. There holds an analogue of 2.66 and 2.67.

We finish this considerations with a scheme which makes clear our achievements, where we refer to section 5 for the definition of coarse structures.

One coarse equivalence class

↙ splits into ↘

many GH–components   many L–components,

one L–component = arc component

↙ splits into ↘

L,top,rel–arc components   L,h,rel–arc components

↙   ↘

L,top–components = L,top–arc components   L,h–arc components.

It is now a natural observation that the classification of non-compact proper metric spaces splits into two main tasks

1. "counting" the components at any horizontal level,
2. "counting" the elements inside each component.

A really complete solution to these two problems, i.e. a complete characterization by computable and handy invariants, is now–a–day hopeless. It is a similar utopic goal as the "classification of all topological spaces". Nevertheless stands the task to define series of invariants which at least permit to decide (in good cases) nonequivalence. This will be the topic of the second part of this section, of sections 2 and 3.

Finally we remark that GH–components $(d_{GH}(X,Y) < \infty)$ and L–components $(d_L(x,Y))$ are very different. Roughly speaking, d_{GH} is in the small unsharp and in the large relatively sharp, d_L quite inverse.

In section 5 we define invariants for the components of our uniform structures. It is clear that an invariant of $\text{comp}_{\mathfrak{U}}$ is also an invariant of $\text{comp}_{\mathfrak{U}'} \subset \text{comp}_{\mathfrak{U}}$, $\mathfrak{U}'$ finer than $\mathfrak{U}$.

3 Completed manifolds of maps

Our next class of examples for non–linear Sobolev structures are manifolds of maps and diffeomorphism groups.

Let (M^n, g), $(N^{n'}, h)$ be open, complete, satisfying (I) and (B_k) and let $f \in C^\infty(M, N)$. Then the differential $f_* = df$ is a section of $T^*M \otimes f^*TN$. f^*TM is endowed with the induced connection $f^*\nabla^h$ which is locally given by

$$\Gamma^\nu_{i\mu} = \partial_i f^\alpha(x) \Gamma^{h,\nu}_{\alpha,\mu}(f(x)), \quad \partial_i = \frac{\partial}{\partial x^i}.$$

∇^g and $f^*\nabla^h$ induce metric connections ∇ in all tensor bundles $T^q_s(M) \otimes f^*T^u_v(N)$. Therefore $\nabla^m df$ is well defined. Since (I) and (B_0) imply the boundedness of the g_{ij}, g^{km}, $h_{\mu\nu}$ in normal coordinates, the conditions df to be bounded and $\partial_i f$ to be bounded are equivalent.

In local coordinates

$$\sup_{x \in M} |df|_x = \sup \operatorname{tr}^g(f^*h) = \sup g^{ij} h_{\mu\nu} \partial_j f^\mu \partial_i f^\nu.$$

For (M^n, g), $(N^{n'}, h)$ of bounded geometry up to order k and $m \le k$ we denote by $C^{\infty,m}(M, N)$ the set of all $f \in C^\infty(M, N)$ satisfying

$$^{b,m}|df| := \sum_{\mu=0}^{m-1} \sup_{x \in M} |\nabla^\mu df|_x < \infty.$$

Assume (M^n, g), $(N^{n'}, h)$ are open, complete, and of bounded geometry up to order k, $r \le m \le k$, $1 \le p < \infty$, $r > \frac{n}{p} + 1$. Consider $f \in C^{\infty,m}(M, N)$. According to theorem 1.16 b) for $r > \frac{n}{p} + s$

$$\Omega^{p,r}(f^*TN) \hookrightarrow {}^{b,s}\Omega(f^*TN), \quad (3.1)$$
$$^{b,s}|Y| \le D \cdot |Y|_{p,r}, \quad (3.2)$$

where $|Y|_{p,r} = \left(\int \sum_{i=0}^{r} |\nabla^i Y|^p \, \mathrm{dvol} \right)^{\frac{1}{p}}$. Set for $\delta > 0$, $\delta \cdot D \le \delta_N < r_{\mathrm{inj}}(N)/2$, $1 \le p < \infty$, $V_\delta = \{(f, g) \in C^{\infty,m}(M, N) \times C^{\infty,m}(M, N) \,|\, \exists Y \in \Omega^p_r(f^*TN)$ such that $g = g_Y = \exp Y$ and $|Y|_{p,r} < \delta\}$.

Proposition 3.1 $\mathfrak{B} = \{V_\delta\}_{0 < \delta < r_{\mathrm{inj}}(N)/2D}$ *is a basis for uniform structure* $\mathfrak{U}^{p,r}(C^{\infty,m}(M, N))$.

We refer to [31], [27] for the very long proof. □

$\mathfrak{U}^{p,r}(C^{\infty,m}(M, N))$ is metrizable. Let ${}^m\Omega^{p,r}(M, N)$ be the completion of $C^{\infty,m}(M, N)$. From now on we assume $r = m$ and denote ${}^r\Omega^{p,r}(M, N) = \Omega^{p,r}(M, N)$.

Theorem 3.2 *Let (M^n, g), (N^n, h) be open, complete, of bounded geometry up to order k, $1 \le p < \infty$, $r \le k$, $r > \frac{n}{p} + 1$. Then each component of $\Omega^{p,r}(M, N)$ is a C^{1+k-r}-Banach manifold, and for $p = 2$ it is a Hilbert manifold.*

We refer to [31], [27] for the proof. □

The elements of $\Omega^{p,r}(M, N)$ are characterized by the following property. $f \in \Omega^{p,r}(M, N)$ if and only if for every $\varepsilon > 0$ there exist $\tilde{f} \in C^{\infty,r}(M, N)$ and a Sobolev vector field X along $\tilde{f}$, $X \in \Omega^{p,r}(\tilde{f}^*TN)$, $|X|_{p,r} < \varepsilon$, such that

$$f(x) = (\exp X)(x) = \exp_{\tilde{f}(x)} X_{\tilde{f}(x)} = (\exp_{\tilde{f}} X \circ \tilde{f})(x). \quad (3.3)$$

In particular $C^{\infty,r}(M, N)$ is dense in $\Omega^{p,r}(M, N)$. A special case is given if we restrict to diffeomorphisms. Let (M^n, g), (N^n, h)

as in the hypothesis of theorem 3.2. Define

$$\mathcal{D}^{p,r}(M,N) = \{f \in \Omega^{p,r}(M,N) \mid f \text{ is a diffeomorphism} \\ \text{ and there exists constants } c, C > 0 \text{ such} \\ \text{ that } c \leq \inf_{x \in M} |df|_x \leq \sup_x |df|_x \leq C\}. \quad (3.4)$$

(3.4) automatically implies the existence of constants $c_1, C_1 > 0$ such that

$$c_1 \leq \inf_x |df^{-1}|_x \leq \sup_x |df^{-1}|_x \leq C_1. \quad (3.5)$$

In fact, for diffeomorphisms (3.4) and (3.5) are equivalent. Moreover, (3.4) is an open condition in $\Omega^{p,r}(M,N)$. Hence we have

Theorem 3.3 *Suppose the hypothesis of 3.2. Then each component of $\mathcal{D}^{p,r}(M,N)$ is a C^{1+k-r}-Banach manifold and for $p = 2$ it is a Hilbert manifold.* □

Denote $\mathcal{D}^{p,r}(M) = \mathcal{D}^{p,r}(M,M)$.

Corollary 3.4 *Suppose for $M = N$ the hypothesis of 3.2. Then each component of $\mathcal{D}^{p,r}(M)$ is a C^{1+k-r}-Banach manifold and for $p = 2$ it is a Hilbert manifold.* □

In the sequel, we need still a relative version of these manifolds of maps.

Suppose (M^n, g), (N^n, h), k, r, p as in 3.2 and that there exist compact submanifolds $K_M \subset M$, $K_N \subset N$ such that there exists $f \in \Omega^{p,r}(M,N)$ with the following properties.

a) $f|_{M \setminus K_M}$ maps $M \setminus K_M$ diffeomorphic onto $N \setminus K_N$ and
b) there exist constants $c, C > 0$ such that

$$c \leq \inf_{x \in M \setminus K_M} |df|_{M \setminus K}|_x \leq \sup_{x \in M \setminus K_M} |df|_{M \setminus K}|_x \leq C. \quad (3.6)$$

Then, automatically,

$$c_1 \leq \inf_{y \in N \setminus K_N} |d(f|_{M \setminus K_M})^{-1}|_y \leq \sup_{y \in N \setminus K_N} |d(f|_{M \setminus K_M})^{-1}|_y \leq C_1. \quad (3.7)$$

We denote for fixed K_M, K_N the subset $\subset \Omega^{p,r}(M,N)$ of these f by $\mathcal{D}_{rel}^{p,r}(M,N)$. Clearly, $\mathcal{D}_{rel}^{p,r}(M,N)$ depends on the choice of K_M, K_N.

Finally, we need this construction still for Riemannian vector bundles. Let $(E_i, h_i, \nabla^{h_i}) \longrightarrow (M_i^n, g_i)$, $i = 1, 2$ be Riemannian vector bundles satisfying (I), $(B_k(M_i, g_i))$, $(B_k(E_i, h_i, \nabla_i))$, $k \geq r > \frac{n}{p} + 1$. If we endow the total spaces E_i with the Kaluza–Klein metric $g_E(X, Y) = h(X^\nu, Y^\nu) + g_M(\pi_* X, \pi_* Y)$, X^ν, Y^ν vertical components, then E_1, E_2 are again manifolds with bounded geometry (cf. [30]) and $\Omega^{p,r}(E_1, E_2)$ is well defined. If we restrict to bundle maps $(f_E, f_M = \pi \circ f_E \circ \pi^{-1}$ then we obtain a subset $\Omega_{vb}^{p,r}(E_1, E_2) \subset \Omega^{p,r}(E_1, E_2)$. Quite analogously to above we define $\mathcal{D}_{vb}^{p,r}(E_1, E_2) \subset \Omega_{vb}^{p,r}(E_1, E_2)$ if the bundles are isomorphic and $\mathcal{D}_{vb,rel}^{p,r}(E_1, E_2)$ if they are isomorphic over $M \setminus K_M$ and $N \setminus K_N$. Here we require (3.3) and (3.5) both for f_E and f_M. We apply this notations in the next section.

4 Uniform structures of manifolds and Clifford bundles

We introduce in chapters IV – VI relative index theory, relative eta and zeta functions, relative determinants and relative analytic torsion. The whole approach relies on the following construction. We endow the set of isometry classes of Clifford bundles (of bounded geometry) with a metrizable uniform structure, define generalized components $\mathrm{gen\,comp}(E)$ (= set of Clifford bundles E' with finite Sobolev distance from a given E), associate the corresponding generalized Dirac operators D, D' and make all constructions for the pair D, D', where D' is running through $\mathrm{gen\,comp}(E)$. The first step for doing this is the introduction of the corresponding uniform structure(s). This is the content of this section. The applications will be performed in chapters IV – VI.

Denote by $\mathfrak{M}^n(mf, I, B_k)$ the set of isometry classes of n–dimensional Riemannian manifolds (M^n, g) satisfying the conditions (I) and (B_k). We defined for $(M_1^n, g_1), (M_2^n, g_2) \in \mathfrak{M}^n(mf, I, B_k)$ and $k > \frac{n}{p}+1$ the diffeomorphisms $\mathcal{D}^{p,r}(M_1, M_2)$ and for appropriate compact submanifolds $K_i \subset M_i$, $M_1 \setminus K_1 \cong M_2 \setminus K_2$ the maps $\mathcal{D}_{rel}^{p,r}(M_1, M_2) \equiv \mathcal{D}_{rel}^{p,r}(M_1, M_2; K_1, K_2) \subset \Omega^{p,r}(M_1, M_2)$ as at the end of section 3. Recall $^b|df| = \sup_x |df|_x$. The elements f of $\mathcal{D}_{rel}^{p,r}, \Omega^{p,r}(M_1, M_2)$ are not smooth. For $k \geq r > \frac{n}{p} + 2$ they are C^2. Hence f^*g is a C^1 metric. This would cause some troubles if we would consider in the sequel only classical derivatives which would disappear if we work with distributional derivatives. Another way to work with the non–smoothness of f^*g is to work with smooth approximations of f. We decide to go this way and define $C^{r,p,r}(M_1, M_2) = \{f \in \Omega^{p,r}(M_1, M_2) | f \in C^{r+1}(M_1, M_2)$ and $^b|\nabla^i df| < \infty, i = 1, \ldots, r\}$. Completing the uniform structure below, we end up with fs $\in \mathcal{D}_{rel}^{p,r}, \Omega^{p,r}$, i.e. the restriction at the beginning to C^k fs implies at the end no further restriction. We restrict in the sequel to $k \geq r > \frac{n}{p} + 2$. Further we remark that the conditions $c \leq {}^b|f_*| \leq C$ and $c_1 \leq {}^b|f_*^{-1}| \leq C_1$ are equivalent: $^b|f_*^{-1}| \geq \frac{1}{C} = c_1$ follows from $^b|f_* f_*^{-1}| = 1$ and $^b|f_*| \leq C$ and $^b|f_*^{-1}| \leq C_1(c, C)$ follows from elementary matrix calculus. If f^* is the induced map between twofold covariant tensors then $c_2 = c^2 \leq {}^b|f^*| \leq C^2 = C_2$, similarly $c_3 \leq {}^b|f^{*-1}| \leq C_3$. Under these conditions, $|\nabla^i f_*| \leq d, 1 \leq i \leq \nu$ implies $|\nabla^i f^*| \leq d_1$, $1 \leq i \leq \nu$, and $|\nabla^i f^{*-1}| \leq d_2, 1 \leq i \leq \nu$, where d_1, d_2 are continuous functions in c, C, d. All this follows from $f_*^{-1} f_* = \mathrm{id}_*$, $f^{*-1} f^* = \mathrm{id}^*$ and $0 = \nabla \mathrm{id}_* = \nabla \mathrm{id}^*$.

Consider now pairs $(M_1^n, g_1), (M_2^n, g_2) \in \mathfrak{M}^n(mf, I, B_k)$ with this property: There exist compact submanifolds $K_1^n \subset M_1^n$, $K_2^n \subset M_2^n$ and an $f \in \mathcal{D}_{rel}^{p,r}(M_1, M_2, K_1, K_2)$. For such pairs define

$$d^{p,r}_{L,diff,rel}((M_1,g_1),(M_2,g_2)) := \inf \Big\{ \max\{0, \log^b |df|\}$$
$$+ \max\{0, \log^b |dh|\} + \sup_{x \in M_1} \text{dist}(x, hfx)$$
$$+ \sup_{y \in M_2} \text{dist}(y, hfy) + \sup_{\substack{x \in M_1 \\ 1 \le i \le r}} |\nabla^i df| + \sup_{\substack{x \in M_2 \\ 1 \le i \le r}} |\nabla^i dh|$$
$$+ |(f|_{M_1 \setminus K_1})^* g_2 - g_1|_{M_1 \setminus K_1}|_{g_1, p, r}$$
$$\Big| f \in C^{r,p,r}(M_1, M_2), h \in C^{r,p,r}(M_1, M_2)$$
$$\text{and for some } K_1 \subset M_1, K_2 \subset M_2 \text{ holds}$$
$$f|_{M_1 \setminus K_1} \in \mathcal{D}^{p,r}(M_1 \setminus K_1, M_2 \setminus K_2)$$
$$\text{and } h_{M_2 \setminus K_2} = (f|_{M_1 \setminus K_1})^{-1} \Big\}, \qquad (4.1)$$

if $\{\ldots\} \ne \emptyset$ and $\inf\{\ldots\} < \infty$. In the other case set $d^{p,r}_{L,diff,rel}((M_1,g_1),(M_2,g_2)) = \infty$. Set

$$V_\delta = \{((M_1,g_1),(M_2,g_2)) \in (\mathfrak{M}^n(mf,I,B_k))^2 \,|\,$$
$$d^{p,r}_{L,diff,rel}((M_1,g_1),(M_2,g_2)) < \delta\}.$$

Proposition 4.1 *Suppose $r > \frac{n}{p} + 2$. Then $\mathfrak{B} = \{V_\delta\}_{\delta > 0}$ is a basis for a metrizable uniform structure on $\mathfrak{M}^n(mf, I, B_k)/ \sim$, where $(M_1, g_1) \sim (M_2, g_2)$ if $d^{p,r}_{L,diff,rel}((M_1,g_1),(M_2,g_2)) = 0$.*

Proof. We have to verify (U_2') and (U_3'), i.e. the symmetry and transitivity of the basis (not of $d^{p,r}_{L,diff,rel}$) and start with (U_2'). For this it is sufficient that

$$d^{p,r}_{L,diff,rel}((M_1,g_1),(M_2,g_2)) < \delta \qquad (4.2)$$

implies

$$d^{p,r}_{L,diff,rel}((M_1,g_1),(M_2,g_2)) < \delta'(\delta) \qquad (4.3)$$

such that

$$\delta'(\delta) \xrightarrow[\delta \to 0]{} 0. \qquad (4.4)$$

We consider the single numbers in the set (4.1). The first number, sum of 4 terms is symmetric in f and h. The second number $\sup_{\substack{y \in M_1 \\ 1 \leq i \leq r}} |\nabla^i df| + \sup_{\substack{y \in M_2 \\ 1 \leq i \leq r}} |\nabla^i dh|$ is symmetric in f and h too. Suppose

$$|(f|_{M_1 \setminus K_1})^* g_2 - g_1|_{M_1 \setminus K_1}|_{g_1,p,r}$$
$$\equiv \Big\{ \int_{M_1 \setminus K_1} |(f|_{M_1 \setminus K_1})^* g_2 - g_1|^p_{g_1,x}$$
$$+ \sum_{i=0}^{r-1} ||(\nabla^{g_1})^i (\nabla^{g_1} - \nabla^{f^* g_2})|^p_{g_1,x} \, \mathrm{dvol}_x(g_1) \Big\}^{\frac{1}{p}} < \delta_1. \quad (4.5)$$

Now we have to estimate

$$|(f|_{M_1 \setminus K_1})^{*-1} g_1 - g_2|_{M_2 \setminus K_2}|_{g_2,p,r}.$$

We omit in the notation $|_{M_i \setminus K_i}$ since in the remaining part of the proof we restrict to this. Then

$$|f^{*-1} g_1 - g_2|_{g_2,p,r} = |(f^{*-1})(g_1 - f^* g_2)|_{g_2,p,r}. \quad (4.6)$$

Now

$$|f^{*-1}(g_1 - f^* g_2)|_{g_2, y=f(x)} \leq |f^{*-1}|_x |g_1 - f^* g_2|_{g_1,x} \leq C_3 |g_1 - f^* g_2|_{g_1,x}.$$

Hence, in the case $r = 0$,

$$|f^{*-1}(g_1 - f^* g_2)|_{g_2,p,0} \leq {}^b|f^{*-1}| \, |g_1 - f^* g_2|_{g_1,p,0} = P_0(|g_1 - f^* g_2|_{g,p,0}). \quad (4.7)$$

Consider now the case $r = 1$. Then

$$\begin{aligned}
|\nabla(f^{*-1}(g_1 - f^* g_2))|_y &= |\nabla(f^{*-1})(g_1 - f^* g_2) \\
&\quad + f^{*-1} \nabla(g_1 - f^* g_2)|_y \\
&\leq {}^b|\nabla(f^{*-1})| \cdot |g_1 - f^* g_2| \\
&\quad + {}^b|f^{*-1}| \cdot |\nabla(g_1 - f^* g_2)|. \quad (4.8)
\end{aligned}$$

We briefly discuss the case $r = 2$, to indicate the general rule.

$$|\nabla[\nabla(f^{*-1}(g_1 - f^*g_2))]|_{g_2,y}$$
$$= |\nabla[\nabla(f^{*-1})(g_1 - f^*g_2) + f^{*-1}\nabla(g_1 - f^*g_2)]|_{g_2,y}$$
$$= |\nabla^2(f^{*-1})(g_1 - f^*g_2) + \nabla(f^{*-1})\nabla(g_1 - f^*g_2)$$
$$+ \nabla f^{*-1}\nabla(g_1 - f^*g_2) + f^{*-1}\nabla^2(g_1 - f^*g_2)|_{g_2,y}$$
$$\leq {}^b|\nabla^2 f^{*-1}| \cdot |g_1 - f^*g_2| + {}^b|\nabla f^{*-1}||\nabla(g_1 - f^*g_2)|$$
$$+ {}^b|\nabla f^{*-1}||\nabla(g_1 - f^*g_2)| + {}^b|f^{*-1}||\nabla^2(g_1 - f^*g_2)|. \quad (4.9)$$

Continuing in this manner, we obtain on the right hand side linear polynomials in $|\nabla^i(g_1 - f^*g_2)|$ without constant term and where the coefficients can be estimated by δ. Summing up (4.7), (4.8), (4.9) and integrating over $M_1 \setminus K_1$, we obtain

$$|(f|_{M_1 \setminus K_1})^{*-1} g_1 - g_2|_{M_2 \setminus K_2}|_{g_2,p,r} \leq C \cdot |g_1 - f^*g_2|_{g_1,p,r}. \quad (4.10)$$

In particular,
$$|g_1 - f^*g_2|_{g_1,p,r} < \delta \longrightarrow 0 \quad (4.11)$$

implies
$$|f^{*-1}g_1 - g_2|_{g_2,p,r} < \delta'(\delta) \xrightarrow[\delta \to 0]{} 0. \quad (4.12)$$

This proves (U_2'). Completely similar is the proof of (U_3'), i.e. the transitivity of the basis.

We assume $(M_1, g_1) \underset{h_1}{\overset{f_1}{\rightleftarrows}} (M_2, g_2) \underset{h_2}{\overset{f_2}{\rightleftarrows}} (M_3, g_3)$, $f_i : M_i \setminus K_i \cong M_{i+1} \setminus K_{i+1}$, $i = 1, 2$, with the desired properties. The triangle inequality for the sum of the first 4 terms in the set (4.1) is just proposition 2.52. Consider the next to numbers in the set (4.1). Applying the Leibniz rule, immediately yields

$$\sup_{\substack{x \in M_1 \\ 1 \leq i \leq r}} |\nabla^i(f_{2*}f_{1*})| + \sup_{\substack{z \in M_3 \\ 1 \leq i \leq r}} |\nabla^i(h_{1*}h_{2*})|$$
$$\leq C\Big[\sup_{\substack{x \in M \\ 1 \leq i \leq r}} |\nabla^i f_{1*}| \cdot \sup_{\substack{y \in M_2 \\ 1 \leq i \leq r}} |\nabla^i f_{2*}|$$
$$+ \sup_{\substack{z \in M_3 \\ 1 \leq i \leq r}} |\nabla^i h_{2*}| \cdot \sup_{\substack{y \in M_2 \\ 1 \leq i \leq r}} |\nabla^i h_{1*}|\Big], \quad (4.13)$$

where C essentially is an expression in binomial coefficients. (4.13) expresses the desired transitivity of the basis. The desired transitivity of the last number in the set (4.1) would be established if

$$|f_1^* g_2 - g_1|_{g_1,p,r} < \delta, \quad |f_2^* g_3 - g_2|_{g_2,p,r} < \delta \qquad (4.14)$$

would imply

$$|(f_2 f_1)^* g_3 - g_1|_{g_1,p,r} < \delta'(\delta) \text{ with } \delta'(\delta) \xrightarrow[\delta \to 0]{} 0. \qquad (4.15)$$

We estimate by the triangle inequality for Sobolev norms

$$\begin{aligned}
&|(f_2 f_1)^* g_3 - g_1|_{M_1 \setminus K_1, g_1, p, r} \\
&\equiv |f_1^*(f_2^* g_3 - f_1^{*-1} g_1)|_{g_1,p,r} \\
&\leq |f_1^*(f_2^* g_3 - g_2)|_{g_1,p,r} + |f_1^*(g_2 - f_1^{*-1} g_1)|_{g_1,p,r} \\
&= |f_1^*(f_2^* g_3 - g_2)|_{g_1,p,r} + |f_1^* g_2 - g_1)|_{g_1,p,r} \\
&\leq \sup_{\substack{x \in M_1 \\ 0 \leq i \leq r}} |\nabla^i f_{1*}| \cdot |f_2^* g_3 - g_2|_{g_2,p,r} + |f_1^* g_2 - g_1|_{g_1,p,r} \\
&< \delta^3 + \delta \xrightarrow[\delta \to 0]{} 0. \qquad (4.16)
\end{aligned}$$

$\square$

Denote the corresponding uniform structure with $\mathfrak{U}^{p,r}_{L,diff,rel}$ and $\mathfrak{M}^{n,p,r}_{L,diff,rel}$ for the completion of $\mathfrak{M}^n(mf, I, B_k)$ with respect to this uniform structure.

It follows again from the definition that $d^{p,r}_{L,diff,rel}((M_1, g_1), (M_2, g_2)) < \infty$ implies $d_L((M_1, g_1), (M_2, g_2)) < \infty$, where d_L is the Lipschitz distance of section 2. Hence $(M_2, g_2) \in \text{comp}_L (M_1, g_1)$, where comp_L denotes the corresponding Lipschitz component, i.e.

$$\{(M_2, g_2) \in \mathfrak{M}^{n,p,r}_{L,diff,rel} \mid d^{p,r}_{L,diff,rel}(M_1, M_2) < \infty\}$$
$$\subseteq \text{comp}_L(M_1, g_1).$$

For this reason we denote the left hand side $\{\ldots\}$ by $\text{gen comp}^{p,r}_{L,diff,rel}(M_1, g_1) = \{\ldots\} = \{\ldots\} \cap \text{comp}_L(M_1, g_1)$

keeping in mind that this is not an arc component but a subset (of manifolds) of a Lipschitz arc component, endowed with the induced topology.

We extend all this to Riemannian vector bundles $(E, h, \nabla^h) \longrightarrow (M^n, g)$ of bounded geometry. First we have to define $\mathcal{D}^{p,r}(E \to M)$. For this we consider as at the end of section 3 the total space E as open Riemannian manifold of bounded geometry with respect to the Kaluza–Klein metric and restrict the uniform structure to bundle maps $f = (f_E, f_M)$. Quite similar we define for $E_i = ((E_i, h_i, \nabla^{h_i}) \longrightarrow (M_i^n, g_i))$, $i = 1, 2$, $\mathcal{D}^{p,r}(E_1, E_2)$ by corresponding bundle isomorphisms and $\mathcal{D}^{r,p,r}(E_1, E_2) = \mathcal{D}^{p,r}(E_1, E_2) \cap C^{r,p,r}(E_1, E_2) \subset \mathcal{D}^{p,r}(E_1, E_2)$ as the C^{r+1} elements with r–bounded differential, i.e. for $f \in (f_E, f_M) \in \mathcal{D}^{r,p,r}(E_1, E_2)$ there holds

$$\sup_{\substack{x \in M_1 \\ 1 \leq i \leq r}} |\nabla^i df_M| \leq d, \sup_{\substack{e \in E_1 \\ 1 \leq i \leq r}} |\nabla^i df_E| \leq d.$$

Quite analogously to $\mathfrak{M}^n(mf, I, B_k)$ we denote the bundle isometry classes of Riemannian vector bundles $(E, h, \nabla) \longrightarrow (M^n, g)$ with $(I), (B_k(g)), (B_k(\nabla))$ of rkN over n–manifolds by $\mathcal{B}^{N,n}(I, B_k)$. Set for $k \geq r > \frac{n}{p} + 2$, $E_i = ((E_i, h_i, \nabla^{h_i}) \longrightarrow (M_i^n, g_i)) \in \mathcal{B}^{N,n}(I, B_k)$, $i = 1, 2$

$$\begin{aligned} d^{p,r}_{L,diff}(E_1, E_2) &= \inf\{ \max\{0, \log{}^b|df_E|\} + \max\{0, \log{}^b|df_E^{-1}|\} \\ &+ \max\{0, \log{}^b|df_M|\} + \max\{0, \log{}^b|df_M^{-1}|\} \\ &+ \sup_{\substack{x \in M_1 \\ 1 \leq i \leq r}} |\nabla^i df_M| + \sup_{\substack{e \in E_1 \\ 1 \leq i \leq r}} |\nabla^i df_E| \\ &+ |g_1 - f_M^* g_2|_{g_1,p,r} + |h_1 - f_E^* h_2|_{g_1,h_1,\nabla^{h_1},p,r} \\ &+ |\nabla^{h_1} - f_E^* \nabla^{h_2}|_{g_1,h_1,\nabla^{h_1},p,r} | \\ & |f = (f_E, f_M) \in \mathcal{D}^{r,p,r}(E_1, E_2)\}, \end{aligned} \qquad (4.17)$$

if $\{\ldots\} \neq \emptyset$ and $\inf\{\ldots\} < \infty$. In the other case set $d^{p,r}_{L,diff}(E_1, E_2) = \infty$. Here we remark that ${}^b|df_E|$, ${}^b|df_E^{-1}|$, ${}^b|df_M|$, ${}^b|df_M^{-1}| < \infty$ automatically imply the quasi isometry of $h_1, f_E^* h_2$ or $g_1, f_M^* g_2$, respectively. A simple consideration shows that $d(E_1, E_2) = 0$ is an equivalence relation $\sim$. Set $\mathcal{B}^{N,n}_{L,diff}(I, B_k) := \mathcal{B}^{N,n}(I, B_k)/\sim$ and for $\delta > 0$

$$V_\delta = \{(E_1, E_2) \in (\mathcal{B}^{N,n}_{L,diff}(I, B_k))^2 \mid d^{p,r}_{L,diff}(E_1, E_2) < \delta\}.$$

Proposition 4.2 $\mathfrak{B} = \{V_\delta\}_{\delta>0}$ *is a basis for a metrizable uniform structure* $\mathfrak{U}^{p,r}_{L,diff}$.

The proof is quite analogous to that of proposition 4.1 with $K_1 = K_2 = \emptyset$. □

The set (4.17) contains some more terms as the set (4.1). The new terms are $|h_1 - f_E^* h_2|_{g_1,h_1 \nabla^{h_1},p,r}$ and $|\nabla^{h_1} - f_E^* \nabla^{h_2}|_{g_1,h_1 \nabla^{h_1},p,r}$. For the symmetry we consider

$$\begin{aligned}
|h_2 - f_E^{*-1} h_1|_{g_2,h_2 \nabla^{h_2},p,r} &= |f_E^{*-1}(f_E^* h_2 - h_1)|_{g_2,h_2 \nabla^{h_2},p,r} \\
&\leq {}^{b,r}|f_E^{*-1}| \cdot |f_E^* h_2 - h_1|_{g_1,h_1 \nabla^{h_1},p,r} \\
&\leq k_1(\delta) \cdot \delta \xrightarrow[\delta \to 0]{} 0
\end{aligned}$$

and

$$\begin{aligned}
|f_E^{*-1} \nabla^{h_1} \nabla^{h_2}|_{g_2,h_2 \nabla^{h_2},p,r} &= |f_E^{*-1}(\nabla^{h_1} - f_E^* \nabla^{h_2})|_{g_2,h_2 \nabla^{h_2},p,r} \\
&\leq {}^{b,r}|f_E^{*-1}| \cdot |\nabla^{h_1} - f_E^* \nabla^{h_2}|_{g_1,h_1 \nabla^{h_1},p,r} \\
&\leq k_2(\delta) \cdot \delta \xrightarrow[\delta \to 0]{} 0
\end{aligned}$$

The proof of the transitivity is completely parallel to (4.14) – (4.16).

Denote $\mathcal{B}^{N,n,p}_{L,diff,r}(I, B_k)$ for the pair $(\mathcal{B}^{N,n}(I, B_k), \mathfrak{U}^{p,r}_{L,diff})$ and $\mathcal{B}^{N,n,p,r}_{L,diff}(I, B_k)$ for the completion.

The next task would be to prove the locally arcwise connectedness of $\mathcal{B}^{N,n,p,r}_{L,diff}$. If we restrict to $(E, h) \longrightarrow (M^n, g)$, i.e. we forget the metric connection ∇^k, then the corresponding space is locally arcwise connected according to 5.19 of [33]. Taking into account the metric connection ∇^h, the situation becomes much worse. Given $(g, h, \nabla^h), (g', h', \nabla^{h'})$ sufficiently neighboured, we have to prove that they could be connected by a (sufficiently short) arc $\{(g_t, h_t, \nabla^{h_t})\}$. Here ∇^{h_t} must be metric w. r. t. h_t.

We were not able to construct the arc $\{\nabla^{h_t}\}_t$ for given $\{h_t\}_t$. One could also try to set $\nabla^t = (1-t)\nabla + t\nabla$ and to construct h_t from ∇^t s. t. ∇^t is metric w. r. t. h_t. In local bases $e_1, \ldots, e_n$, $\Phi_1, \ldots, \Phi_N$ this would lead to the system

$$\nabla^t_{e_i} h_{t,\alpha\beta} = \Gamma^\gamma_{t,i\alpha} h_{t,\gamma\beta} + \Gamma^\gamma_{t,i\beta} h_{t,\alpha\gamma}, \quad i = 1, \ldots, n, \alpha, \beta = 1, \ldots, N,$$

where $h_{t,\alpha\beta} = h_t(\Phi_\alpha, \Phi_\beta)$, $\nabla^t_i \Phi_\alpha = \Gamma^\gamma_{t,i\alpha} \Phi_\gamma$. This is a system of $n\frac{N(N+1)}{2}$ equations for the $\frac{N(N+1)}{2}$ components $h_{\alpha\beta}$, i.e. it is overdetermined. With other words, we don't see a comparetively simple and natural proof for locally arcwise connectedness. $\mathcal{B}^{N,n,p,r}_{L,diff}(I, B_k)$ is a complete metric space. Hence locally and locally arcwise connectedness coincide. But to prove locally connectedness amounts very soon to similar questions just discussed.

Consider for $E = ((E, h, \nabla^h) \longrightarrow (M, g)) \in \mathcal{B}^{N,n}(I, B_k)$

$$\{E' \in \mathcal{B}^{N,n,p,r}_{L,diff}(I, B_k) \mid d^{p,r}_{L,diff}(E, E') < \infty\}. \quad (4.18)$$

The set is open and contains the arc component of E. If $\mathcal{B}^{N,n,p,r}_{L,diff}(I, B_k)$ would be locally arcwise connected = locally connected then we would have

$$\mathrm{arccomp}(E) = \mathrm{comp}(E). \quad (4.19)$$

If we endow the total spaces E with the Kaluza–Klein metric

$$g_E(X, Y) = h(X^V, Y^V) + g_M(\pi_* X, \pi_* Y),$$
$$X^V, Y^V \text{ vertical components,}$$

then (E, g_E) becomes a Riemannian manifold of bounded geometry, hence a proper metric space. It follows from the definition that

$$\{E' \in \mathcal{B}^{N,n,p,r}_{L,diff}(I, B_k) \mid d^{p,r}_{L,diff}(E, E') < \infty\} \subseteq \mathrm{comp}_L(E). \quad (4.20)$$

(4.18), (4.19) and the foregoing considerations are for us motivation enough to define the generalized component gen comp(E) by

$$\mathrm{gen\,comp}^{p,r}_{L,diff}(E) := \{E' \in \mathcal{B}^{N,n,p,r}_{L,diff}(I, B_k) \mid d^{p,r}_{L,diff}(E, E') < \infty\}. \quad (4.21)$$

In particular gen comp(E) is a subset of a Lipschitz component and is endowed with a well defined topology coming from $\mathfrak{U}_{L,diff}^{p,r}$. The next step in this section consists of the additional admission of compact topological perturbations, quite similar to the case above of manifolds. We consider pairs $E_i = ((E_i, h_i, \nabla^{h_i}) \longrightarrow (M_i^n, g_i)) \in \mathcal{B}^{N,n}(I, B_k)$, $i = 1, 2$, with the following property. There exist compact submanifolds $K_i^n \subset M_i^n$ and $f = (f_E, f_M) \in C^{r,p,r}(E_1, E_2)$, $f|_{M_1 \setminus K_1} \in \mathcal{D}^{r,p,r}(E_1|_{M_1 \setminus K_1}, E_2|_{M_2 \setminus K_2})$. For such pairs define

$$d_{L,diff,rel}^{p,r}(E_1, E_2) = \inf\{\max\{0, \log^b|df_E|\}$$
$$+ \max\{0, \log^b|dh_E|\} + \max\{0, \log^b|df_M|\}$$
$$+ \max\{0, \log^b|dh_M|\} + \sup_{e_1} d(h_E f_E e_1, e_1)$$
$$+ \sup_{e_2} d(f_E h_E e_2, e_2) + \sup_{x_1} d(h_M f_M x_1, x_1)$$
$$+ \sup_{x_2} d(f_M h_M x_2, x_2) + \sup_{\substack{x \in M_1 \\ 1 \le i \le r}} |\nabla^i df_M|$$
$$+ \sup_{\substack{e \in E_1 \\ 1 \le i \le r}} |\nabla^i df_E| + |(f_M|_{M_1 \setminus K_1})^* g_2 - g_1|_{M_1 \setminus K_1}|_{g_1, p, r}$$
$$+ |(f_E|_{E|_{M_1 \setminus K_1}})^* h_2 - h_1|_{E_1|_{M_1 \setminus K_1}}|_{g_1, h_1, \nabla^{h_1}, p, r}$$
$$+ |(f_E|_{E|_{M_1 \setminus K_1}})^* \nabla^{h_2} - \nabla^{h_1}|_{E_1|_{M_1 \setminus K_1}}|_{g_1, h_1, \nabla^{h_1}, p, r}$$
$$| \; f = (f_E, f_M) \in C^{r,p,r}(E_1, E_2),$$
$$h = (h_E, h_M) \in C^{r,p,r}(E_2, E_1)\}$$

bundle maps and for some $K_1 \subset M_1$ holds

$$f|_{E_1|_{M_1 \setminus K_1}} \in \mathcal{D}_{rel}^{r,p,r}(E_1|_{M_1 \setminus K_1}, E_2|_{f_M(M_1 \setminus K_1)}) \text{ and}$$
$$h|_{E_2|_{f(M_1 \setminus K_1)}} = (f|_{E|_{M_1 \setminus K_1}})^{-1}\} \tag{4.22}$$

if $\{\ldots\} \ne \emptyset$ and $\inf\{\ldots\} < \infty$. In the other case set $d_{L,diff,rel}^{p,r}(E_1, E_2) = \infty$. This definition seems to be quite lengthy but it is quite natural. It measures outside a compact set the distinction of "shape" and the geometric objects in question. Set

$$V_\delta = \{(E_1, E_2) \in (\mathcal{B}^{N,n}(I, B_k))^2 \; | \; d_{L,diff,rel}^{p,r}(E_1, E_2) < \delta\}.$$

Proposition 4.3 $\mathfrak{B} = \{V_\delta\}_{\delta>0}$ *is a basis for a metrizable uniform structure* $\mathfrak{U}^{p,r}_{L,diff,rel}$ *on* $\mathcal{B}^{N,n}(I, B_k)/\sim$ *where* $E_1 \sim E_2$ *if* $d^{p,r}_{L,diff,rel}(E_1, E_2) = 0$.

The proof is completely parallel to that of 4.1 combined with that of 4.2. □

Denote $\mathcal{B}^{N,n,p,r}_{L,diff,rel}(I, B_k)$ for the completed $\mathcal{B}^{N,n}(I, B_k)$ endowed with this uniform structure. We have again that $d^{p,r}_{L,diff,rel}(E_1, E_2) < \infty$ implies $d_L(E_1, E_2) < \infty$ (here we consider E_1, E_2 as proper metric spaces). Hence $E_2 \in \text{comp}_L(E_1)$, i.e.

$$\{E_2 \in \mathcal{B}^{N,n,p,r}_{L,diff,rel}(I, B_k) \mid d^{p,r}_{L,diff,rel}(E_1, E_2) < \infty\} \subseteq \text{comp}_L(E_1). \quad (4.23)$$

For this reason we denote again the left hand side $\{\ldots\}$ of (4.23) by gen $\text{comp}^{p,r}_{L,diff,rel}(E_1)$ keeping in mind that this is not an arc component but a subset of a Lipschitz component endowed with the induced topology from $\mathfrak{U}^{p,r}_{L,diff,rel}$.

In our later applications we prove and thereafter use the trace class property of $e^{-tD^2} - e^{-tD'^2}$. Here essentially enter estimates for $D - D'$, coming from the explicit expression for $D - D'$. But in this expression only $g, \nabla, \cdot, g', \nabla', \cdot'$ enter. This is the reason why we consider in some of our applications smaller generalized components, which are in fact arc components. Exactly speaking, we restrict in some of our applications to those uniform structures and components where $h_1 = f_E^* h_2$, i.e. the fibre metric does not vary. Nevertheless, the generalized components play the more important role as appropriate equivalence classes in classification theory. We prove the trace class property of $e^{-tD^2} - e^{-tD'^2}$ also for generalized components.

Set now

$$d^{p,r}_{L,diff,F}(E_1, E_2) = \inf\{\max\{0, \log^b |df_M|\}$$
$$+ \max\{0, \log^b |df_M^{-1}|\} + \sup_{\substack{x \in M_1 \\ 1 \le i \le r}} |\nabla^i df_M|$$
$$+ \sup_{\substack{y \in M_2 \\ 1 \le i \le r}} |\nabla^i df_M^{-1}| + |g_1 - f_M^* g_2|_{g_1,p,r}$$
$$+ |\nabla^{h_1} - f_M^* \nabla^{h_2}|_{g_1,h_1,\nabla^{h_1},p,r}$$
$$| \; f = (f_E, f_M) \in \mathcal{D}^{r,p,r}(E_1, E_2),$$
$$f_E \text{ fibrewise an isometry } \} \qquad (4.24)$$

if $\{\ldots\} \ne \emptyset$ and $\inf\{\ldots\} < \infty$. In the other case set $d^{p,r}_{L,diff,F}(E_1, E_2) = \infty$. $d^{p,r}_{L,diff,F}(\cdot, \cdot) = 0$ is an equivalence relation $\sim$. Set $\mathcal{B}^{N,n}_{L,diff,F}(I, B_k) = \mathcal{B}^{N,n}(I, B_k)/\sim$ and for $\delta > 0$

$$V_\delta = \{(E_1, E_2) \in (\mathcal{B}^{N,n}_{L,diff,F}(I, B_k))^2 \; | \; d^{p,r}_{L,diff,F}(E_1, E_2) < \delta\}.$$

Proposition 4.4 $\mathfrak{B} = \{V_\delta\}_{\delta > 0}$ *is a basis for a metrizable uniform structure* $\mathfrak{U}^{p,r}_{L,diff,F}$. □

Denote $\mathcal{B}^{N,n,p,r}_{L,diff,F}(I, B_k)$ for the corresponding completion.

Proposition 4.5 a) $\mathcal{B}^{N,n,p,r}_{L,diff,F}(I, B_k)$ *is locally arcwise connected.*
b) *In* $\mathcal{B}^{N,n,p,r}_{L,diff,F}(I, B_k)$ *components coincide with arc components.*
c) $\mathcal{B}^{N,n,p,r}_{L,diff,F}(I, B_k) = \sum_{i \in I} \operatorname{comp}^{p,r}_{L,diff,F}(E_i)$ *as topological sum.*
d) *For* $E \in \mathcal{B}^{N,n}$

$$\operatorname{gen comp}^{p,r}_{L,diff,F}(E) = \{E' \in \mathcal{B}^{N,n,p,r}_{L,diff,F}(I, B_k) \; | $$
$$d^{p,r}_{L,diff,F}(E, E') < \infty\}.$$

Proof. a) $g_t = (1-t)g_1 + t f_M^* g_2$, $\nabla_t = (1-t)\nabla^{h_1} + t f_E^* \nabla^{h_2}$ yield an arc between E_1 and $f^* E_2$. Here we use $h_1 = f_E^* h_2$ and that $\nabla^{h_1}, \nabla^{h_2}$ are metric. □

Quite analogously we define – based on $h_1|_{E|_{M\setminus K}} = f_E^*(h_2|_{E'|_{M'\setminus K'}})$ – the uniform space $\mathcal{B}_{L,diff,F,rel}^{N,n,p,r}(I,B_k)$ and its generalized components

$$\text{gen comp}_{L,diff,F,rel}^{p,r}(E) = \{E' \mid d_{L,diff,F,rel}^{p,r}(E,E') < \infty\}.$$

Here $d_{L,diff,F,rel}^{p,r}(E,E')$ is defined as $d_{L,diff,F,rel}^{p,r}(\cdot,\cdot)$ with the additional condition $h|_{E|_{M\setminus K}} = f_E^*(h'|_{E'|_{M'\setminus K'}}$. $\mathcal{B}_{L,diff,F,rel}^{N,n,p,r}$ is not locally arcwise connected. Now the classification of $\mathcal{B}^{N,n}(I,B_k)$ amounts to two tasks.

1) Classification (i. e. "counting") the (generalized) components gen comp(E) by invariants,

2) Classification of the elements inside a component by invariants, where number valued invariants should be relative invariants.

Until now g, h, ∇^h could be fixed independently, keeping in mind that ∇^h should be a metric connection with respect to h. The situation rapidly changes if we restrict to Clifford bundles. The new ingredient is the Clifford multiplication $\cdot$ which relates g, h, ∇^h.

As we know from the definition, a Clifford bundle $(E, h, \nabla^h, \cdot) \longrightarrow (M^n, g)$ has as additional ingredient the module structure of E_m over $\text{CL}_m(g) = \text{CL}(T_mM, g_m)$. A change of g, $g \longrightarrow g'$, changes point by point the Clifford algebra, $\text{CL}_m(g) \longrightarrow \text{CL}_m(g')$. Locally they are isomorphic since by radial parallel transport of orthonormal bases in a normal neighbourhood $U(m_0)$ always

$$\text{CL}(M,g)|_{U(m_0)} \cong U(m_0) \times \text{CL}(\mathbb{R}^n) \cong U(m_0) \times \text{CL}(M,g')|_{U(m_0)}. \tag{4.25}$$

The same holds for bundles of Clifford modules if we fix the typical fibre, i. e.

$$E|_U \cong E'|_U \tag{4.26}$$

but globally (4.26) in general does not hold although as vector bundles E and E' can be isomorphic. The point is the module structure which includes g (in $\text{CL}_m(g)$) as operating algebra and

$\cdot : T_m M \otimes E_m \longrightarrow E_m$. Therefore for a moment we consider the following admitted deformations. Let $(E, h, \nabla^h, \cdot) \longrightarrow (M^n, g)$ be a Clifford bundle of rkN. The vector bundle structure $E \longrightarrow M$ (of rkN) shall remain fixed. We admit variation of g, hence of $CL(TM)$, variation of $\cdot \in \mathrm{Hom}\,(TM \otimes E, E)$, hence variation of the structure of E as bundle of Clifford modules, compatible variation of h, ∇^h. $\mathrm{Hom}\,(TM \otimes E, E) \cong T^*M \otimes E^* \otimes E$ is for given g, h a Riemannian vector bundle. Including ∇^h, the notion of Sobolev sections is well defined. For fixed g, h, ∇^h the space $\Gamma(\mathrm{mult}, g, h, \nabla^h)$ of Clifford multiplications $\cdot$ is a well defined subspace of $\Gamma(\mathrm{Hom}\,(TM \otimes E, E))$ described invariantly by the conditions

$$\langle X \cdot \Phi, \Psi \rangle = -\langle \Phi, \Psi \cdot Y \rangle, \qquad (4.27)$$
$$\nabla_X(Y \cdot \Phi) = (\nabla_X Y) \cdot \Phi + Y \cdot \nabla_Y \Phi, \qquad (4.28)$$

where here $\langle \cdot, \cdot \rangle = h(\cdot, \cdot)$. We describe this space locally as follows. Let $U(m_0) \subset M$ and $e_1(m), \ldots, e_n(m)$ be a field of orthonormal bases obtained from $e_1(m_0), \ldots, e_n(m_0)$ by radial parallel transport, similarly $\Phi_1, \ldots, \Phi_N$ a field of orthonormal bases in $E|_{U(m_0)}$, obtained also by radial parallel transport. Then (4.27), (4.28) mean for the attachment $e_i \otimes \Phi_j \longrightarrow e_i \cdot \Phi_j$

$$\langle e_i \cdot \Phi_j, \Phi_k \rangle = -\langle \Phi_j, e_i \cdot \Phi_k \rangle,$$
$$i = 1, \ldots, n, \quad j = 1, \ldots, N, \qquad (4.29)$$
$$\nabla_{e_i}(e_j \cdot \Phi_k) = (\nabla_{e_i} e_j) \cdot \Phi_k + e_i \cdot \nabla_{e_j} \Phi_k,$$
$$i, j = 1, \ldots, n, \quad k = 1, \ldots, N. \qquad (4.30)$$

If we write in the linear space $\mathrm{Hom}\,(T_m \otimes E_m, E_m)$ $e_i \cdot \Phi_j = \sum_{k=1}^{N} a_{ij}^k \Phi_k$ then (4.29) reduces to $\frac{nN(N+1)}{2}$ independent linear equations between the a_{ij}^k, $a_{ij}^k = -a_{ik}^j$. Hence the fibre $\mathrm{mult}_m(g, h, \nabla^h)$ is an $n \cdot N^2 - \frac{nN(N+1)}{2} = \frac{nN}{2}(N-1)$-dimensional affine subspace at any point m. This establishes a locally trivial fibre bundle $\mathrm{mult}(g, h, \nabla^h)$. The charts for local trivialization arise from radial parallel transport P of $\cdot$ from m to m_0 since $\nabla_{\frac{\partial}{\partial r}}(P(e_i \cdot \Phi_j) - (Pe_i) \cdot (P\Phi_j)) = 0$. $CLM(g, h, \nabla^h) \subset$

$\Gamma(\text{mult}(g, h, \nabla_h))$ now are those sections of $\text{mult}(g, h, \nabla^h)$ which additionally satisfy (4.28) or (4.30). Consider the Riemannian vector bundle $(\mathcal{H} = \text{Hom}(TM \otimes E, E) = (T^*M \otimes E^* \otimes E, h_\mathcal{H}) \longrightarrow (M^n, g))$. Assume as always (M^n, g) with $(I), (B_k)$, (E, h, ∇^h) with (B_k), $k \geq \frac{n}{p} + 2$ and $(\pi : E \longrightarrow M) \in C^{\infty,k+1}(E, M)$. Then with respect to the Kaluza–Klein metric $g_\mathcal{H}(X, Y) = h_\mathcal{H}(X^V, Y^V) + g_M(\pi_*X, \pi_*Y)$, X^V, Y^V vertical components, the total space of $\mathcal{H}$ becomes a Riemannian manifold satisfying $(I), (B_k)$. The fibres $\mathcal{H}_m$ are totally geodesic submanifolds, moreover they are flat. The latter also holds for the affine fibres $\text{mult}_m(g, h, \nabla^h)$ of $\text{mult}(g, h, \nabla^h)$. If $\cdot$ and $\cdot'$ are sections of $\text{mult}(g, h, \nabla^h)$ satisfying (4.28), i. e. $\cdot, \cdot' \in \text{CLM}(g, h, \nabla^h)$ then

$$(1 - t) \cdot + t \cdot' \in \text{CLM}(g, h, \nabla^h), \qquad (4.31)$$

Let $k \geq r > \frac{n}{p} + 2$, $\delta > 0$. Set

$$V_\delta = \{(\cdot, \cdot') \in \text{CLM}(g, h, \nabla^h)^2 \mid |\cdot - \cdot'|_{g,h,\nabla^h,p,r} < \delta\}.$$

Lemma 4.6 $\mathfrak{B} = \{V_\delta\}_{\delta>0}$ *is a basis for a metrizable uniform structure* $\mathfrak{U}^{p,r}(\text{CLM}(g, h, \nabla^h))$. □

Denote by $\text{CLM}^{p,r}(g, h, \nabla^h)$ the completion.

Proposition 4.7 a) $\text{CLM}^{p,r}(g, h, \nabla^h)$ *is locally arcwise connected.*
b) *In* $\text{CLM}^{p,r}(g, h, \nabla^h)$ *components and arc components coincide.*
c) $\text{CLM}^{p,r}(g, h, \nabla^h) = \sum_{i \in I} \text{comp}^{p,r}(\cdot_i)$.
d) $\text{comp}^{p,r}(\cdot) = \{\cdot' \mid |\cdot - \cdot'|_{g,h,\nabla^h,p,r} < \infty\}$.

Proof. a) follows from (4.31), b) and c) follow from a), d) is a simple calculation. □

Remark 4.8 In the language of the intrinsic Riemannian geometry of $\text{mult}(g,h,\nabla^h)$ and of $\Gamma(\text{mult}(g,h,\nabla^h))$ we can rewrite $|\cdot - \cdot'|_{g,h,\nabla^h,p,r} < \delta$ as $\cdot' = \exp X \circ \cdot$, $X \in \Gamma(T(\text{mult}(g,h,\nabla^h)))$, $|X|_{g,h,\nabla^h,p,r} < \delta$. Here X_m lies in the affine subspace mult_m. □

Denote by $\text{CL}\mathcal{B}^{N,n}(I,B_k)$ the set of (Clifford isometry classes) of all Clifford bundles $(E,h,\nabla^h,\cdot) \longrightarrow (M^n,g)$ of (module) rank N over n–manifolds, all with (I) and (B_k).

Lemma 4.9 Let $E_i = ((E_i, h_i, \nabla^{h_i}, \cdot_i) \longrightarrow (M_i^n, g_i)) \in \text{CL}\mathcal{B}^{N,n}(I,B_k)$, $i = 1,2$ and $f = (f_E, f_M) \in \mathcal{D}^{p,r+1}(E_1,E_2) \cap C^{\infty,k+1}(E_1,E_2)$ be a vector bundle isomorphism between bundles of Clifford modules satisfying $f_E(X \cdot_1 \Phi) = (f_M)_* X \cdot_2 f_E \Phi$. Then $f^*E_2 := ((E_1, f_E^* h_2, f_E^* \nabla^{h_2}, f_E^* \cdot_2) \longrightarrow (M_1, f_M^* g_2)) \in \text{CL}\mathcal{B}^{N,n}(I,B_k)$.

Proof. The definitions of $f_E^* h_2$, $f_E^* \nabla^{h_2}$, $f_M^* g_2$ are clear. $f_E^* \cdot_2$ is defined by $X(f_E^* \cdot_2)\Phi = f_E^{-1}(f_* X \cdot_2 f_E \Phi)$. It is now an easy calculation that $f^*E_2 \in \text{CL}\mathcal{B}^{N,n}(I,B_k)$. □

Let $k \geq r > \frac{n}{p} + 2$ and define for $E_1, E_2 \in \text{CL}\mathcal{B}^{N,n}(I,B_k)$

$$d^{p,r}_{L,diff}(E_1,E_2) = \inf\{\max\{0, \log{}^b|df_E|\}$$
$$+ \max\{0, \log{}^b|df_E^{-1}|\} + \max\{0, \log{}^b|df_M|\}$$
$$+ \max\{0, \log{}^b|df_M^{-1}|\} + \sup_{\substack{m \in M_1 \\ 1 \leq i \leq r}} |\nabla^i df_M|$$
$$+ |g_1 - f_M^* g_2|_{g_1,p,r} + |h_1 - f_E^* h_2|_{g_1,h_1,\nabla^{h_1},p,r}$$
$$+ |\nabla^{h_1} - f_E^* \nabla^{h_2}|_{g_1,h_1,\nabla^{h_1},p,r} + |\cdot_1 - f_E^* \cdot_2|_{g_1,h_1,\nabla^{h_1},p,r} |$$
$$f = (f_E, f_M) \in \mathcal{D}^{r,p,r}(E_1,E_2) \text{ is a } (r+1)$$
$$-\text{bounded isomorphism of Clifford bundles}\}$$

if $\{\ldots\} \neq \emptyset$ and $\inf\{\ldots\} < \infty$. In the other case set $d^{p,r}_{L,diff}(E_1,E_2) = \infty$. $d^{p,r}_{L,diff}$ is numerically not symmetric but nevertheless it defines a uniform structure which is by definition symmetric. Set for $\delta > 0$

$$V_\delta = \{(E_1,E_2) \in \text{CL}\mathcal{B}^{N,n}(I,B_k))^2\} \mid d^{p,r}_{L,diff}(E_1,E_2) < \delta\}.$$

Proposition 4.10 $\mathfrak{B} = \{V_\delta\}_{\delta>0}$ *is a basis for a metrizable uniform structure* $\mathfrak{U}^{p,r}_{L,diff}(\mathrm{CL}\mathcal{B}^{N,n}(I, B_k))$. □

Denote $\mathrm{CL}\mathcal{B}^{N,n,p}_{L,diff,r}(I, B_k)$ for the pair $(\mathrm{CL}\mathcal{B}^{N,n}(I, B_k), \mathcal{U}^{p,r})$ and $\mathrm{CL}\mathcal{B}^{N,n,p,r}_{L,diff}(I, B_k)$ for the completion. By the same motivation as above we introduce again the generalized component $\operatorname{gen comp}(E) = \operatorname{gen comp}^{p,r}_{L,diff}((E, h, \nabla^h) \longrightarrow (M, g)) \subset \mathrm{CL}\mathcal{B}^{N,n,p,r}_{L,diff}(I, B_k)$ by

$$\operatorname{gen comp}^{p,r}_{L,diff}(E) = \{E' \in \mathrm{CL}\mathcal{B}^{N,n,p,r}_{L,diff}(I, B_k) \mid d^{p,r}_{L,diff}(E, E') < \infty\}. \quad (4.32)$$

$\operatorname{gen comp}(E)$ contains $\operatorname{arccomp}(E)$ and is endowed with a Sobolev topology induced from $\mathcal{U}^{p,r}_{L,diff}$.

The absolutely last step in our uniform structures approach is the additional admission of compact topological perturbations. We proceed assuming additionally $E_i = ((E_i, h_i, \nabla^{h_i}, \cdot_i) \longrightarrow (M^n_i, g_i)) \in \mathrm{CL}\mathcal{B}^{N,n}(I, B_k)$, adding still $|(f_E|_{E|_{M_1 \setminus K_1}})^* \cdot_2 - \cdot_1 |_{E|_{M_1 \setminus K_1}}|_{g_1, h_1, \nabla^{h_1}, p, r}$ and assuming $f = (f_E, f_M)|_{M_1 \setminus K_1}$, $h = (h_E, h_M)|_{M_2 \setminus K_2 = f_M(M_1 \setminus K_1)}$ vector bundle isomorphisms (not necessary Clifford isometries). Then we get $d^{p,r}_{L,diff,rel}(E_1, E_2)$, define V_δ, $\mathfrak{B} = \{v_\delta\}_{\delta > 0}$, obtain the metrizable uniform structure $\mathfrak{U}^{p,r}_{L,diff,rel}(\mathrm{CL}\mathcal{B}^{N,n}(I, B_k))$ and finally the completion $\mathrm{CL}\mathcal{B}^{N,n,p,r}_{L,diff,rel}$. We set again

$$\begin{aligned}\operatorname{gen comp}(E) &= \operatorname{gen comp}^{p,r}_{L,diff,rel}(E) \\ &= \{E' \in \mathrm{CL}\mathcal{B}^{N,n,p,r}_{L,diff,rel}(I, B_k)) \mid \\ & \quad d^{p,r}_{L,diff,rel}(E, E') < \infty\}\end{aligned}$$

which contains the arc component and inherits a Sobolev topology from $\mathfrak{U}^{p,r}_{L,diff,rel}$.

As in the preceding considerations we obtain by requiring additionally $h_1 = f_E^* h_2$ or $h_1|_{E_1|_{M_1 \setminus K_1}} = f_E^*(h_2|_{E_2|_{M_2 \setminus K_2}})$ local distances $d^{p,r}_{L,diff,F}(\cdot, \cdot)$ or $d^{p,r}_{L,diff,F,rel}(\cdot, \cdot)$ and corresponding uniform spaces $\mathrm{CL}\mathcal{B}^{N,n,p,r}_{L,diff,F}(I, B_k)$ or $\mathrm{CL}\mathcal{B}^{N,n,p,r}_{L,diff,F,rel}(I, B_k)$ respectively. We obtain generalized components

$$\operatorname{gen comp}^{p,r}_{L,diff,F}(E) \quad (4.33)$$

and
$$\operatorname{gen\,comp}^{p,r}_{L,diff,F,rel}(E) \tag{4.34}$$

as before. One of our main technical results in chapter IV will be that E and E' being in the same generalized component implies that after transforming $e^{-tD'^2}$ into the Hilbert space $L_2((M,E),g,h)$, $e^{-tD^2}-e^{-tD'^2}$ and $e^{-tD^2}D-e^{-tD'^2}D'$ are of trace class and their trace norm is uniformly bounded on compact t–Intervalls $[a_0, a_1]$, $a_0 > 0$. For our later applications the components (4.33), (4.34) are most important, excluded one case, the case $D^2 = \Delta(g)$, $D'^2 = \nabla(g')$. In this case variation of g automatically induces variation of the fibre metric and we have to consider (4.32) and $\operatorname{gen\,comp}^{p,r}_{L,diff,rel}(E)$.

Perhaps, for the reader the definitions for the $\operatorname{gen\,comp}(E)$ look very involved. We recall, roughly speaking, the main points are as follows. The distance which defines $\operatorname{gen\,comp}\ldots(E)$ measures step by step the distance between the main ingredients of a Clifford bundle: the smooth Lipschitz distance between the diffeomorphic parts of the manifolds and the bundles and the Sobolev distance between the manifold metrics, the fibre metrics, the fibre connections and the Clifford multiplications.

Remark 4.11 The $\operatorname{gen\,comp}^{\cdot\cdot}_{L,diff,rel}(\cdot)$–definition can be extended canonically to structures with boundary. □

5 The classification problem, new (co–)homologies and relative characteristic numbers

As we already indicated, we understand this treatise as a contribution to the classification problem for open manifolds. We proved in chapter I that meaningful number valued invariants for all open manifold do not exist. The way out from this situation is to introduce relative number valued invariants or to give up the claim for number valued invariants and to admit group valued invariants as e.g. in classical algebraic topology. We go both ways. The heart of this treatise are new number valued relative invariants like relative determinants, relative analytic torsion, relative eta invariants, relative indices. This will be the content of chapters IV – VI. Our general approach consists in two steps,

1. to decompose the class of manifolds/bundles under consideration into generalized components and to try to "count", to "classify" them,

2. to "count", to "classify" the elements inside a generalized component.

Chapters IV – VI are exclusively devoted to the second step. Concerning the first step, we developed in [34] some new (co–) homologies which are invariants of the corresponding generalized component and hence represent steps within the first task above. In this section, we give a brief review of these (co–) homologies. In the second part, we give an outline of bordism theory for open manifolds and corresponding relative characteristic numbers.

Let X and Y be proper metric spaces. We call a map $\Phi : X \to Y$ *coarse* if it is

1. metrically proper, i.e. for each bounded subset $B \subseteq Y$ the inverse image $\Phi^{-1}(B)$ is bounded in X, and

2. uniformly expansive, i.e. for $R > 0$ there is $S > 0$ such that $d(x_1, x_2) \leq R$ implies $d(\Phi x_1, \Phi x_2) \leq S$.

A coarse map is called *rough* if it is additionally uniformly metrically proper. X and Y are called *coarsely* or *roughly equivalent* if there exist coarse or rough maps $\Phi : X \to Y$, $\Psi : Y \to X$, respectively, such that there exist constants D_X, D_Y satisfying

$$d(\Psi\Phi x, x) \leq D_X, \quad d(\Phi\Psi y, y) \leq D_Y.$$

Proposition 5.1 *X and Y are coarsely equivalent if and only if they are roughly equivalent.*

We refer to [33] for the proof. □

The equivalence class of X under coarse equivalence is called the coarse type of X.

Let X be a proper metric space. Then we have sequences of inclusions

$$\text{coarse type } (X) \supset \text{comp}_{GH}(X), \qquad (5.1)$$

$$\text{coarse type } (X) \supset \text{comp}_{GH}(X) \supset \text{arccomp}_{L,h,rel}(X) \supset$$
$$\supset \text{arccomp}_{L,h}(X) \supset \text{comp}_{L,top}(X), \quad (5.2)$$

$$\text{coarse type } (X) \supset \text{comp}_L(X) \supset \text{arccomp}_{L,h,rel}(X) \supset$$
$$\supset \text{arccomp}_{L,top,rel}(X) \supset \text{comp}_{L,top}(X), \quad (5.3)$$

The arising task is to define for any sequence of inclusions invariants depending only on the component and becoming sharper and sharper if we move from the left to the right. Start with the coarse type which has been extensively studied by J. Roe.

Given $X = (X,d)$, X^{q+1} becomes a proper metric space by $d((x_0,\ldots,x_q),(y_0,\ldots,y_q)) = \max\{d(x_0,y_0),\ldots,d(x_q,y_q)\}$. Let $\Delta = \Delta_q \subset X^{q+1}$ be the multidiagonal and set

$$Pen(\Delta, R) = \{y \in X^{q+1} | d(\Delta, y) \le R\}.$$

Then J. Roe defines in [63] the coarse complex $(CX^*(X), \delta) = (CX^q(X), \delta)_q$ by

$CX^q(X) := \{f : X^{q+1} \longrightarrow \mathbb{R} \mid f$ is locally bounded Borel function and for each $R > 0$ is supp $f \cap Pen(\Delta, R)$ relatively compact in $X^{q+1}\}$,

$$\delta f(x_0,\ldots,x_{q+1}) := \sum_{i=0}^{q+1}(-1)^i f(x_0,\ldots,\hat{x}_i,\ldots,x_{q+1}) \qquad (5.4)$$

The coarse cohomology $HX^*(X)$ of X is then defined as

$$HX^*(X) := H^*(CX^*(X)).$$

Theorem 5.2 $HX^*(X)$ *is an invariant of the coarse type, i.e. coarse equivalences* $\Phi : X \longrightarrow Y$, $\Psi : Y \longrightarrow X$ *induce isomorphisms.*

We refer to [63] for a proof. □

Corollary 5.3 *$HX^*(X)$ is an invariant for all components right from the coarse type.*

Remark 5.4 It is well known that without the support condition

$$\text{supp } f \cap Pen(\Delta, R) \text{ relatively compact} \tag{5.5}$$

the complex $CX^*(X)$ would be contractible. After fixing a base point $\bar{x} \in X$ the map $D : C^q \longrightarrow C^{q-1}$,

$$Df(x_1, \ldots, x_q) := f(\bar{x}, x_1, \ldots, x_q) \tag{5.6}$$

would be a contracting homotopy. □

It is now possible to define in a canonical way a cohomology theory which is an invariant of $\text{comp}_L(\cdot)$. One only has to choose the "right category". Let

$$C_L^q(X) = \{f : X^{q+1} \longrightarrow \mathbb{R} \mid f \text{ is Lipschitz continuous and}$$
$$\text{supp } f \cap Pen(\Delta, R) \text{ is relatively compact for all } R\}. \tag{5.7}$$

Then, with δ from (3.4), $C_L^*(X) = (C_L^q(X), \delta)_q$ is a complex and we define

$$H_L^*(X) := H^*(C_L^*(X)).$$

If $\Phi : X \longrightarrow Y$ is (u.p.) Lipschitz then Φ induces $\Phi_L^\# : C_L^q(Y) \longrightarrow C_L^q(X)$ by $(\Phi_L^\#(X)f)(x_0, \ldots, x_q) := f(\Phi x_0, \ldots, \Phi x_q)$, $f \in C_L^q(Y)$, and $\Phi_L^* : H_L^*(Y) \longrightarrow H_L^*(X)$.

Using Roe's anti Cech systems and uniqueness of the cohomology of uniform resolutions by appropriate sheaves as in [63], one easily obtains

Theorem 5.5 *If $Y \in \text{comp}_L(X)$ then there exist $\Phi : X \longrightarrow Y$, $\Psi : Y \longrightarrow X$ wich induce inverse to each other isomorphisms*

$$H_L^*(X) \underset{\Phi_L^*}{\overset{\Psi_L^*}{\rightleftarrows}} H_L^*(Y).$$

But this approach is very unsatisfactory since we did in fact not define a really new invariant but the categorial restriction of a coarse invariant. The situation rapidly changes if we factorize or impose decay conditions. Let $C_L^q(X)$ as above, ${}^bC_L^q(X)$ the subspace of bounded functions in $C_L^q(X)$ and $C_{L,b}^q(X) = C_L^q(X)/{}^bC_L^q(X)$. Then δ maps ${}^bC_L^q(X)$ into ${}^bC_L^{q+1}(X)$, i.e. ${}^bC_L^*$ is a subcomplex and we obtain a complex $C_{L,b}^*(X) = (C_{L,b}^q(X), \delta)_q$. Define

$$H_{L,b}^*(X) := H^*(C_{L,b}^*(X)).$$

Any (u.p.) Lipschitz map $\Phi : X \longrightarrow Y$ induces $\Phi^\# : {}^bC_L^*(Y) \longrightarrow {}^bC_L^*(X)$, hence $\Phi^\# : {}^bC_{L,b}^*(Y) \longrightarrow {}^bC_{L,b}^*(X)$ and

$$\Phi^* : H_{L,b}^*(Y) \longrightarrow H_{L,b}^*(X).$$

Theorem 5.6 $H_{L,b}^*(X)$ *is an invariant of* $\mathrm{comp}_L(X)$.

Proof. Let $Y \in \mathrm{comp}_L(X)$, $d_L(X,Y) < \varepsilon$, $\Phi : X \longrightarrow Y$, $\Psi : Y \longrightarrow X$, $d(\Psi\Phi x, x) < \varepsilon$, $d(\Phi\Psi y, y) < \varepsilon$ and let $[f] \in H_{L,b}^q(X)$. Then $(\Psi\Phi)^*[f] = [f]$ since $|((\Psi\Phi)^\# f - f)(x)| = |f(\Psi\Phi x) - f(x)| \leq C \cdot d(\Psi\Phi x, x) < C \cdot \varepsilon$, i.e. $(\Psi\Phi)^\# f - f \in {}^bC_L^q(X)$. Here $x = (x_0, \ldots, x_q)$. Similiarly $(\Phi\Psi)^*[g] = [g]$ for $[g] \in H_{L,b}^q(Y)$. Φ^* and Ψ^* are inverse to each other isomorphisms. □

Remark 5.7 For compact X, $H_{L,b}^*(X) = 0$ as it should be. □

Let Z be a proper metric space, $f : Z \longrightarrow \mathbb{R}$ Borel and locally bounded. We say f is uniformly locally bounded if for every $D > 0$ their exists $\delta = \delta_f(D) > 0$ s.t. $|f(z) - f(z')| < \delta$ for all $z' \in B_D(z)$, δ independent of z. Let $C_{ulb}^q(X) \subset CX^q(X)$ the subspace of all u.l.b. functions $f \in CX^q(X)$. δ maps $C_{ulb}^q(X)$ into $C_{ulb}^{q+1}(X)$. We obtain hence a complex $C_{GH}^*(X) = (C_{GH}^q(X) = CX^q(X)/C_{ulb}^q(X), \delta)_q$ and define

$$H_{GH}^*(X) := H^*(C_{GH}^*(X)).$$

If $\Phi : X \longrightarrow Y$ is Borel and metrically semilinear then Φ induces $\Phi^\# : CX^q(Y) \longrightarrow CX^q(X)$, $\Phi^\# : C^q_{ulb}(Y) \longrightarrow C^q_{ulb}(X)$ (since $|(\Phi^\# f)(x) - (\Phi^\# f)(x')| = |f(\Phi x) - f(\Phi x')| < \delta_f(D + C_\Phi)$, $d(\Phi x, \Phi x') \leq d(x, x') + C_\Phi$, $x = (x_0, \ldots, x_q)$), $\Phi^\# : C^*_{GH}(Y) \longrightarrow C^*_{GH}(X)$ and
$$\Phi^* : H^*_{GH}(Y) \longrightarrow H^*_{GH}(X).$$

Theorem 5.8 $H^*_{GH}(X)$ is an invariant of $\text{comp}_{GH}(X)$.

Proof. The proof is completely parallel to that of 5.6, $Y \in \text{comp}_{GH}(X)$, $d_{GH}(X, Y) < \varepsilon$, $\Phi : X \longrightarrow Y$, $\Psi : Y \longrightarrow X$ Borel and metrically semilinear, for $[f] \in H^q_{GH}(X)$
$$(\Psi\Phi)^\# f - f \in C^q_{ulb}(X)$$
since $|f(\Psi\Phi x) - f(x)| \leq \delta_f(d_{GH}(X, Y))$. □

Remark 5.9 For X compact, $H^*_{GH}(X) = 0$. □

Next we consider $\text{comp}_{L,h,rel}(X)$. Denote by $C_*(X) = C_*(S(X))$ the finite singular chains and by $C_{*,b,ulf}(X)$ the complex of all uniformly locally finite bounded singular chains. Then $C_*(X) \subset C_{*,b,ulf}$ is a subcomplex and we define
$$C_{*,b,ulf,\infty}(X) = C_{*,b,ulf}(X)/C_*(X)$$
and
$$H_{*,b,ulf,\infty}(X) = H_*(C_{*,b,ulf,\infty}(X)).$$
Similarly let $C^{*,b,ulf}(X)$ be the complex of uniformly locally finite bounded singular cochains, $C_c^*(X)$ the subcomplex of cochains with compact support,
$$C^{*,b,ulf,\infty}(X) = C^{*,b,ulf}(X)/C_c^*(X)$$
and
$$H^{*,b,ulf,\infty}(X) = H^*(C^{*,b,ulf,\infty}(X)).$$

Theorem 5.10 *If $Y \in \text{arccomp}_{L,h,rel}(X)$ then there exist uniformly proper Lipschitz maps $\Phi : X \longrightarrow Y$, $\Psi : Y \longrightarrow X$ which induce inverse to each other isomorphisms*

$$H_{*,b,ulf,\infty}(X) \underset{\Psi_*}{\overset{\Phi_*}{\underset{\cong}{\overset{\cong}{\rightleftarrows}}}} H_{*,b,ulf,\infty}(Y)$$

and

$$H^{*,b,ulf,\infty}(Y) \underset{\Psi^*}{\overset{\Phi^*}{\underset{\cong}{\overset{\cong}{\rightleftarrows}}}} H^{*,b,ulf,\infty}(X).$$

Proof. By assumption there exists u.p. Lipschitz maps $\Phi : X \longrightarrow Y$, $\Psi : Y \longrightarrow X$ and compact subsets $K_X \subset X$, $K_Y \subset Y$ s.t. $\Phi|_{X \setminus K_X} : X \setminus K_X \longrightarrow Y \setminus K_Y$, $\Psi|_{Y \setminus K_Y} : Y \setminus K_Y \longrightarrow X \setminus K_X$ are inverse to each other Lipschitz homotopy equivalences. It follows immediately from the definition that Φ, Ψ induce chain maps

$$\Phi_\# : C_{*,b,ulf}(X) \longrightarrow C_{*,b,ulf}(Y), \quad \Phi_\# : C_*(X) \longrightarrow C_*(Y),$$
$$\Psi_\# : C_{*,b,ulf}(Y) \longrightarrow C_{*,b,ulf}(X), \quad \Psi_\# : C_*(Y) \longrightarrow C_*(X),$$
$$\Phi_\# : C_{*,b,ulf,\infty}(X) \longrightarrow C_{*,b,ulf,\infty}(Y),$$
$$\Psi_\# : C_{*,b,ulf,\infty}(Y) \longrightarrow C_{*,b,ulf,\infty}(X),$$

and hence morphisms

$$\Phi_* : H_{*,b,ulf,\infty}(X) \longrightarrow H_{*,b,ulf,\infty}(Y),$$
$$\Psi_* : H_{*,b,ulf,\infty}(Y) \longrightarrow H_{*,b,ulf,\infty}(X).$$

Any singular chain $c \in C_{q,b,ulf}(X)$ can be decomposed as $c = c_\infty + c_{K_x}$, where $\text{supp } c_\infty \subset X \setminus \text{Pen}(K_x, R)$, $R = R(c)$. In particular every homology class $[z] \in H_{q,b,ulf,\infty}(X)$ has a representation $z_\infty + B_{q,b,ulf,\infty}(X)$, $\partial z_\infty \in C_{q-1}(X)$. The refined prism construction in [6], p. 512 for uniformly proper Lipschitz homotopy equivalences now shows that $\Psi_\# \Phi_\# + B_{q,b,ulf,\infty} = z_\infty + B_{q,b,ulf,\infty}$, similarly for $\Phi_\# \Psi_\#$. Hence $\Psi_* \Phi_* = id$, $\Phi_* \Psi_* = id$. Similarly for cohomology. □

Remark 5.11 It is also clear that Φ, Ψ induce isomorphisms of the usual end (co–) homology. □

Omitting the factorization by $C_*(X)$ or $C_c^*(X)$, respectively, we obtain the uniformly locally finite bounded homology $H_{*,b,ulf}(X)$ or cohomology $H^{*,b,ulf}(X)$, respectively.

Theorem 5.12 *If $Y \in \mathrm{arccomp}_{L,h}(X)$ then there exist u.p. Lipschitz maps $\Phi : X \longrightarrow Y$, $\Psi : Y \longrightarrow X$ which induce inverse to each other isomorphisms*

$$H_{*,b,ulf}(X) \underset{\Psi_*}{\overset{\Phi_*}{\underset{\cong}{\overset{\cong}{\rightleftarrows}}}} H_{*,b,ulf}(Y)$$

and

$$H^{*,b,ulf}(Y) \underset{\Psi^*}{\overset{\Phi^*}{\underset{\cong}{\overset{\cong}{\rightleftarrows}}}} H^{*,b,ulf}(X).$$

□

Remark 5.13 It is completely clear that the classical homotopy invariants like singular (co–) homology etc. are invariants of $\mathrm{arccomp}_{L,h}(\cdot)$ too. □

We can generalize 5.10 to $1 \leq p < \infty$.
Denote

$C_{q,b,ulf}(N) := \{c \in C_{q,b,ulf} \mid \text{For all } x \in X \text{ and } R > 0 \text{ is } (\#\{ \text{ singular simplexes } \sigma^q \text{ of } c \mid \mathrm{supp}\, \sigma^q \subset B_R(x)\})/R \leq N\}$.

Roughly speaking for all singular chains $\in C_{q,b,ulf}(N)$ simultanously holds that every metric ball of radius R contains at most $R \cdot N$ singular simplexes.

From the definition
$$C_{q,b,ulf} \supseteq \varinjlim_N C_{q,b,ulf}(N) = \bigcup_N C_{q,b,ulf}(N).$$

Set
$$C_{q,p,ulf}(N) = \Big\{ c = \sum_{\sigma^q} c_\sigma \sigma \in C_{q,b,ulf}(N) \;\Big|$$
$$|c|_p := \Big(\sum_\sigma |c_\sigma|^p\Big)^{\frac{1}{p}} < \infty \Big\}.$$

$(C_{q,p,ulf}(N), |\;|_p)$ is a normed space (nonseparable) and we have $\cdots \subseteq C_{q,p,ulf}(N) \subseteq C_{q,p,ulf}(N+1) \subseteq \cdots$. Denote $C_{q,p,ulf}(\infty) = \varinjlim C_{q,p,ulf}(N)$ with the inductive limit topology. Then $\partial : C_{q,p,ulf}(\infty) \longrightarrow C_{q-1,p,ulf}(\infty)$ is continuous since $\partial : C_{q,p,ulf}(N) \longrightarrow C_{q-1,p,ulf}(N)$ is norm–continuous. We obtain

$$H_{q,p,ulf}(\infty)(X), \quad \overline{H}_{q,p,ulf}(\infty)(X),$$
$$H^{q,p,ulf}(\infty)(X), \quad \overline{H}^{q,p,ulf}(\infty)(X), \qquad (5.8)$$

where $\overline{H}$ denotes the reduced (co)homology.

Theorem 5.14 *The (co–)homologies of (5.8) are invariants of* $\mathrm{arccomp}_{L,h}(\cdot)$. $\qquad\square$

Corollary 5.15 (a) $H_{*,b,ulf,\infty}$ *and* $H^{*,b,ulf,\infty}$ *are invariants of* $\mathrm{arccomp}_{L,top,rel}(\cdot)$.
(b) $H_{*,b,ulf}$, $H^{*,b,ulf}$ *and the (co–)homologies of (5.8) are invariants of* $\mathrm{comp}_{L,top}(\cdot)$.

$\qquad\square$

The proof of 5.14, 5.15 follows from the fact that the admitted maps induce chain maps and chain homotopy equivalences between the corresponding complexes. There are many other classes of invariants which we did not consider explicitely until now. These include the K–theory of C^*–algebras, $K_*(C^*X)$, and Kasparovs K–homology for locally compact spaces, K_*X.

We conclude this section with a brief review of bordism for open manifolds and relative characteristic numbers.

We consider as before oriented open manifolds (M^n, g) satisfying

$$|\nabla^i R| \leq C_i, i = 0, 1, 2, \ldots, k \qquad (B_k)$$

and
$$r_{\text{inj}}(M, g) \equiv \inf r_{\text{inj}}(g, x) > 0. \qquad (I)$$

(B^{n+1}, g_B) is a bordism between (M_1^n, g_1) and (M_2^n, g_2) if it satisfies the following conditions.

1) $(\partial B, g_B|_{\partial B}) \cong (M_1, g_1) \cup (-M_2, g_2)$,
2) there exists $\delta > 0$ such that $g_B|_{U_\delta(\partial B)} \cong g_{\partial B} + dt^2$,
3) (B, g_B) satisfies (B_k) and $\inf_{x \in B \setminus U_\delta(\partial B)} r_{\text{inj}}(g_B, x) > 0$,
4) there exists $R > 0$ such that $B \subset U_R(M_1)$, $B \subset U_R(M_2)$.

We denote $(M_1, g_1) \underset{b}{\sim} (M_2, g_2)$. (B^{n+1}, g_B) is called a bordism. Sometimes we denote additionally $\underset{b, b_g}{\sim}$, b_g stands for bounded geometry, i.e. (I) and (B_k).

Lemma 5.16 a) $\underset{b}{\sim}$ *is an equivalence relation. Denote by $[M^n, g]$ the bordism class.*
b) $[M \cup M', g \cup g'] = [M \# M', g \# g']$.
c) *Set* $[M, g] + [M', g'] := [M \cup M', g \cup g'] = [M \# M', g \# g']$. *Then $+$ is well defined and the set of all $[M^n, g]$ becomes an abelian semigroup.* □

Denote by $\Omega_n^{nc} = \Omega_n^{nc}(I, B_k)$ the corresponding Grothendieck group. Similarly one defines $\Omega_n^{nc}(X)$ generated by pairs $((M^n, g), f : M^n \longrightarrow X)$, f bounded and uniformly proper.

Remarks 5.17 1) Condition 4) above looks like $d_{GH}(M, M') \leq R$, where d_{GH} is the Gromov–Hausdorff distance. But this is wrong.

2) There is no chance to calculate Ω_n^{nc}.

3) One would like to have a geometric representative for 0 and for $-[M, g]$. □

The way out from this is to establish bordism theory for special classes of open manifolds or/and further restrictions to bordism. Our first example is bordism with compact support. Here condition 1) above remains but one replaces 2) – 4) by the condition

There exists a compact submanifold $C^{n+1} \subset B^{n+1}$ such that $(\overline{B \setminus C}, g_B|_{\overline{B \setminus C}})$ is a product bordism, i.e.
$$(\overline{B \setminus C}, g_{\overline{B \setminus C}}) \cong (\overline{M \setminus C} \times [0, 1], g_{\overline{M \setminus C}} + dt^2). \quad (cs)$$

We write $\underset{b,cs}{\sim}$. Then one gets a bordism group $\Omega_n^{nc}(cs)$ (= Grothendieck group).

At the first glance, the calculation of $\Omega_n^{nc}(cs)$ or at least the characterization of the bordism classes seems to be very difficult. But we will see, that this is not the case. For this, we introduce still some uniform, structures. Denote by $\mathfrak{M}^n(mf) := \mathfrak{M}^n(mf, nc) \subset \mathfrak{M}_L$ the set of isometry classes of complete, open, oriented Riemannian manifolds. Consider pairs (M_1^n, g_1), $(M_2^n, g_2) \in \mathfrak{M}^n(mf)$ with the following property:

There exist compact submanifolds $K_1^n \subset M_1^n$ and $K_2^n \subset M_2^n$ and an isometry $M_1 \setminus K_1 \xrightarrow{\Phi} M_2 \setminus K_2$. (5.9)

For such pairs, we define in analogy to sections 2 and 4

$${}^b d_{L,iso,rel}((M_1, g_1), (M_2, g_2)) :=$$
$$\inf \{ \max\{0, \log {}^b|df|\} + \max\{0, \log {}^b|dh|\}$$
$$+ \sup_{x \in M_1} \mathrm{dist}(x, hfx) + \sup_{y \in M_2} \mathrm{dist}(y, fhy) \mid$$
$$f \in C^\infty(M_1, M_2), g \in C^\infty(M_2, M_1), \text{ and for some}$$
$$K_1 \subset K, f|_{M_1 \setminus K_1} \text{ is an isometry and } g|_{f(M_1 \setminus K)} = f^{-1} \}.$$

If (M_1, g_1) and (M_2, g_2) do not satisfy (5.9), then we define $^b d_{L,iso,rel}((M_1, g_1), (M_2, g_2)) = \infty$. We have $^b d_{L,iso,rel}((M_1, g_1), (M_2, g_2)) = 0$ if (M_1, g_1) and (M_2, g_2) are isometric.

Remarks 5.18 1) The notions Riemannian isometry and distance isometry coincide for Riemannian manifolds. Furthermore, if f is an isometry f, then we have $^b|df| = 1$.

2) Any f that occurs in the definition of $d_{L,iso,rel}$ is automatically an element of $C^{\infty,m}(M_1, M_2)$ for all m. The same holds true for g. □

We write $\mathfrak{M}^n_{L,iso,rel}(mf) = \mathfrak{M}^n(mf)/\sim$ where $(M_1, g_1) \sim (M_2, g_2)$ if $^b d_{L,iso,rel}((M_1, g_1), (M_2, g_2)) = 0$. Set

$$V_\delta = \{((M_1, g_1), (M_2, g_2)) \in (\mathfrak{M})^2_{L,iso,rel}(mf))^2 \mid$$
$$^b d_{L,iso,rel}((M_1, g_1), (M_2, g_2)) < \delta\}.$$

Proposition 5.19 $\mathcal{L} = \{V_\delta\}_{\delta>0}$ *is a basis for a metrizable uniform structure* $^b\mathcal{U}_{L,iso,rel}$. □

Denote by $^b\mathfrak{M}^n_{L,iso,rel}(mf)$ the corresponding uniform space.

Proposition 5.20 *If* $r_{inj}(M_i, g_i) = r_i > 0$, $r = \min\{r_1, r_2\}$ *and* $^b d_{L,iso,rel}((M_1, g_1), (M_2, g_2)) < r$ *then* M_1 *and* M_2 *are (uniformly proper) bi–Lipschitz homotopy equivalent.* □

Corollary 5.21 *If we restrict ourselves to open manifolds with injectivity radius* $\geq r$, *then manifolds* (M_1, g_1) *and* (M_2, g_2) *with* $^b d_{L,iso,rel}$-*distance less than* r *are automatically (uniformly proper) bi–Lipschitz homotopy equivalent.* □

Remark 5.22 If (M_1, g_1) satisfies (I) or (I) and (B_k) and $^b d_{L,iso,rel}(M_1, g_1), (M_2, g_2)) < \infty$ then (M_2, g_2) also satifies (I) or (I) and (B_k). □

We cannot show that $^b\mathfrak{M}^n_{L,iso,rel}$ is locally arcwise connected, that components are arc components and $^b\text{comp}_{L,iso,rel}(M,g) = \{(M',g')|{}^b d_{L,iso,rel}((M,g),(M',g')) < \infty\}$ is wrong. The reason is that we cannot connect non–homotopy–equivalent manifolds by a continuous family of manifolds. A parametrization of non-trivial surgery always contains bifurcation levels where we leave the category of manifolds. A very simple case comes from corollary 5.21.

Corollary 5.23 *If we restrict $^b\mathcal{U}_{L,iso,rel}$ to open manifolds with injectivity radius $\geq r > 0$, then the manifolds in each arc component of this subspace are bi–Lipschitz homotopy equivalent.*

Proof. This subspace is locally arcwise connected and components are arc components. Consider an (arc) component and two elements (M_1, g_1) and (M_2, g_2) of it, connect them by an arc, cover this arc by sufficiently small balls, and apply 5.21. □

By definition, we have

$$^b d_{L,iso,rel}((M_1,g_1),(M_2,g_2)) < \infty \implies d_L((M_1,g_1),(M_2,g_2)) < \infty,$$

where d_L is the Lipschitz distance of section 2. Hence, $(M_2, g_2) \in \text{comp}_L(M_1, g_1)$, i. e.

$$\{(M_2,g_2) \in \mathfrak{M}^n(mf)|{}^b d_{L,iso,rel}((M_1,g_1),(M_2,g_2)) < \infty\}$$
$$\subseteq \text{comp}_L(M_1,g_1). \tag{5.10}$$

For this reason, we denote the left hand side $\{\ldots\}$ of (5.10) by $\text{gen}^b\text{comp}_{L,iso,rel}(M_1,g_1) = \{\ldots\} = \{\ldots\} \cap \text{comp}_L(M_1,g_1)$, keeping in mind that this is not an arc component, but a subset of (manifolds in) a Lipschitz arc component.

If we fix (M_1, g_1), then in a special case, we have a good overview of the elements in $\text{gen}^b\text{comp}_{L,iso,rel}(M_1, g_1)$.

Example 5.24 *Let $(M_1, g_1) = (\mathbb{R}^n, g_{\text{standard}})$. Then $\text{gen}^b\text{comp}_{L,iso,rel}(M_1,g_1)$ is in a 1-1-correspondence with $\{(M^n, g)| M^n$ is a closed manifold and g is flat in an annulus contained in a disk neighbourhood of a point $\}$.* □

This can be generalized as follows.

Theorem 5.25 *Any component* $\text{gen}^b\text{comp}_{L,iso,rel}(M,g)$ *contains at most countably many diffeomorphism types.*

Proof. Fix $(M,g) \in \text{gen}^b\text{comp}_{L,iso,rel}(M,g)$ and an exhaustion $K_1 \subset K_2 \subset \ldots$, $\bigcup K_i = M$, of M by compact submanifolds, and let $(M',g') \in {}^b\text{comp}_{L,iso,rel}(M,g)$. Then there are $K' \subset M'$ and $K_i \subset M$ such that $M \setminus K_i$ and $M' \setminus K'$ are isometric. The diffeomorphism type of M' is completely determined by that of the pair $(K_1 \underset{\partial K_1 \cong \partial K'}{\cup} K', K_1)$, but the set of types of such pairs (after fixing M and $K_1 \subset K_2 \subset \ldots$) is at most countable. □

Thus, after fixing (M,g), the diffeomorphism classification of the elements in ${}^b\text{comp}_{L,iso,rel}(M,g)$ seems to be reduced to a "handy" countable discrete problem. This is in fact the case in a sense which is parallel to the classification of compact manifolds. Now we connect the calculation of $\Omega_n^{nc}(cs)$ with the generalized components $\text{gen}^b\text{comp}_{L,iso,rel}(\cdot) \subset {}^b\mathfrak{M}_{L,iso,rel}^n(mf)$.

Remark 5.26 If $(M_1,g_1), (M_2,g_2) \in \text{gen}^b\text{comp}_{L,iso,rel}(M,g)$, then, in general, $(M_1,g_1)\#(M_2,g_2) \notin \text{gen}^b\text{comp}_{L,iso,rel}(M,g)$. □

Let $\Omega_n^{nc}(cs, \text{gen}^b\text{comp}_{L,iso,rel}(M,g)) \subset \Omega_n^{nc}(cs)$ be the subgroup generated by

$$\{[M',g']_{cs} | (M',g') \in \text{gen}^b\text{comp}_{L,iso,rel}(M,g)\}.$$

We know $\Omega_n^{nc}(cs)$ completely if we know all $\Omega_n^{nc}(cs, \text{gen}^b\text{comp}_{L,iso,rel}(M,g))$, and we know

$$\Omega_n^{nc}(cs, {}^b\text{comp}_{L,iso,rel}(M,g))$$

completely if we know a corresponding generating set. However, the elements of such a set are completely determined by their (relative) characteristic numbers.

Fix $(M^n, g) \in \text{gen}^b\text{comp}_{L,iso,rel}(M^n, g)$, where M is oriented. Assume that $(M_1, g_1) \in \text{gen}^b\text{comp}_{L,iso,rel}(M^n, g)$, and let $\Phi : M \setminus K \longrightarrow M_1 \setminus K_1$ be an orientation preserving isometry. Define the (relative) Stiefel–Whitney numbers of the pair (M_1, M) by

$$w_1^{r_1} \ldots w_n^{r_n}(M_1, M) := \langle w_1^{r_1} \ldots w_n^{r_n}, [K_1]\rangle + \langle w_1^{r_1} \ldots w_n^{r_n}, [K]\rangle. \tag{5.11}$$

Similarly, for (M_1, M) and $n = 4k$, we define the (relative) Pontrjagin numbers

$$p_1^{r_1} \ldots p_k^{r_k}(M_1, M) := \int_{K_1} p_1^{r_1} \ldots p_k^{r_k}(M_1) - \int_K p_1^{r_1} \ldots p_k^{r_k}(M) \tag{5.12}$$

and the (relative) signature by

$$\sigma(M_1, M) := \sigma(K_1) + \sigma(-K). \tag{5.13}$$

Lemma 5.27 *The numbers* $w_1^{r_1} \ldots w_n^{r_n}(M_1, M)$, $p_1^{r_1} \ldots p_k^{r_k}(M_1, M)$, *and* $\sigma(M_1, M)$ *are well defined, and we have*

$$w_1^{r_1} \ldots w_n^{r_n}(M_1, M) = \langle w_1^{r_1} \ldots w_n^{r_n}(K_1 \cup K), [K_1 \cup K]\rangle, \tag{5.14}$$

$$p_1^{r_1} \ldots p_k^{r_k}(M_1, M) = \langle p_1^{r_1} \ldots p_k^{r_k}(K_1 \cup -K), [K_1 \cup -K]\rangle, \tag{5.15}$$

and

$$\sigma(M_1, M) = \sigma(K_1 \cup -K). \tag{5.16}$$

Proof. The equations (5.11), (5.12) are clear. (5.13) comes from Novikov additivity of σ. Hence we have only to show the well definedness, i.e. the independence of the choice of $K \subset M$, $K_1 \subset M_1$. Start with (5.12). If $K_1' \supset K_1$, $K' \supset K$, $\Phi|_{M \setminus K'} : M \setminus K' \xrightarrow{\cong} M_1 \setminus K_1'$ orientation preserving isometric, then

$$\int_{K_1'} \cdots - \int_{K'} \cdots = \int_{K_1' \setminus K_1} \cdots + \int_{K_1} \cdots - \left(\int_{K' \setminus \overset{\circ}{K}} \cdots + \int_K \cdots \right). \tag{5.17}$$

But $\int_{K'_1\setminus \mathring{K}_1} \cdots - \int_{K'\setminus \mathring{K}} \cdots = 0$ since $K'_1 \setminus \mathring{K}_1$ and $K' \setminus \mathring{K}$ are isometric under Φ by assumption. The analogous conclusion can be done for $K''_1 \subset K_1$, $K'' \subset K$, $\Phi : M \setminus \mathring{K}'' \longrightarrow M_1 \setminus \mathring{K}''_1$ already an isometry. In the general case K_1, K'_1, K, K' one considers $K'_1 \cap K_1$, $K \cap K'$ and reduces to the first two considerations after smoothing out $K'_1 \cap K_1$, $K' \cap K$ by arbitrarily small perturbations. The proof for (5.11) is quite similar replacing integrations in (5.17) by application of cocycles to cycles. The independence of (5.13) comes again from Novikov additivity, applying it several times. □

Theorem 5.28 *Fix* $(M_1, g_1), (M_2, g_2) \in \text{gen}^b\text{comp}_{L,iso,rel}(M, g)$. *Then* $(M_1, g_1) \underset{b,cs}{\sim} (M_2, g_2)$ *if and only if all characteristic numbers of* (M_1, M) *coincide with the corresponding characteristic numbers of* (M_2, M).

Proof. Assume $(M_1, g_1) \underset{b,cs}{\sim} (M_2, g_2)$. Choose $C^{n+1} \subset B^{n+1}$ large enough such that with $\partial C^{n+1} = \partial_1 C \cup \partial_2 C \cup \partial_3 C$, $\partial_1 C = K_1 \subset M_1$, $\partial_2 C = K_2 \subset M_2$, $\partial_3 C \cong \partial K_1 \times [0,1]$ there holds $M_1 \setminus \mathring{K}_1 \underset{\Phi_1}{\cong} M \setminus \mathring{K}$, $M_2 \setminus \mathring{K}_2 \underset{\Phi_2}{\cong} M \setminus \mathring{K}$. Then after smoothing out by arbitrary small perturbations, ∂C is diffeomorphic ot $K_1 \underset{\Phi}{\cup} -K_2$, $\Phi = \partial(\Phi_2^{-1}\Phi_1)$. Hence char.n.$(K_1 \underset{\Phi}{\cup} -K_2) =$ char.n.$(K_1 \underset{\Phi_1}{\cup} -K) +$ char.n.$(K \underset{\Phi_2^{-1}}{\cup} -K_2) =$ char.n.$(M_1, M) +$ char.n.$(M, M_2) =$ char.n.$(M_1, M) -$ char.n.(M_2, M), i.e. char.n.$(M_1, M) =$ char.n.(M_2, M). Conversely, if char.n.$(M_1, M) =$ char.n.(M_2, M) then char.n.$(K_1 \underset{\Phi}{\cup} -K_2) = 0$, $K_1 \underset{\Phi}{\cup} -K_2$ is 0–bordant, $K_1 \underset{\Phi}{\cup} -K_2 = \partial C^{n+1}$. Form $K_1 \cup (\partial K_1 \times [0,1]) \underset{\Phi}{\cup} -K_2$ which equals to ∂C^{n+1}, glue $(M_1 \setminus \mathring{K}_1) \times [0,1] \cong (M \setminus \mathring{K}) \times [0,1] \cong (M_2 \setminus \mathring{K}_2) \times [0,1]$ and smooth out (the topology and the metrics). The result is a bordism (B^{n+1}, g_B) with compact support between (M_1, g) and (M_2, g). □

Corollary 5.29 *Description of all elements of $\Omega_n^{nc}(cs)$ reduces to "counting" the generalized components of $^b\mathfrak{M}_{L,iso,rel}^n(mf)$.* □

Example 5.30 *Consider $M_i = (M_i' \cup \partial M_i' \times [0, \infty[, g_i), i = 1, 2$, M_i' compact, $\partial M_1' = \partial M_2'$, $(\partial M_1' \times [0, \infty[, g_{1,a,\infty})$ isometric to $(\partial M_2' \times [0, \infty[, g_{2,a,\infty}), g_{i,a,\infty} = g_i|_{\partial M_i' \times [a, \infty[}$. Let $M = (D^n \cup S^n \times [0, \infty[, g_{standard})$. If $\sigma(M_1') \neq \sigma(M_2')$ then $(M_1, g), (M_2, g) \in \text{gen}^b\text{comp}_{L,iso,rel}(M, g)$ are not cs-bordant.* □

There is a simple approach to calculate the local algebraic structure of $\Omega_n^{nc}(cs)$. Consider as above $\Omega_n^{nc}(cs, M) := \Omega_n^{nc}(cs, \text{gen}^b\text{comp}_{L,iso,rel}(M))$ and let Ω_n be the usual bordism group of closed oriented n–manifolds. Then there exists a map $\Phi = \Phi_M : \Omega_n \longrightarrow \Omega_n^{nc}(cs, M)$, $\Phi([N]) := [M \# N, g_{M \# N}]_{cs}$. Here the bordism class in $\Omega_n^{nc}(cs, M)$ is independent of the metric of N. Moreover, we have a map $\Psi = \Psi_M : \Omega_n^{nc}(cs, M) \longrightarrow \Omega_n$, $\Psi([M', g']) := [K' \cup -K]$, where $(M' \setminus \mathring{K}', g'|_{M' \setminus \mathring{K}'})$ is isometric to $(M \setminus \mathring{K}, g|_{M \setminus \mathring{K}})$. It is very easy to see that Ψ_M is well defined: Let $(M'', g'') \in [M', g']_{cs}$. Then there exist $K_1'' \subset M''$, $K_1' \subset M'$, $K_1 \subset M$ such that $M'' \setminus \mathring{K}_1''$, $M' \setminus \mathring{K}_1'$, $M \setminus \mathring{K}_1$ are isometric. By assumption and according to theorem 5.27 there holds char.n.$(M'', M) = $ char.n.(M', M), i.e. char.n.$(K_1'' \cup -K_1) = $ char.n.$(K_1' \cup -K_1) = $ char.n.$(K' \cup -K)$, $[K_1'' \cup -K_1]_{\Omega_n} = [K' \cup -K]_{\Omega_n}$. There holds $\Psi_M \Phi_M = $ id $: (\Psi\Phi)[N] = \Psi([M\#N]) = [N]$ and $\Phi_M \Psi_M = $ id $: (\Phi\Psi)([M', g']) = \Phi([K' \cup K]) = [M\#(K' \cup -K), g_{M\#(K' \cup -K)}] = [M', g']$ since char.n.$(M', M) = $ char.n.$(M\#(K' \cup -K), M)$. Here Φ_M and Ψ_M are 1-1 maps. Moreover we have maps

$\Phi_M \times \Phi_{M'} : \Omega_n \times \Omega_n \longrightarrow \Omega_n^{nc}(cs, M) \times \Omega_n^{nc}(cs, M'),$

$\Phi_M \times \Phi_{M'}([N_1], [N_2]) = ([M\#N_1, g_{M\#N_1}], [M'\#N_2, g_{M'\#N_2}]),$

$\Phi_{M\#M'} : \Omega_n \longrightarrow \Omega_n^{nc}(cs, M\#M'),$

$[N]_{\Omega_n} \longrightarrow [M\#M'\#N, g_{M\#M'\#N}]_{cs},$

$$\begin{array}{ccc} \Omega_n \times \Omega_n, & ([N_1], [N_2]) \\ \downarrow & \downarrow \\ \Omega_n & [N_1 \# N_2] \end{array}$$

and

$$\begin{array}{ccc} \Omega_n^{nc}(cs, M) \times \Omega_n^{nc}(cs, M'), & ([M_1, g_1], [M_1', g_1']) \\ \downarrow & \downarrow \\ \Omega_n^{nc}(cs, M \# M') & [M_1 \# M_1', g_1 \# g_1'] \end{array}$$

Proposition 5.31 *The diagrams*

$$\begin{array}{ccc} \Omega_n \times \Omega_n & \xrightarrow{\Phi_M \times \Phi_{M'}} & \Omega_n^{nc}(cs, M) \times \Omega_n^{nc}(cs, M') \\ \downarrow & & \downarrow \\ \Omega_n & \xrightarrow{\Phi_{M \# M'}} & \Omega_n^{nc}(cs, M \# M') \end{array} \qquad (5.18)$$

and

$$\begin{array}{ccc} \Omega_n^{nc}(cs, M) \times \Omega_n^{nc}(cs, M') & \xrightarrow{\Phi_M \times \Phi_{M'}} & \Omega_n \times \Omega_n \\ \downarrow & & \downarrow \\ \Omega_n^{nc}(cs, M \# M') & \xrightarrow{\Phi_{M \# M'}} & \Omega_n \end{array} \qquad (5.19)$$

commute. □

Remark 5.32 It is important that we consider (5.18), (5.19) at bordism class level. In (5.19) e.g. $(K_1 \cup -K) \# (K_1' \cup -K') \neq K_1 \# K_1' \cup -(K \# K')$, but their bordism classes coincide. □

The 1–1 property of Φ_M, Ψ_M moreover implies:

Proposition 5.33 *Modulo torsion (which is well defined at the Ω_n-level) is $\Omega_n^{nc}(cs, M^n) = 0$ for $n \neq 4k$ and for $n = 4k$ are $[M \# P^{2i_1}(\mathbb{C}) \times \cdots \times P^{2i_k}(\mathbb{C}), g_{M \# P^{2i_1}(\mathbb{C}) \times \cdots \times P^{2i_k}(\mathbb{C})}]$ independent generators for $\Omega_n^{nc}(cs, M^n)$ over $\mathbb{Q}$, $i_1 + \cdots + i_k = k$.* □

Remarks 5.34 1) 5.28, 5.29, 5.33 provide sufficient means to characterize cs–bordism classes and to calculate $\Omega_n^{nc}(cs)$.

2) An analogous procedure can be applied to calculate e.g. $\Omega_n^{nc,spin}(cs, M)$, where $^b\text{comp}_{L,iso,rel}(M)$ is now a component consisting of Spin–manifolds. □

As we pointed out, a contentful theory should be developed under three aspects.

1) A convenient characterization of bordism classes should be desirable.

2) It should be possible to exhibit sets of independent generators, at least for the intersections with gen–components.

3) A geometric realization of zero and the inverse should be desirable.

The general bordism group Ω_n^{nc} did not satisfy any of these three wishes. $\Omega_n^{nc}(cs)$ satisfies the first two wishes. We develop below a bordism theory which satisfies the second and the third wish. This will be the bordism theory for manifolds with a finite number of ends, each of them nonexpanding.

Let ε be an isolated end of (M^n, g). A ray in ε is a geodesic γ defined on $[0, \infty[$ which is a shortest geodesic between any two of its points and such that some neighbourhood of ε contains up to a finite segment the whole of $|\gamma|$. Then the latter holds for any neighbourhood of ε.

Lemma 5.35 *Let ε be an isolated end of (M^n, g).*
a) *Then there exists a ray in ε.*
b) *If (M^n, g) additionally satisfies (I) then there exists a ray in ε with a uniformly thick neighbourhood.*

Proof. A proof of a) is e.g. contained in [37], p. 43. b) follows immediately from a) and (I).

□

We call an end ε of (M^n, g) nonexpanding if there exist a ray γ in ε and an $R = R_M > 0$ and an element $G \in \varepsilon$ such that $G \subseteq U_R(|\gamma|)$, roughly written $\varepsilon \subseteq U_R(|\gamma|)$.

In the sequel we restrict to open manifolds satisfying (I), $B(\infty)$, with finitely many ends, each of them nonexpanding.

Examples 5.36 1) Consider the sphere $S_r^{n-1} \subset \mathbb{R}^n \subset \mathbb{R}^{n+1}$ of radius r and

$$\operatorname{chc}^n(r) := (S_r^{n-1} \times [0, \infty[\cup D_r^n, g_{st}),$$

i.e. the closed half cylinder of radius r with a standard metric g_{st} which should be the product metric of $S_r^{n-1} \times [0, \infty[$ smoothly extended to the glued bottom D_r^n and with standard orientation. Then $\operatorname{chc}^n(r)$ is an open manifold with one nonexpanding end, satisfying (I), (B_∞).

2) $\overset{k}{\underset{i=1}{\#}} \operatorname{chc}(r_i)$ has finitely many nonexpanding ends.

3) Any manifold $(M^n = M' \cup \partial M' \times [0, \infty[, g_M)$ where M' is compact and $g_M|_{\partial M' \times [a, \infty[} = dt^2 + g_{\partial M'}$ satisfying (I), (B_∞) and has finitely many nonexpanding ends.

4) The same is true if we allow g_M of 3) to vary in $\operatorname{comp}^{p,r}(g_M) \cap C^\infty$.

5) If we consider M of smooth type of 3) and $g_M|_{\partial M' \times [a, \infty[} = dt^2 + f(t)^2 g_{\partial M'}$ with $c_2 \geq f(t) \geq c_1 > 0$, $\frac{f^{(\nu)}}{f}$ bounded for all ν, $t \geq a$ then M has finitely many nonexpanding ends. $\square$

We define now a slightly sharpened bordism relation.

Let (M^n, g), (M'^n, g') be as above, each with finitely many nonexpanding ends $\varepsilon_1, \ldots, \varepsilon_s$ or $\varepsilon'_1, \ldots, \varepsilon'_{s'}$, respectively. Let $\gamma_{M,1}, \ldots, \gamma_{M,s}$ or $\gamma_{M',1}, \ldots, \gamma_{M',s'}$ corresponding rays as above. From $(M, g) \underset{bg}{\sim} (M', g')$ and all ends nonexpanding follows in particular that for all sufficiently large compact $C^{n+1} \subset B^{n+1}$

there exists $R = R_B > 0$ s.t.

$$B^{n+1} \setminus C^{n+1} \subset \bigcup_1^s U_R(|\gamma_{M,\sigma}|),$$

$$B^{n+1} \setminus C^{n+1} \subset \bigcup_1^{s'} U_R(|\gamma_{M',\sigma'}|).$$

We require additionally to this condition the additive compability of the inner γ–distance and the $(B \setminus C)$–distance for points x_γ, y_γ on the γ's.
There exist $C^{n+1} \subset B^{n+1}$ and $c' > 0$ s.t. for $x_\gamma, y_\gamma \in |\gamma| \setminus C$ holds

$$d_\gamma(x_\gamma, y_\gamma) - c' \leq d_{B \setminus C}(x_\gamma, y_\gamma) \leq d_\gamma(x_\gamma, y_\gamma) + c'. (GH)$$

Here γ stands for $\gamma_{M,1}, \ldots, \gamma_{M,s}, \gamma_{M',1}, \ldots, \gamma_{M',s'}$, respectively and $d(\cdot, \cdot) \equiv \text{dist}(\cdot, \cdot)$.
We denote $(M, g) \underset{ne}{\sim} (M', g')$ if they are bg–bordant by means (B, g_B) satisfying (GH).

Remarks 5.37 1) The right hand inequality of (GH) trivially holds. We added it only for symmetry reasons.
2) It was essentially Thomas Schick who pointed out to the author the meaning of the condition (GH) or (GH_1) and who proposed to include them into the definition of bordism. □

We consider instead of (GH) the condition
There exist $C^{n+1} \subset B^{n+1}$ and $c' > 0$ s.t. for all $x, y \in U(\varepsilon)$ holds

$$d_{U_\varepsilon}(x, y) - c \leq d_{B \setminus C}(x, y) \leq d_{U(\varepsilon)}(x, y) + c. (GH_1)$$

Here ε stands for $\varepsilon_1, \ldots, \varepsilon_s, \varepsilon'_1, \ldots, \varepsilon'_{s'}$ and $U(\varepsilon)$ for a neighbourhood of ε, $U(\varepsilon) \cap C = \emptyset$.

Lemma 5.38 (GH) and (GH_1) are equivalent.

Proof. Assume (GH_1). Then (GH) holds since for $x_\gamma, y_\gamma \in |\gamma| \subset U(\varepsilon)$, $U(\varepsilon) \cap C = \emptyset$, $d_{U(\varepsilon)}(x_\gamma, y_\gamma) = d_\gamma(x_\gamma, y_\gamma)$. If conversely $x, y \in U(\varepsilon)$ then there exists $x_\gamma, y_\gamma \in |\gamma| \subset U(\varepsilon)$ s.t. $d_{U(\varepsilon)}(x, x_\gamma) \leq R_M$, $d_{U(\varepsilon)}(y, y_\gamma) \leq R_M$. Then the assertion follows from

$$d_{U(\varepsilon)}(x,y) - d_{U(\varepsilon)}(x_\gamma, y_\gamma) \leq d_{U(\varepsilon)}(x, x_\gamma) + d_{U(\varepsilon)}(y, y_\gamma),$$
$$d_{U(\varepsilon)}(x,y) - d_{U(\varepsilon)}(x, x_\gamma) - d_{U(\varepsilon)}(y, y_\gamma) \leq d_\gamma(x_\gamma, y_\gamma)$$
$$= d_\gamma(x_\gamma, y_\gamma) - c' + c' \leq d_{B\setminus C}(x_\gamma, y_\gamma) + c',$$
$$d_{U(\varepsilon)} - 2R_M - c' \leq d_{B\setminus C}(x_\gamma, y_\gamma),$$
$$d_{U(\varepsilon)} - 4R_M - c' \leq d_{B\setminus C}(x, y).$$

□

Remark 5.39 (GH_1) immediately implies that $d_{GH}(\overline{B\setminus C}, \overline{U(\bigcup_\sigma \varepsilon_\sigma)}) < \infty$, where $d_{G,H}(\cdot, \cdot)$ is the Gromov–Hausdorff distance between proper metric spaces. This follows from the following facts. $d_{GH}(\overline{B\setminus C}, \overline{U(\bigcup_\sigma \varepsilon_\sigma)}) < \infty$ if we endow $\overline{U(\bigcup_\sigma \varepsilon_\sigma)}$ with the induced lengths metric and use $(\overline{B\setminus C} \subset U_R(U(\bigcup_\sigma \varepsilon_\sigma))$. Then we use $d_{GH}(U(\varepsilon))$, its own lengths metric, $U(\varepsilon)$, induced lengths metric $< \infty$, which follows from (GH_1). As a matter of fact, we introduced (GH) to assure $d_{GH}(\overline{B\setminus C}, \overline{U(\varepsilon)}) < \infty$. □

Proposition 5.40 $\underset{ne}{\sim}$ *is an equivalence relation.*

We refer to [26] for the proof. □

$\Omega_n^{nc}(ne) \equiv \Omega^{nc}(f_e, ne, b_g)$ is again defined as Grothendieck group. Next we develop geometric realizations for 0 and $-[M, g]_{ne}$ in $\Omega_n^{nc}(ne)$.

Let (M^n, g) be as above, i.e. oriented, with (I), (B_∞), finitely many ends $\varepsilon_1, \ldots, \varepsilon_s$, each of them nonexpanding. Let ε be one of them, $C \subset M$ compact and so large that ε is defined by one of the components of $M \setminus C$, $U_\varepsilon \subset M \setminus C$ a neighbourhood, γ a ray in $U(\varepsilon)$. γ admits a tubular neighbourhood of radius $\delta_3 > 0$. Consider $(B, g_B) = (M \times I, g_M + dr^2)$. Then $\varepsilon \times I = \{U_j(\varepsilon) \times I\}_{j \in J}$ is an end of $M \times I$, $U(\varepsilon \times I) = U(\varepsilon) \times I$ a neighbourhood disjoint to $C_{M \times I} = C \times I$, and for $0 < \delta_1 < 1$, the curve $\gamma_{\delta_1} = \gamma \times \{\delta_1\} = (\gamma, \delta_1)$ is a ray in $U(\varepsilon \times I)$. $\varepsilon \times I$ is nonexpanding. γ_{δ_1} admits a tubular neighbourhood with a radius $\delta_2 > 0$, $T_{\delta_2}(\gamma_{\delta_1})$.

Theorem 5.41 $\partial T_{\delta_2}(\gamma_{\delta_1})$ *has bounded geometry, one nonexpanding end and there holds*

$$\partial T_{\delta_2}(\gamma_{\delta_1}) \underset{ne}{\sim} \mathrm{chc}^n(\delta_2), \quad \delta_2 > 0.$$

We refer to [26] for the proof. □

Next we shall see, $(\mathrm{chc}^n(\delta), g_{st})$ will play the role of our zero in $\Omega_n^{nc}(ne)$.

Lemma 5.42

a)   For $r_1 < r_2$ is $\mathrm{chc}^n(r_1) \underset{ne}{\sim} \mathrm{chc}^n(r_2)$.   (5.20)

b) $\left[\overset{k}{\underset{i=1}{\#}} \mathrm{chc}^n(r_i)\right]_{ne} = [\mathrm{chc}^n(r)]_{ne}$ *for* $r > r_1 + \cdots + r_k$.

(5.21)

Proof. a) is immediately clear (or follows from b)). Set for b) $r = r_1 + \cdots + r_k + \delta$, place $\mathrm{chc}^n(r_1) \cup \cdots \cup \mathrm{chc}^n(r_k)$ all with parallel $[0, \infty[$ direction into $\mathrm{int}(\mathrm{chc}^n(r))$, where $\mathrm{int}(\mathrm{chc}^n(r))$ coresponds to $\mathring{D}_r^n \times]0, \infty[$. Then $\mathrm{CL}(\mathrm{int}(\mathrm{chc}^n(r)) \setminus \mathrm{int}(\mathrm{chc}^n(r_1) \cup \cdots \cup \mathrm{chc}^n(r_k)))$ defines the desired ne–bordism. □

Theorem 5.43 *For any oriented manifold* (M^n, g) *of bounded geometry and a finite number of ends, each of them nonexpanding, there holds*

$$[M^n, g]_{ne} = [(M^n, g) \cup (\mathrm{chc}^n(r), g_{st})]. \quad (5.22)$$

Proof. We must construct a ne–bordism between (M^n, g) and $-((M^n, g) \cup (\mathrm{chc}^n(r), g_{st}))$. Let $(B^{n+1}, g_B) = (M \times [0,1], g_M + dt^2)$, ε be an end of M, γ a ray in ε, form $\gamma_{\delta_1} = (\gamma, \delta_1) \subset M \times [0,1]$, $T_{\delta_2}(\gamma_{\delta_1})$, $\delta_2 < \inf\{\frac{\delta_1}{2}, r_{\mathrm{inj}}(M)/2\}$ and set $B_\gamma = B^{n+1} \setminus \mathrm{int} T_{\delta_2}(\gamma_{\delta_1})$ with the induced metric. From our assumption $r_{\mathrm{inj}} > 0$ follows easily that $\partial T_{\delta_2}(\gamma_{\delta_1})$ has a smooth collar $U_\delta(\partial T)$. Endow $U_{\frac{\delta}{2}}$ with the product metric $g_{\frac{\delta}{2}}$ and form on $U_\delta - U_{\frac{\delta}{2}}$ the smooth bg–convex combination of $g_{\frac{\delta}{2}}$ and g_B getting g_{B_γ}. Endow $\partial T_{\gamma_2}(\gamma_{\delta_1})$ with the induced orientation. Then (B_γ, g_{B_γ}) is a bg, ne–bordism between (M^n, g) and $(M^n, g) \cup (\partial T_{\delta_2}(\gamma_{\delta_1}), g_{\partial T})$. Theorem 5.41 yields

$$(M^n, g) \cup (\partial T, g_{\partial T}) \underset{ne}{\sim} (M^n, g) \cup (\mathrm{chc}^n(\delta_2), g_{st}).$$

□

Theorem 5.44 $\Omega_n^{nc}(ne) \equiv \Omega_n^{nc}(bg, ne)$ *is an abelian group with* $-[M^n, g] = [(-M^n, g)]$ *and* $0 = [\mathrm{chc}^n(r), g_{st}]$. □

Our next goal is to produce independent generators of $\Omega_n^{nc}(ne)$. As we shall see in the sequel, infinite connected sums of complex projective spaces (or their cartesian products) supply such elements. We prepare this by several assertions

Lemma 5.45 *Let* (M_i^n, g_i), $i = 1, 2$, *be open, oriented of bounded geometry and with a finite number of ends, each of them nonexpanding. Let further* (B^{n+1}, g_B) *be a ne–bordism between them and* $K \subset B$ *compact such that the ends of* B *coincide with the components of* $B \setminus K$. *Let* $C_\varepsilon \subset B \setminus K$ *a component of* $B \setminus K$ *and* $x_0 \in C_\varepsilon$. *Then there exists a constant* $C_1 > 0$ *such that the diameter of any metric sphere*

$$S_\varrho(x_0) = \{x \in C_\varepsilon | d_B(x, x_0) = \varrho\}$$

is $\leq C_1$. *Here we understand the diameter with respect to the induced length metric* d_B *of* B.

We refer to [26] for the proof. □

Now we recall once again the chopping theorem of Cheeger/Gromov (cf. [17]) which is a consequence of Abresch's habilitation (cf. [1]) and was our I 1.33.

Theorem 5.46 *Suppose (M^n, g) open, complete with bounded sectional curvature $|K| \leq C$. Given a closed set $X \subset M^n$ and $0 < r \leq 1$, there is a submanifold, U^n, with smooth boundary, ∂U^n, such that for some constant $c(n, C)$*

$$X \subset U \subset T_r(X),$$
$$\text{vol}(\partial U) \leq c(n, C)\text{vol}(T_r(X) \setminus X)r^{-1},$$
$$|II(\partial U)| \leq c(n, C)r^{-1}.$$

Moreover, U can be chosen to be invariant under $\mathcal{I}(r, X) =$ group of isometries of $T_r(X)$ which fix X. □

In our case, $X = X_\varrho = \overline{B_\varrho(x_0)} \subset B^{n+1}$. To apply 5.46, we form $(V^{n+1}, g_V) = (B^{n+1} \cup B^{n+1}, g_B \cup g_B)$ which is well defined and smooth since we assumed the Riemannian collar $g_B|_{\text{collar}} = g_{\partial B} + dt^2$. Now we set $X_V = X \cup X$ and apply 5.46. Fix $0 < r \leq 1$. Then we get U_V, $H_{\varrho,V,r} = \partial U_V$.

$$X_V \subset U_V \subset T_r(X_V)(= \{x \in V | d_V(x, X_V) \leq r\}), \quad (5.23)$$
$$\text{vol}(H_{\varrho,V,r}) = \text{vol}(\partial U_V)$$
$$\leq c(n+1, C)\text{vol}(T_r(X_V) \setminus X_V)r^{-1} \quad (5.24)$$
$$|II(\partial U_V)| \leq c(n+1, C)r^{-1} \quad (5.25)$$

and U_V is invariant under $\mathcal{I}(r, X_V)$.

The main idea of the proof consists in considering the distance function $F = d(\cdot, X_V)$ where for points $\in V \setminus X_V$, $d(\cdot, X_V) = d(\cdot, X_\varrho) = d(\cdot, S_\varrho)$. Then one applies Yomdin's theorem to F in Abresch's smoothed out metric. All constructions are invariant under the metric involution and this involution remains an isometry also with respect to Abresch's smoothed out metric.

Restricting the obtained U_V, ∂U_V to B, we obtain the desired result for $X = \overline{B_\varrho(x_0)} \subset B$. Restricting for ϱ large to C_ε and using the construction of U as preimage under the smoothed F, we obtain in $\overline{C}_\varepsilon$ a hypersurface $H = H_\varrho$ which decomposes $\overline{C}_\varepsilon$ into a compact and noncompact part $\overline{C}_{\varepsilon,c}$ and $\overline{C}_{\varepsilon,nc}$, respectively. Under our assumptions (∂B is totally geodesic) it is possible to arrange that H^n intersects ∂B transversally under an angle $> \delta$ and that there exists a constant C_1 independent of ϱ such that

$$|II(\partial H_\varrho^n)| \leq C_1. \tag{5.26}$$

We infer from (5.23), bounded curvature and lemma 5.45 that for fixed $0 < r \leq 1$ there is a constant $C_2 > 0$ such that

$$\mathrm{vol}(H_\varrho^n) \leq C_2 \tag{5.27}$$

for all ϱ. Moreover, H_ϱ^n has bounded geometry (at least of order 0) according to (5.24) and to the bounded geometry of B.

Now we are able to present independent generators of $\Omega_{4k}^{nc}(ne)$. Let $P^{2k}(\mathbb{C})$ be the complex projective space with its standard orientation and with its Fubini–Study metric, fix two points z_1, z_2 and form by means of fixed spheres about z_1, z_2 the infinite connected sum

$$M^{4k} = (M^{4k}, g) = \overset{\infty}{\underset{1}{\#}} P^{2k}(\mathbb{C}), \tag{5.28}$$

always with same glueing metric. Then (M^{4k}, g) is oriented, has bounded geometry, one end which is nonexpanding.

Theorem 5.47 $M^{4k} = \overset{\infty}{\underset{1}{\#}} P^{2k}(\mathbb{C})$ *defines a non zero bordism class in* $\Omega_{4k}^{nc}(ne)$.

Proof. Suppose $[M^{4k}] = 0$. Then there exists a bordism (B^{n+1}, g_B), $\partial B = M^{4k} \cup -\mathrm{chc}^{4k}(r)$, $g_B|_{U_\delta(\partial B)} = g_{\partial B} + dt^2$, $U_R(M^{4k}) \supseteq B$, $U_R(\mathrm{chc}^{4k}(r)) \supseteq B$ and $d_B \geq d_M - c$, $d_B \geq d_{\mathrm{chc}} - c$. We choose $z_0 \in P_1^{2k}(\mathbb{C})$, $K = \emptyset$ and obtain for any $\varrho > 0$ a compact hypersurface $H_\varrho^{4k} \subset B = B \setminus \emptyset = C_\varepsilon$ which decomposes B into a compact and noncompact part $\overline{B}_c$ and $\overline{B}_{nc}$, respectively, and which satisfies (5.26), (5.27) and has bounded geometry at least of order 0 with constants independent of ϱ. Then $\partial \overline{B_c^{4k+1}} = (\partial \overline{B_c^{4k+1}} \cap M^{4k}) \cup H_\varrho \cup (\partial \overline{B_c^{4k+1}} \cap \mathrm{chc}^{4k})$. Here $\sigma(\partial \overline{B_c^{4k+1}} \cap \mathrm{chc}^{4k}) = 0$. $\sigma(\partial \overline{B_c^{4k+1}})$ must be zero since it is 0-bordant (if one wants, after smoothing out). Hence

$$0 = \sigma(\partial \overline{B_c^{4k+1}} \cap M^{4k}) - \sigma(H_\varrho^{4k}). \tag{5.29}$$

But

$$\sigma(H_\varrho^{4k}) = \int_{H_\varrho^{4k}} L + \eta(\partial H_\varrho^{4k}) + \int_{\partial H_\varrho^{4k}} \mathrm{expression}(II(\partial H_\varrho^{4k})). \tag{5.30}$$

The first expression on the r.h.s. of (5.30) is bounded by a bound independent of ϱ according to (5.27) and (B_0) for H_ϱ^{4k}. The same holds for the second expression according to

$$|\eta(\partial H_\varrho^{4k})| \leq C_3 \mathrm{vol}(\partial H_\varrho^{4k})$$

and for the third expression according to (5.26), (5.27). On the other hand, choosing ϱ sufficiently large, $\sigma(\partial \overline{B^{4k+1}} \cap M^{4k})$ can be made arbitrarily large. This contradicts (5.29). □

Looking at the proof of theorem 5.47, we immediately infer

Theorem 5.48 *Let (M^{4k}, g) be open, oriented, of bounded geometry and with a finite number of ends, each of them nonexpanding. If for any exhaustion $M_1 \subset M_2 \subset \cdots$ by compact submanifolds, $\bigcup M_i = M$, there holds*

$$\lim_{i \to \infty} \sigma(M_i^{4k}) = \infty$$

then $[M^{4k}, g] \neq 0$ in $\Omega_{4k}^{nc}(ne)$. □

Corollary 5.49 $\#_1^\infty P^{2k}(\mathbb{C})$, *or, more general,* $P^{2i_1}(\mathbb{C}) \times \cdots \times P^{2i_{r_1}} \# P^{2j_1}(\mathbb{C}) \times \cdots \times P^{2j_{r_2}} \# \cdots$, $i_1 + \cdots i_{r_1} = k$, $j_1 + \cdots + j_{r_2} = k, \ldots$ *are not torsion elements in $\Omega_{4k}^{nc}(ne)$.* □

A special case for theorem 5.48 is given by manifolds M^{4k} of the type
$$M^{4k} = \#_{1}^{\infty} M_i^{4k},$$
$\text{vol}(M_i^{4k}) \leq C_1$, $|K(g_i)| \leq C_2$, $r_{\text{inj}}(g_i) \geq C_3 > 0$, $\sigma(M_i^{4k}) \geq 0$ for $i \geq i_0$ and > 0 for infinitely many $i \geq i_0$. Then, in particular, $\mathcal{H}_{2k,2}(M^{4k})$ is infinitedimensional and $[M^{4k}, g] \neq 0$ in $\Omega_{4k}^{nc}(ne)$, i.e. adding a finite number of closed manifolds with negative signature and an infinite number of closed manifolds with zero signature (such that the bg, ne–end struture remains preserved) does not transform a nonzero element into zero in $\Omega_{4k}^{nc}(ne)$. A finer characterization of nonzero elements in $\Omega_{4k}^{nc}(ne)$ will be presented at another place. Moreover there are very interesting specializations of the theory developed until now and generalizations, e.g. the restriction to manifolds with warped product structure at infinity or with prescribed volume growth of the ends etc.. This will be the topic of another investigation.

III   The heat kernel of generalized Dirac operators

Substantial estimates for the operator e^{-tD^2} are more or less equivalent to estimates for the corresponding heat kernel. We present in the first section those estimates which are needed in the sequel and establish some invariance properties of the spectrum which we apply in chapters IV, V and VI.

1   Invariance properties of the spectrum and the heat kernel

We start with an absolutely fundamental theorem.

Theorem 1.1 *Let* $(E, h, \nabla_i) \longrightarrow (M^n, g)$ *be a Clifford bundle,* (M^n, g) *complete and D the generalized Dirac operator. Then all powers D^n, $n \geq 0$, are essential self-adjoint.*

We refer to [20] for the proof. □

Corollary 1.2 *Let* $(E, h, \nabla) \longrightarrow (M^n, g)$ *be a Riemannian vector bundle, (M^n, g) complete and Δ_q the Laplace operator acting on q–forms with values in E. Then $(\Delta_q)^n$, $n = 1, 2, \ldots$ are essentially self–adjoint. In particular this holds for the Laplace operator acting on ordinary q–forms.*

Proof. $\Delta_q = D^2$ for the Clifford bundle $\Lambda^* T^* M \otimes E$. □

In what follows, we always consider the self–adjoint closure $\overline{D^n}$ and write $\overline{D^n} \equiv D^n$.

Corollary 1.3 *There is a spectral decomposition*

$$\sigma(D^n) = \sigma_e(D^n) \cup \sigma_{pd}(D^n), \quad \sigma_e(D^n) \cap \sigma_{pd}(D^n) = \emptyset, \quad (1.1)$$

where σ_e denotes the essential and σ_{pd} the purely discrete point spectrum. In particular,

$$\sigma(\Delta_q) = \sigma_e(\Delta_q) \cup \sigma_{pd}(\Delta_q), \quad \sigma_e(\Delta_q) \cap \sigma_{pd}(\Delta_q) = \emptyset. \quad (1.2)$$

□

$\lambda \in \sigma_e$ if and only if there exists a Weyl sequence for λ. Properties of Weyl sequences imply very important invariance properties for the spectrum.

Proposition 1.4 Let $(E, h, \nabla^h, \cdot) \longrightarrow (M^n, g)$ be a Clifford bundle, M^n open and complete, $K \subset M$ a compact subset, $D_F(E|_{M\setminus K})$ Friedrichs' extension of $D|_{C_c^\infty(E|_{M\setminus K})}$. Then there hold

$$\sigma_e(D) = \sigma_e(D_F) = \sigma_e(D_F(E|_{M\setminus K})) \quad (1.3)$$

and

$$\sigma_e(D^2) = \sigma_e(D_F^2) = \sigma_e((D_F(E|_{M\setminus K}))^2). \quad (1.4)$$

Proof. We start with (1.3) and $\sigma_e(D) \subseteq \sigma_e(D_F(E|_{M\setminus K}))$. Let $\lambda \in \sigma_e(D)$, $(\psi_\nu)_\nu$ be an orthonormal Weyl sequence for λ, $D\psi_\nu - \lambda\psi_\nu \longrightarrow 0$. Then $(\omega_\nu)_\nu$, $\omega_\nu = \psi_{2\nu+1} - \psi_{2\nu}$ is still a Weyl sequence for λ. Let $\Phi \in C_c^\infty(M)$, $0 \leq \Phi \leq 1$, $\Phi = 1$ on a neighbourhood $U = U(K)$ of K. According to the Rellich chain property of Sobolev spaces (with real index) on compact manifolds, $(\Phi\psi_\nu)_\nu$ contains an L_2–convergent subsequence which we denote again by $(\Phi\psi_\nu)_\nu$. This yields $\Phi\omega_\nu \longrightarrow 0$ and $\text{grad } \Phi \cdot \omega_\nu \longrightarrow 0$ in L_2. $((1-\Phi)\omega_\nu)_\nu$ is a Weyl sequence for $\lambda \in \sigma_e(D_F(E|_{M\setminus K}))$:

$$D_F(E|_{M\setminus K})(1-\Phi)\omega_\nu - \lambda(1-\Phi)\omega_\nu$$
$$= D\omega_\nu - \lambda_\nu\omega_\nu - (D(\Phi\omega_\nu)) - \lambda\Phi\omega_\nu$$
$$= D\omega_\nu - \lambda\omega_\nu - \Phi(D\omega_\nu - \lambda\omega_\nu) - \text{grad } \Phi \cdot \omega_\nu \xrightarrow[\nu\to\infty]{} 0$$

since by assumption and construction $D\omega_\nu - \lambda\omega_\nu \longrightarrow 0$, $\text{grad } \Phi \cdot \omega_\nu \longrightarrow 0$. Hence $\sigma_e(D) \subseteq \sigma_e(D_F(E|_{M\setminus K}))$. $\mathcal{D}_{D_F(E|_{M\setminus K})} \subseteq \mathcal{D}_{D_F}$ and every Weyl sequence for $\lambda \in \sigma_e(D_F(E|_{M\setminus K}))$ is also a Weyl sequence for $\lambda \in \sigma_e(D)$. This finishes the proof of (1.3). (1.4) is an immediate consequence of (1.3) by means of the spectral theorem but it can also similarly be proven. □

Corollary 1.5 *The essential spectrum of D and D^2 remains invariant under compact perturbations of the topology and the metric. In particular this holds for the Laplace operators acting on forms with values in a vector bundle.* □

As for compact manifolds, we can define the Riemannian connected sum for open Riemannian manifolds, even for Riemannian vector bundles $(E_i, h_i, \nabla^{h_i}) \longrightarrow (M_i^n, g_i)$, where at the compact glueing domain the metric and connection are not uniquely determined. Another corollary is then given by

Proposition 1.6 *Suppose $(E_i, h_i, \nabla^{h_i}) \longrightarrow (M_i^n, g_i)$, $i = 1, \ldots, r$ Riemannian vector bundles of the same rank, (M_i^n, g_i) complete, and let $\Delta = \Delta_q$ be the Laplace operator acting on q–forms with values in E_i (resp. E). Then*

$$\sigma_e \Delta_q \left(\overset{r}{\underset{i=1}{\#}} (E_i \longrightarrow M_i) \right) = \bigcup_{i=1}^{r} \sigma_e(\Delta_q(E_i \longrightarrow M_i)). \quad (1.5)$$

□

1.4 can be reformulated as the statement that the essential spectrum for an isolated end ε is well defined. We denote it by $\sigma_e(D_F(\varepsilon))$, $\sigma_e(D_F^2(\varepsilon))$.

Proposition 1.7 *If (M^n, g) is complete and has finitely many ends $\varepsilon_1, \ldots, \varepsilon_r$ then*

$$\sigma_e(D) = \bigcup_{i=1}^{r} \sigma_e(D_F(\varepsilon_i)), \quad \sigma_e(D^2) = \bigcup_{i=1}^{r} \sigma_e(D_F^2(\varepsilon_i)). \quad (1.6)$$

Proposition 1.8 *Assume the hypothesis of 1.4. Suppose $\lambda \in \sigma_e(D)$. Then there exists a Weyl sequence $(\varphi_\nu)_\nu$ for λ such that for any compact subset $K \subset M$*

$$\lim_{\nu \to \infty} |\varphi_\nu|_{L_2(E|_K)} = 0. \tag{1.7}$$

For every $\lambda \in \sigma_e(D^2)$ there exists a Weyl sequence $(\varphi_\nu)_\nu$ satisfying (1.7) and

$$\lim_{\nu \to \infty} |D\varphi_\nu|_{L_2(E|_K)} = 0. \tag{1.8}$$

Proof. Start with (1.7). Let $(\psi_\nu)_\nu$ be a Weyl sequence for $\lambda \in \sigma_e(D)$, $K_1 \subset K_2 \subset \cdots \subset K_i \subset K_{i+1} \subset \cdots$, $\bigcup K_i = M$, an exhaustion by compact submanifolds. By a Rellich compactness argument there exists a subsequence $(\psi_\nu^{(1)})_\nu$ of $(\psi_\nu)_\nu$ $(\psi_\nu^{(0)})_\nu$ converging on K_1. Inductively, there exists a subsequence $(\psi_\nu^{(i+1)})_\nu$ of $(\psi_\nu^{(i)})_\nu$ converging on K_{i+1}. Set $(\varphi_\nu)_\nu = ((\psi_{2\nu+1}^{(2\nu+1)} - \psi_{2\nu}^{(2\nu)})/\sqrt{2})_\nu$. Then $(\varphi_\nu)_\nu$ is a Weyl sequence for $\lambda \in \sigma_e(D)$ satisfying (1.7). For $\lambda \in \sigma_e(D^2)$ with Weyl sequence $(\psi_\nu)_\nu$, we choose the subsequence $(\psi_\nu^{(i+1)})_\nu$ of $(\psi_\nu^{(i)})_\nu$ such that $(\psi_\nu^{(i+1)})_\nu$ and $(D\psi_\nu^{(i+1)})_\nu$ converge on K_{i+1} (in L_2, as always). □

1.8 means that w.l.o.g. Weyl sequences should "leave" (in the sense of the L_2-norm) any compact subset, i.e. there must be "place enough at infinity".

Proposition 1.9 *Let $(E, h, \nabla, \cdot) \longrightarrow (M^n, g)$ be a Clifford bundle with (I), $(B_{r-3}(M, g))$, $(B_{r-3}(E, \nabla))$, $r > \frac{n}{2} + 1$ and ∇' a second Clifford connection satisfying $|\nabla' - \nabla|_{\nabla, 2, r-1} < \infty$. Then for $D = D(\nabla)$ and $D' = D(\nabla')$ there holds*

$$\mathcal{D}_D = \mathcal{D}_{D'} \tag{1.9}$$

and

$$\sigma_e(D) = \sigma_e(D'). \tag{1.10}$$

Proof. (M^n, g) is complete, D and D' are self–adjoint. $\mathcal{D}_D = \Omega^{2,1}(E, D) = \Omega^{2,1}(E, \nabla) = \Omega^{2,1}(E, \nabla') = \Omega^{2,1}(E, D') = \mathcal{D}_{D'}$

according to II 1.25 and II 1.32. We write $\nabla' = \nabla + \eta$. Then $D' = \sum_i e_i \cdot \nabla'_{e_i} = \sum_i e_i \cdot (\nabla_{e_i} + \eta_{e_i}(\cdot)) = D + \eta^{op}$, where the operator η^{op} acts as $\eta^{op}(\varphi)_x = \sum_i e_i \cdot \eta_{e_i}(\varphi)_x$. We can apply the Sobolev embedding theorem since $r - 1 > \frac{n}{2}$. Then, pointwise, $|\eta^{op}|_x \leq C_1 \cdot |\eta|_x$, C_1 independent of x. Given $\varepsilon' > 0$, there exists a compact set $K = K(\varepsilon') \subset M$ such that

$$\sup_{x \in M \setminus K} |\eta|_x < \frac{\varepsilon'}{C_1}, \text{ i. e. } \sup_{x \in M \setminus K} |\eta^{op}|_x < \varepsilon'. \tag{1.11}$$

Assume now $\lambda \in \sigma_e(D)$, $(\varphi_\nu)_\nu$ a Weyl sequence as in (1.7), $|\varphi_\nu|_{L_2(M,E)} \leq C_2$. According to (1.9), $\varphi_\nu \in \mathcal{D}_{D'}$. Then

$$(D' - \lambda)\varphi_\nu = (D' - D)\varphi_\nu + (D - \lambda)\varphi_\nu.$$

By assumption, $(D - \lambda)\varphi_\nu \longrightarrow 0$. Let $\varepsilon > 0$ arbitrarily be given, choose $\varepsilon' < \frac{\varepsilon}{C_2}$ and $K = K(\varepsilon')$ such that (1.11) is satisfied. Then

$$|(D' - D)\varphi_\nu|_{L_2(M,E)} = |\eta^{op}\varphi_\nu|_{L_2(M,E)}$$
$$\leq C_1(|\eta\varphi_\nu|_{L_2(K,E)} + |\eta\varphi_\nu|_{L_2(M \setminus K,E)}).$$

By (1.7)
$$|\eta\varphi_\nu|_{L_2(K,E)} \longrightarrow 0$$

and

$$C_1|\eta\varphi_\nu|_{L_2(M \setminus K,E)} \leq C_1 \sup_{x \in M \setminus K} |\eta|_x \cdot |\varphi_\nu|_{L_2(M \setminus K,E)} < \varepsilon' \cdot C_2 < \varepsilon.$$

Hence $(D' - \lambda)\varphi_\nu \longrightarrow 0$, $\lambda \in \sigma_e(D')$, $\sigma_e(D) \subseteq \sigma_e(D')$. Exchanging the role of D and D', we obtain $\sigma_e(D') \subseteq \sigma_e(D)$. $\square$

Corollary 1.10 *Assume the hypotheses of 1.9. Then for all $i \in \mathbb{N}$*

$$\mathcal{D}_{D^i} = \mathcal{D}_{D'^i} \tag{1.12}$$

and

$$\sigma_e(D^i) = \sigma_e(D'^i). \tag{1.13}$$

Proof. This follows from the self–adjointness of D on $\mathcal{D}_D$. □

Remark 1.11 For $\Omega^{2,1}(E, \nabla) = \Omega^{2,1}(E, D)$ (B_0) would be sufficient which easily follows from the Weitzenboeck formula, but we need moreover the Sobolev embedding theorem for $\Omega^{2,r-1}$, in particular $\overset{\circ}{\Omega}{}^{2,r-1} = \Omega^{2,r-1}$, hence (B_{r-3}). One could try to estimate $||\eta|_x \cdot |\varphi|_x|_{L_2}$ instead of $\sup_x |\eta|_x \cdot |\varphi|_{L_2}$, i. e. one could try to work without the Sobolev embedding theorem, but in this case we would need some module structure. Hence an assumption (B_i), $i > 0$, seems to be inavoidable in any case. □

For the Laplace operator on forms, 1.9 is not immediately applicable since if we replace for (M^n, g) the Levi–Civita connection ∇^g by another metric connection then we lose many of the standard formulas, i. e. we should consider a change $g \longrightarrow g'$.

Proposition 1.12 *Let (M^n, g) be open, complete, with (I) and $(B_r(M^n, g))$, $r > \frac{n}{2} + 1$, g' another metric satisfying the same conditions and suppose g, g' quasi isometric and $|g' - g|_{g,2,r} = (\int (|g' - g|^2_{g,x} + \sum_{i=0}^{r-1} |(\nabla^g)^i - \nabla^g|^2_{g,x}) \operatorname{dvol}_x(g))^{\frac{1}{2}} < \infty$. Then*

$$\mathcal{D}_{\Delta_q(g)} = \mathcal{D}_{\Delta_q(g')} \text{ as equivalent Hilbert spaces} \quad (1.14)$$

and

$$\sigma_e(\Delta_q(g)) = \sigma_e(\Delta_q(g')), \quad q = 0, 1, \ldots, n. \quad (1.15)$$

Proof. We write $\Delta = \Delta_q(g)$, $\Delta' = \Delta_q(g')$, $\nabla = \nabla^g$, $\nabla' = \nabla^{g'}$. Then, according to II 1, $\mathcal{D}_\Delta = \Omega^{q,2,2}(\Delta, g) = \Omega^{q,2,2}(\Delta_{\text{Bochner}}, g) = \Omega^{q,2,2}(\nabla, g) \cong \Omega^{q,2,2}(\nabla', g') = \Omega^{q,2,2}(\Delta'_{\text{Bochner}}, g') = \Omega^{q,2,2}(\Delta', g') = \mathcal{D}_{\Delta'}$. Denote by $\mathcal{R}$ the Weitzenboeck endomorphism,

$$\Delta = \nabla^* \nabla + \mathcal{R}. \quad (1.16)$$

We write $\Delta' - \Delta = \nabla'^*\nabla' - \nabla^*\nabla + \mathcal{R}' - \mathcal{R}$. Let $\lambda \in \sigma_e(\Delta)$ and $(\omega_\nu)_\nu$ be a Weyl sequence for λ as in proposition 1.8, for any K

$$\lim_{\nu \to \infty} |\omega_\nu|_{L_2(\Lambda^q|_K)} = 0 \tag{1.17}$$

and

$$\lim_{\nu \to \infty} |\nabla \omega_\nu|_{L_2((T^* \otimes \Lambda^q)|_K)} = 0 \tag{1.18}$$

We infer as above from $(\omega_\nu)_\nu$ bounded, $\Delta \omega_\nu - \lambda \omega_\nu \longrightarrow 0$ that $(\omega_\nu)_\nu, (\Delta \omega_\nu)_\nu$ are bounded, hence

$$(\omega_\nu)_\nu \text{ is a bounded sequence in } \Omega^{q,2,2}(\Delta, g). \tag{1.19}$$

But under our assumption $r > \frac{n}{2} + 1$, according to II 1.27,

$$\Omega^{q,2,2}(\Delta, g) = \Omega^{q,2,2}(\nabla, g) \cong \Omega^{q,2,2}(\nabla', g') = \Omega^{q,2,2}(\Delta', g') \tag{1.20}$$

as equivalent Sobolev spaces. Hence

$$(\nabla \omega_\nu)_\nu, (\nabla^2 \omega_\nu)_{\nu'}, (\nabla' \omega_\nu)_{\nu'}, (\nabla'^2 \omega_\nu)_{\nu'}, (\Delta' \omega_\nu)_\nu \text{ are bounded} \tag{1.21}$$

and

$$(\nabla' \nabla \omega_\nu)_{\nu'} = ((\nabla' - \nabla)(\nabla \omega_\nu)_\nu + (\nabla^2 \omega_\nu)_\nu \text{ too.} \tag{1.22}$$

To be very explicit, we choose a u. l. f. cover of (M^n, g) by normal charts with respect to g, $\mathcal{U} = \{(U_\alpha, u_\alpha)\}_\alpha$ and an associated partition $\{\varphi_\alpha\}_\alpha$ of unity with bounded derivatives as in I 1.
Then for a q-form $\omega|_{U_\alpha} = \sum_{i_1 < \cdots < i_q} \omega_{i_1 \ldots i_q} du_\alpha^{i_1} \wedge \cdots \wedge du_\alpha^{i_q}$

$$-(\nabla'^*\nabla' - \nabla^*\nabla)\omega_{i_1 \ldots i_q}$$
$$= (g'^{ij} \nabla'_j \nabla'_i - g^{ij} \nabla_j \nabla_i) \omega_{i_1 \ldots i_q} \tag{1.23}$$
$$= g'^{ij} \nabla'_j (\nabla'_i - \nabla_i) \omega_{i_1 \ldots i_q} + \tag{1.24}$$
$$+ (g'^{ij} - g^{ij}) \nabla'_j \nabla'_i \omega_{i_1 \ldots i_q} + \tag{1.25}$$
$$+ g^{ij} (\nabla'_j - \nabla_j) \nabla_i \omega_{i_1 \ldots i_q}. \tag{1.26}$$

If we write $\nabla_i \omega_{i_1...i_q} = \frac{\partial}{\partial u^i}\omega_{i_1...i_q} + \Gamma_i^{op}\omega_{i_1...i_q}$ then we can rewrite (1.24) as

$$g'^{ij}\nabla'_j(\Gamma'^{op}_i - \Gamma^{op}_i)\omega_{i_1...i_q} \tag{1.27}$$

Insert now $\omega_\nu = \sum_\alpha \varphi_\alpha \omega_\nu$. Then

$$g'^{ij}\nabla'_j(\Gamma'^{op}_i - \Gamma^{op}_i)\varphi_\alpha \omega_{\nu,i_1...i_q} = \tag{1.28}$$
$$= g'^{ij}(\nabla_j \varphi_\alpha)(\Gamma'^{op}_i - \Gamma^{op}_i)\omega_{\nu,i_1...i_q} + \tag{1.29}$$
$$+\varphi_\alpha g'^{ij}\nabla'_j(\Gamma'^{op}_i - \Gamma^{op}_i)\omega_{\nu,i_1...i_q}. \tag{1.30}$$

We obtain from $\nabla'_j \varphi_\alpha$, $\nabla_j(\Gamma'^{op}_i - \Gamma^{op}_i)$ bounded, $\nabla'_j = \nabla'_j - \nabla_j + \nabla_j$, $\nabla'_j - \nabla_j = \Gamma'^{op}_i - \Gamma^{op}_i$, $g'^{ij} = (g'^{ij} - g^{ij}) + g^{ij}$ bounded, (1.17) and (1.18)

$$\lim_{\nu \to \infty} \left| \sum_\alpha g'^{ij}\nabla'_j(\nabla'_i - \nabla_i)\varphi_\alpha \omega_{\nu,i_1,...,i_q} \right|_K = 0 \tag{1.31}$$

where $||_K = ||_{L_2(\Lambda^q|_K)}$, $K \subset M$ compact. Here we used that in normal coordinates w. r. t. g, (g_{ij}) and (g^{ij}) are bounded. Moreover, for $K = K(\varepsilon)$ sufficiently large,

$$\sup_{x \in M \setminus K} |\nabla'(\Gamma'^{op} - \Gamma^{op})|, \quad \sup_{x \in M \setminus K} |\Gamma'^{op} - \Gamma^{op}|_x \tag{1.32}$$

become arbitrarily small. Together with (1.21) we infer from (1.32) and (1.31)

$$\lim_{\nu \to \infty} \left| \sum_\alpha g'^{ij}\nabla'_j(\Gamma'^{op}_i - \Gamma^{op}_i)\varphi_\alpha \omega_{\nu,i_1,...,i_q} \right|_{L_2} = 0. \tag{1.33}$$

Quite similar we infer from (1.22)

$$\lim_{\nu \to \infty} \left| \sum_\alpha (g'^{ij} - g^{ij})\nabla'_j \nabla_i \varphi_\alpha \omega_{\nu,i_1,...,i_q} \right|_{L_2} = 0 \tag{1.34}$$

and from (1.21)

$$\lim_{\nu \to \infty} \left| \sum_\alpha g'^{ij}(\Gamma'^{op}_j - \Gamma^{op}_j)\nabla_i \varphi_\alpha \omega_{\nu,i_1,...,i_q} \right|_{L_2} = 0. \tag{1.35}$$

The essential point is that for $K = K(\varepsilon)$ sufficiently large $|g - g'|_{g,x}$, $|g^{-1} - g'^{-1}|_{g,x}$, $|\Gamma'^{op} - \Gamma^{op}|_{g,x}$, $\nabla(\Gamma'^{op} - \Gamma^{op})$, $|\nabla'(\Gamma'^{op} - \Gamma^{op})|_{g,x}$ become uniformly arbitrarily small outside K. But this follows from our assumption and the Sobolev embedding theorem. We conclude from (1.33) – (1.35) $\lim_{\nu \to \infty} (\Delta' - \Delta)\omega_\nu = 0$, $\lambda \in \sigma_e(\Delta')$. Exchanging the role of g and g' yields the other inclusion. □

We state without proof the generalization to forms with values in a vector bundle.

Proposition 1.13 *Let* $(E, h, \nabla^h) \longrightarrow (M^n, g)$ *be a Riemannian vector bundle satisfying* (I), $(B_k(M^n, g))$, $(B_k(E, \nabla))$, $k \geq r > \frac{n}{2} + 1$, *and let* g' *be a second metric,* h' *a second fibre metric with metric connection* $\nabla^{h'}$, g, g' *and* h, h' *quasi isometric, respectively,*

$$|g - g'|_{g,2,r} < \infty, \quad |h - h'|_{h,\nabla^h,g,2,r} < \infty,$$
$$|\nabla^h - \nabla^{h'}|_{h,\nabla^h,g,2,r-1} < \infty.$$

$(E, h', \nabla^{h'}) \longrightarrow (M^n, g')$ *also satisfying* (I) *and* (B_k). *Then there holds for the Laplace operators* $\Delta = \Delta_q(g, h, \nabla^h)$, $\Delta' = \Delta_q(g', h', \nabla^{h'})$ *acting on forms with values in* E

$$\mathcal{D}_\Delta = \mathcal{D}_{\Delta'} \text{ as equivalent Hilbert spaces} \quad (1.36)$$

and

$$\sigma_e(\Delta) = \sigma_e(\Delta'). \quad (1.37)$$

The proof is quite similar to that of 1.12, taking the difference of the Weitzenboeck formulas and proceed as in the proof of proposition 1.12. □

Now we collect some standard facts concerning the heat kernel of e^{-tD^2}. The best references for this are [9], [27].
We consider the self-adjoint closure of D in $L_2(E) = H^0(E)$,
$$D = \int\limits_{-\infty}^{+\infty} \lambda E_\lambda.$$

Lemma 1.14 $\{e^{itD}\}_{t\in\mathbf{R}}$ *defines a unitary group on the spaces* $H^r(E)$, *for* $0 \leq h \leq r$ *holds*

$$|D^h e^{itD}\Psi|_{L_2} \cdot |e^{itD} D^h \Psi|_{L_2} = |D^h \Psi|_{L_2}. \qquad (1.38)$$

□

We can extend this action to $H^{-r}(E)$ by means of duality.

Lemma 1.15 e^{-tD^2} *maps* $L_2(E) \equiv H^0(E) \to H^r(E)$ *for any* $r > 0$ *and*

$$|e^{-tD^2}|_{L_2 \to H^r} \leq C \cdot t^{-\frac{r}{2}}, \quad t \in]0, \infty[, \quad C = C(r). \qquad (1.39)$$

Proof. Insert into $e^{-tD^2} = \int e^{-t\lambda^2}\, dE_\lambda$ the equation

$$e^{-t\lambda^2} = \frac{1}{\sqrt{4\pi t}} \int_{-\infty}^{+\infty} e^{i\lambda s} e^{-\frac{s^2}{4t}}\, ds$$

and use

$$\sup |\lambda^r e^{-t\lambda^2}| \leq C \cdot t^{-\frac{r}{2}}.$$

□

Corollary 1.16 *Let* $r, s \in \mathbf{Z}$ *be arbitrary. Then* $e^{-tD^2} : H^r(E) \to H^s(E)$ *continuously.*

Proof. This follows from 1.15, duality and the semi group property of $\{e^{-tD^2}\}_{t \geq 0}$. □

e^{-tD^2} has a Schwartz kernel $W \in \Gamma(\mathbf{R} \times M \times M, E \boxtimes E)$,

$$W(t, m, p) = \langle \delta(m), e^{-tD^2} \delta(p) \rangle,$$

where $\delta(m) \in H^{-r}(E) \otimes E_m$ is the map $\Psi \in H^r(E) \to \langle \delta(m), \Psi \rangle = \Psi(m)$, $r > \frac{n}{2}$. The main result of this section is the fact that for $t > 0$, $W(t, m, p)$ is a smooth integral kernel in L_2 with good decay properties if we assume bounded geometry. Denote by $C(m)$ the best local Sobolev constant of the map $\Psi \to \Psi(m)$, $r > \frac{n}{2}$, and by $\sigma(D^2)$ the spectrum.

Lemma 1.17 a) $W(t,m,p)$ is for $t > 0$ smooth in all Variables.
b) For any $T > 0$ and sufficiently small $\varepsilon > 0$ there exists $C > 0$ such that

$$|W(t,m,p)| \leq e^{-(t-\varepsilon)\inf \sigma(D^2)} \cdot C \cdot C(m) \cdot C(p) \text{ for all } t \in]T, \infty[. \tag{1.40}$$

c) Similar estimates hold for $(D_m^i D_p^j W)(t,m,p)$.

Proof.
a) First one shows W is continuous, which follows from $\langle \delta(m), \cdot \rangle$ continuous in m and $e^{-tD^2}\delta(p)$ continuous in t and p. Then one applies elliptic regularity.
b) Write
$$|\langle \delta(m), e^{-tD^2}\delta(p)\rangle| = |\langle (1+D^2)^{-\frac{r}{2}}\delta(m), (1+D^2)^r e^{-tD^2}(1+D^2)^{\frac{r}{2}}\delta(p)\rangle|$$
c) Follows similarly as b. □

Lemma 1.18 For any $\varepsilon > 0, T > 0, \delta > 0$ there exists $C > 0$ such that for $r > 0, m \in M, T > t > 0$ holds

$$\int_{M \setminus B_r(m)} |W(t,m,p)|^2 \, dp \leq C \cdot C(m) \cdot e^{-\frac{(r-\varepsilon)^2}{(4+\delta)t}}. \tag{1.41}$$

A similar estimate holds for $D_m^i D_p^j W(t,m,p)$.

We refer to [9] for the proof. □

Lemma 1.19 For any $\varepsilon > 0, T > 0, \delta > 0$ there exists $C > 0$ such that for all $m, p \in M$ with $\text{dist}(m,p) > 2\varepsilon, T > t > 0$ holds

$$|W(t,m,p)|^2 \leq C \cdot C(m) \cdot C(p) \cdot e^{-\frac{(\text{dist}(m,p)-\varepsilon)^2}{(4+\delta)t}}. \tag{1.42}$$

A similar estimate holds for $D_m^i D_p^j W(t,m,p)$.

We refer to [9] for the proof. □

Proposition 1.20 *Assume (M^n, g) with (I) and (B_K), (E, ∇) with (B_K), $k \geq r > \frac{n}{2} + 1$. Then all estimates in (1.40) – (1.42) hold with uniform constants.*

Proof. From the assumptions $H^r(E) \cong W^r(E)$ and $\sup_m C(m) = C$ = global Sobolev constant for $W^r(E)$, according to II 1.4, II 1.6. □

Let $U \subset M$ be precompact, open, (M^+, g^+) closed with $U \subset M^+$ isometrically and $E^+ \to M^+$ a Clifford bundle with $E^+|_U \cong E|_U$ isometrically. Denote by $W^+(t, m, p)$ the heat kernel of $e^{-tD^{+^2}}$.

Lemma 1.21 *Assume $\varepsilon > 0, T > 0, \delta > 0$. Then there exists $C > 0$ such that for all $T > t > 0, m, p \in U$ with $B_{2\varepsilon}(m)$, $B_{2\varepsilon}(p) \subset U$ holds*

$$|W(t, m, p) - W^+(t, m, p)| \leq C \cdot e^{-\frac{\varepsilon^2}{(4+\delta)t}} \qquad (1.43)$$

We refer to [8] for the simple proof. □

Corollary 1.22 *$trW(t, m, m)$ has for $t \to 0^+$ the same asymptotic expansion as for $trW^+(t, m, m)$.* □

2 Duhamel's principle, scattering theory and trace class conditions

We want to prove the trace class property of $e^{-tD^2} - e^{-t\tilde{D}'^2}$, where $\tilde{D}'$ is a perturbation of D. The key to get convenient expressions for $e^{-tD^2} - e^{t\tilde{D}'^2}$ is Duhamel's principle. For closed manifolds, this is a very well known fact. We establish it here for open complete manifolds. In principle, it follows from Stokes' theorem, or, what is the same, from partial integration. Having established Duhamel's principle, the proof of the trace class property amounts to the estimate of a certain number of operator valued integrals. Their estimate occupies the whole 30 pages of chapter IV.

The trace class property is the key for the application of scattering theory. We give an account on those facts of scattering theory which are of great importance in chapters V and VI.

First we establish Duhamel's principle and make the following assumptions: D and D' are generalized Dirac operators acting in the same Hilbert space,

$$\mathcal{D}_D = \mathcal{D}_{D'}, \quad \mathcal{D}_{D^2} = \mathcal{D}_{D'^2}, \quad D' - D = \eta,$$

where $\eta = \eta^{op}$ is an operator acting in the same Hilbert space.

Lemma 2.1 *Assume $t > 0$. Then*

$$e^{-tD^2} - e^{-tD'^2} = \int_0^t e^{-sD^2}(D'^2 - D^2)e^{-(t-s)D'^2}\, ds. \qquad (2.1)$$

Proof. (2.1) means at heat kernel level

$$W(t, m, p) - W'(t, m, p)$$
$$= -\int_0^t \int_M (W(s, m, q), (D^2 - D'^2)W'(t - s, q, p))_q \, dq \, ds,$$
$$\qquad (2.2)$$

where $(,)_q$ means the fibrewise scalar product at q and $dq = dvol_q(g)$. Hence for (2.1) we have to prove (2.2). (2.2) is an immediate consequence of Duhamel's principle. Only for completeness, we present the proof of (2.1), which is the last of the following 7 facts and implications.

1. For $t > 0$ is $W(t, m, p) \in L_2(M, E, dp) \cap \mathcal{D}_D^2$

2. If $\Phi, \Psi \in \mathcal{D}_D^2$ then $\int (D^2\Phi, \Psi) - (\Phi, D^2\Psi)\, dvol = 0$ (Greens formula).

3. $((D^2 + \frac{\partial}{\partial \tau})\Phi(\tau, g)\Psi(t-\tau, q))_q - (\Phi(\tau, g), (D^2 + \frac{\partial}{\partial t})\Psi(t-\tau, q))_q =$
$= (D^2(\Phi(\tau, q), \Psi(t-\tau, q))_q - (\Phi(\tau, q), D^2\Psi(t-\tau, q))_q + \frac{\partial}{\partial \tau}(\Phi(\tau, g), \Psi(t-\tau, q))_q$.

4. $\int_\alpha^\beta \int_M ((D^2 + \frac{\partial}{\partial \tau})\Phi(\tau,q), \Psi(t-\tau,q))_q - (\Phi(\tau,q), (D^2 + \frac{\partial}{\partial t})\Psi(t-\tau,q))_q \, dq \, d\tau =$
$= \int_M [(\Phi(\beta,q), \Psi(t-\beta,q))_q - (\Phi(\alpha,q), \Psi(t-\alpha,q))_q] \, dq.$

5. $\Phi(t,q) = W(t,m,q), \Psi(t,q) = W'(t,q,p)$ yields
$-\int_\alpha^\beta \int_M (W(\tau,m,q), (D^2 + \frac{\partial}{\partial t})W'(t-\tau,q,p) \, dq \, d\tau =$
$= \int_M [(W(\beta,m,q), W'(t-\beta,q,p))_q$
$-(W(\alpha,m,q), W'(t-\alpha,q,p))_q] \, dq \, .$

6. Performing $\alpha \to 0^+, \beta \to t^-$ in 5. yields
$-\int_0^t \int_M (W(s,m,q), (D^2 + \frac{\partial}{\partial t})W'(t-s,q,p))_q \, dq \, ds = W(t,m,p) - W'(t,m,p).$

7. Finally, using $D^2 + \frac{\partial}{\partial t} = D^2 - D'^2 + D'^2 + \frac{\partial}{\partial t}$ and $(D'^2 + \frac{\partial}{\partial t})W' = 0$ we obtain
$W(t,m,p) - W'(t,m,p) = -\int_0^t \int_M (W(s,m,q), (D^2 - D'^2)W'(t-s,q,p))_q \, dq \, ds$
which is (2.2).

If we write $D^2 - D'^2 = D(D - D') + (D - D')D'$ then

$$e^{-tD^2} - e^{-tD'^2} = -\int_0^t e^{-sD^2}(D^2 - D'^2)e^{-(t-s)D'^2}\,ds$$

$$= -\int_0^t e^{-sD^2} D(D - D')e^{-(t-s)D'^2}\,ds$$

$$ -\int_0^t e^{-sD^2}(D - D')D'e^{-(t-s)D'^2}\,ds$$

$$= \int_0^t e^{-sD^2} D\eta e^{-(t-s)D'^2}\,ds$$

$$ +\int_0^t e^{-sD^2}\eta D' e^{-(t-s)D'^2}\,ds,$$

where $\eta = \eta^{op}$ is an operator which will specified in chapter IV.
We split $\int_0^t = \int_0^{\frac{t}{2}} + \int_{\frac{t}{2}}^t$.

$$e^{-tD^2} - e^{-tD'^2} = \int_0^{\frac{t}{2}} e^{-sD^2} D\eta e^{-(t-s)D'^2}\,ds \qquad (I_1)$$

$$+ \int_0^{\frac{t}{2}} e^{-sD^2}\eta D' e^{-(t-s)D'^2}\,ds \qquad (I_2)$$

$$+ \int_{\frac{t}{2}}^t e^{-sD^2} D\eta e^{-(t-s)D'^2}\,ds \qquad (I_3)$$

$$+ \int_{\frac{t}{2}}^t e^{-sD^2}\eta D' e^{-(t-s)D'^2}\,ds. \qquad (I_4)$$

We will show in chapter IV that each integral $(I_1) - (I_4)$ is a product of Hilbert–Schmidt operators and estimate their Hilbert–Schmidt norm. □

Now we introduce scattering theory and recall the standard decomposition of the spectrum $\sigma(A)$ of a self–adjoint operator $A : \mathcal{D}_A \longrightarrow X$.

Every Borel measure m on $\mathbb{R}$ admits a decomposition $m = m_{pp} + m_c = m_{pp} + m_{ac} + m_{sc}$, where m_{pp} is a pure point measure, $m_c = m - m_{pp}$ is characterized by $m_c(\{p\}) = 0$ for all points p, m_{ac} is absolutely continuous with respect to the Lebesgue measure and m_{sc} is singular with respect to the Lebesgue measure, i.e. $m_{sc}(S) = 0$ for a certain set S such that $\mathbb{R} \setminus S$ has zero Lebesgue measure. We consider for $x \in X$ the positive Borel measure $[a,b] \xrightarrow{m_x} \langle E_{[a,b]} x, x \rangle$ and set $X_{pp} = \{x \in X | m_x = m_{x,pp}\}$, $X_{ac} = \{x \in X | m_x = m_{x,ac}\}$, $X_{sc} = \{x \in X | m_x = m_{x,sc}\}$, $X_c = \{x \in X | m_x = m_{x,c}\}$ and $\sigma_{pp}(A) := \sigma(A|X_{pp})$, $\sigma_c(A) := \sigma(A|X_c)$, $\sigma_{ac}(A) := \sigma(A|X_{ac})$, $\sigma_{sc}(A) := \sigma(A|X_{sc})$. Then $\sigma_c(A)$, σ_{ac} and σ_{sc} are called the continuous, the absolutely continuous and the singular continuous spectra. Finally we define $\sigma_{pd}(A) := \{\lambda \in \sigma_p(A) | \mathrm{mult}(\lambda) < \infty$ and λ is an isolated point in $\sigma(A)\}$, called the purely discrete spectrum, and $\sigma_{c,R}(A) = \sigma(A) \setminus \sigma_p(A)$, $\sigma_p(A)$ the point spectrum.

Theorem 2.2 *Let $A : \mathcal{D}_A \longrightarrow X$ be self–adjoint.*
a) $X = X_{pp} \oplus X_{ac} \oplus X_{sc}$.
b) $\sigma(A) = \sigma_{pd}(A) \cup \sigma_e(A)$, $\sigma_{pd}(A) \cap \sigma_e(A) = \emptyset$.
c) $\sigma_e(A) = \sigma_c(A) \cup \sigma_p(A)^1 \cup \{\lambda \in \sigma_p(A) | \mathrm{mult}(\lambda) = \infty\}$.
d) $\sigma_c(A) = \sigma_{ac}(A) \cup \sigma_{sc}(A)$.
e) $\sigma(A) = \sigma_{pp}(A) \cup \sigma_c(A) = \overline{\sigma_p(A)} \cup \sigma_c(A)$.
Here M^1 denotes the set of accumulation points of M.

Remark 2.3 In general, $\sigma_p(A) \subset \sigma_{pp}(A)$, hence $\sigma_p(A) \cup \sigma_{ac}(A) \cup \sigma_{sc}(A) \subset \sigma(A)$, but $\overline{\sigma_p(A)} = \sigma_{pp}(A)$. This shows also that in general $\sigma_c(A) \setminus (\sigma_c(A) \cap \overline{\sigma_p(A)}) \subset \sigma_{c,R}(A) \subseteq \sigma(A) \setminus \sigma_p(A)$. □

One of the main topics are invariance questions concerning the components of the spectrum, in particular the invariance of the absolutely continuous spectrum under perturbations. These questions are essentially settled by the wave operators, their completeness and the scattering matrix. We present them now. Let A and B be self-adjoint operators in a Hilbert space X and let $P_{ac}(B)$ be the projection onto the absolutely continuous subspace of B. The wave operators $W_\pm$ for the pair A, B are defined as strong limits

$$W_\pm = W_{\pm(A,B)} = s - \lim_{t \to \pm\infty} e^{iAt} e^{-iBt} P_{ac}(B)$$

if they exist.

Denote by $E_A(I)$ or $E_B(I)$, I an interval, the spectral projections of A or B, respectively. We collect some standard properties of $W_\pm$:

Proposition 2.4 *Suppose that $W_\pm(A,B)$ exist. Then*
a) $W_\pm$ *are isometric on* $(P_{ac}(B))(X)$,
b) $E_A(I)W_\pm(A,B) = W_\pm(A,B)E_B(I)$,
c) $e^{-iAs}W_\pm(A,B) = W_\pm(A,B)e^{-iBs}$,
d) *if additionally $W_\pm(B,C)$ exist, then $W_\pm(A,C)$ exist and $W_\pm(A,C) = W_\pm(A,B)W_\pm(B,C)$.*

□

2.4 b) immediately implies $\mathrm{Ran} W_\pm \equiv \mathrm{im}\, W_\pm \subseteq (P_{ac}(A))(X)$. We say the $W_\pm$ are complete if $\mathrm{Ran} W_\pm = (P_{ac}(A))(X)$.

Proposition 2.5 *Suppose that $W_\pm(A,B)$ exist. Then they are complete if and only $W_\pm(B,A)$ exist. In this case $W_\pm(B,A) = W_\pm(A,B)^*$.* □

The central question are considerations which assure the existence and completeness of the wave operators. There are several classes of theorems. We restrict here ourselves to the conditions of Birman–Kato–Rosenblum.

Theorem 2.6 *Let A and B self-adjoint operators in X. Then the wave operators $W_\pm(A, B)$ exist and are complete in the following cases.*

a) *A and B are positive operators with $A^2 - B^2$ of trace class,*

b) *A and B positive operators with $(A^2 + 1)^{-1} - (B^2 + 1)^{-1}$ of trace class,*

c) *$A, B \geq -a + I$ and $(A + a)^{-k} - (B + a)^{-k}$ of trace class for some k,*

d) *$e^{-A} - e^{-B}$ of trace class,*

e) *$(A + i)^{-1} - (B + i)^{-1}$ of trace class,*

f) *A and B positive and $e^{-tA}A - e^{tB}B$ of trace class.*

We refer to [58], [72], [73] for the proofs of 2.3 - 2.5. □

All of our efforts in chapter IV will be concentrated to the task to satisfy on of the conditions above. Suppose $W_\pm(A, B)$ exist. Then we define the scattering operator S by

$$S = W_+^* W_-.$$

Remark 2.7 Completeness is not needed to define S. S commutes with B. If $\text{Ran} W_\pm = (P_{ac}(A))(X)$, i.e. if $W_\pm$ are complete then S is unitary on $(P_{ac}(B))(X)$. □

Next we will introduce the spectral decomposition of the scattering operator S. For this we need the notion of a direct integral of Hilbert spaces. The direct integral

$$\int_{-\infty}^{+\infty} \oplus X(\lambda) du(\lambda) \qquad (2.3)$$

is defined as the set of all vector–valued functions f, $f(\lambda) \in X(\lambda)$ which are measurable and square–integrable with respect to the measure m. The scalar product in (2.3) is defined as

$$\langle f, g \rangle := \int_{-\infty}^{+\infty} (f(\lambda), g(\lambda))_{X(\lambda)} dm(\lambda),$$

where $(f(\lambda), g(\lambda))_{X(\lambda)}$ is the scalar product in the Hilbert space $X(\lambda)$.

We say, a Hilbert space X has a decomposition as a direct integral if there is a unitary mapping

$$\mathcal{F}: X \xrightarrow{\cong} \int_{-\infty}^{+\infty} \oplus X(\lambda) du(\lambda).$$

We denote this by

$$X \longleftrightarrow \int_{-\infty}^{+\infty} \oplus X(\lambda) du(\lambda).$$

A special case of such a representation is given by the spectral resolution for a self–adjoint operator A,

$$A = \int_{-\infty}^{+\infty} \lambda dE_A(\lambda)$$

which induces a decomposition of type (2.3) in which the operator $\mathcal{F} A \mathcal{F}^*$ acts as multiplication by λ. Let $\mathcal{B} \subset \mathbb{R}$ be a Borel set. Then $\mathcal{F} E_A(\mathcal{B}) \mathcal{F}^*$ reduces to $\chi_\mathcal{B}$, and we obtain

$$\langle E_A(\mathcal{B}) f, g \rangle = \int_\mathcal{B} ((\mathcal{F} f)(\lambda), (\mathcal{F} g)(\lambda)) dm(\lambda).$$

If we apply these considerations to the self–adjoint operator B (of the pair A, B above) and to $X_{ac} = X_{ac}(B) \equiv (P_{ac}(B))(X)$, then we get

$$X_{ac}(B) \longleftrightarrow \int_{\hat{\sigma}(B)} \oplus X_\lambda(B) d\lambda := X(B)^{(ac)}. \qquad (2.4)$$

Here $\hat{\sigma}(B)$ is a so–called core of $\sigma(B)$, i.a. a Borel set of minimal measure such that $E_B(\mathbb{R} \setminus \hat{\sigma}(B)) = 0$. The right hand side of (2.4) diagonalizes the operator $B^{ac} = B|_{X_{ac}(B)}$. The scattering operator $S = W_+^* W_-$ commutes with B^{ac}, hence under the correspondence (2.4) its action goes over into multiplication by an operator valued function $S(\lambda) : X_\lambda(B) \longrightarrow X_\lambda(B)$. $S(\lambda) = S(\lambda; A, B)$ is called the scattering matrix. If we consider the left hand side of (2.2) then we obtain the version

$$S = \int_{\hat{\sigma}} S(\lambda) dE_B(\lambda) \qquad (2.5)$$

instead of

$$\mathcal{F} S \mathcal{F}^* = \int_{\hat{\sigma}} S(\lambda) d\lambda \qquad (2.6)$$

Both representation are equivalent.

In chapter V and VI, the spectral shift function $(SSF)\xi$ will play a central role. We introduce it now. The main goal is to introduce a function $\xi(\lambda)$ such that

$$\mathrm{tr}(\varphi(A) - \varphi(B)) = \int_{\mathbb{R}} \varphi'(\lambda) \xi(\lambda) d\lambda, \qquad (2.7)$$

where A, B are self–adjoint in X, φ belongs to an appropriate class of C^1–functions and $\varphi(A) - \varphi(B)$ is of trace class. We shall see below that the desired $\xi(\lambda)$ satisfies the equation

$$\det S(\lambda) = e^{-2\pi i \xi(\lambda)}, \text{ a.e. } \lambda \in \sigma(B). \qquad (2.8)$$

First we must show that the left hand side of (2.8) makes sense.

Proposition 2.8 *Let T be a trace class operator in X and $\{u_i\}_i$ a complete orthonormal system (ONS) in X.*

a) *Then*
$$\det(I+T) := \lim_{n\to\infty}(\det\{\delta_{kl} + Tu_k, u_l)\}_{k,l=1}^n)$$
exists and does not depend on the choice of the ONS.

b) *It is a continuous functional on the ideal of trace class operators.*

c) *There exists a representation as absolutely convergent infinite product*
$$\det(I+T) = \prod_k (1 + \lambda_k(T)).$$

d) *There holds*
$$\begin{aligned}\det(I+T^*) &= \overline{\det(I+T)},\\ \det((I+T_1)(I+T_2)) &= \det(I+T_1)\cdot\det(I+T_2),\\ \det(I+T_1T_2) &= \det(I+T_2T_1),\end{aligned}$$

e) *$I+T$ ist invertible if and only if $\det(I+T) \neq 0$, and then holds*
$$\det(I+T)\cdot\det(I+T)^{-1} = 1, \qquad (2.9)$$

f) *if $T(z)$ is a C^1-family of trace class operators then*
$$\frac{d}{dz}\det(I+T(z)) = \det(I+T(z))\,\mathrm{tr}\left((I+T(z))^{-1}\frac{dT(z)}{dz}\right).$$

We refer to [72], [73] for details and proofs. □

For A, B closed operators in X, $\mathcal{D}_A = \mathcal{D}_B$ with $(A-B)R_z(B)$ of trace class, we define the perturbation determinant
$$\Delta(z) \equiv \Delta_{A/B}(z) := \det(I + (A-B)R_z(B)), \quad z \in \varrho(B).$$

The resolvent equations
$$R_z(A) - R_z(B) = -R_z(A)(A-B)R_z(B) = -R_z(B)(A-B)R_z(A)$$

imply
$$\Delta_{A/B}(z) = \det(A - zI)(B - zI)^{-1}), \quad z \in \varrho(B). \quad (2.10)$$

$\Delta(z)$ is holomorphic on $\varrho(B)$ and

$$\Delta'(z)/\Delta(z) = \mathrm{tr}(R_z(B) - R_z(A)), \quad z \in \varrho(B) \cap \varrho(A). \quad (2.11)$$

We infer from (2.9) and (2.10)

$$\Delta_{B/A}(z)\Delta_{A/B}(z) = 1, \quad z \in \varrho(A) \wedge \varrho(B) \quad (2.12)$$

and from 2.8 d)

$$\Delta_{B/A}(z)\Delta_{B/C}(z) = \Delta_{A/C}(z), \quad z \in \varrho(A) \wedge \varrho(C), \quad (2.13)$$

if $(A - B)R_z(B)$ and $(B - C)R_z(C)$ are trace class.
Now we are able to state the first theorem of Klein.

Theorem 2.9 *Suppose A and B self-adjoint, $A - B$ of trace class. Then*

$$\log \Delta_{A/B}(z) = \int_{\mathbb{R}} \frac{\xi(t)}{t - z} dt, \quad \mathrm{im}\, Z \neq 0, \quad (2.14)$$

with $\xi = \xi(\cdot; A, B) = \overline{\xi} \in L_1(\mathbb{R})$

$$\xi(\lambda) = \pi^{-1} \lim_{\varepsilon \to 0^+} \arg \Delta(\lambda + i\varepsilon), \quad a.\,e.\ \lambda \in \mathbb{R},$$

$$\int_{\mathbb{R}} \xi(t; A, B) dt = \mathrm{tr}(A - B)$$

and

$$\int_{\mathbb{R}} |\xi(t; A, B)| dt \leq |A - B|_1, \quad (2.15)$$

where $|T|_1 = \sum_k (\lambda_k(T^*T))^{\frac{1}{2}}$ *is the trace norm.*

We refer to [72], [73] for the proof. □

To state the second theorem of Krein, we introduce the Wiener class $W_1(\mathbb{R})$

$$W_1(\mathbb{R}) = \{\varphi \in C^1_{loc} | \varphi'(\lambda) = \int_{\mathbb{R}} e^{-i\lambda t} d\sigma(t),$$

$\sigma(\cdot)$ a complex–valued finite Borel measure on $\mathbb{R}\}$.

Theorem 2.10 *Suppose the hypthesis of 2.9 and $\varphi \in W_1(\mathbb{R})$. Then $\varphi(A) - \varphi(B)$ is of trace class and*

$$|\varphi(A) - \varphi(B)|_1 \leq |\sigma|(\mathbb{R})|B - A|_1 \qquad (2.16)$$

and

$$\operatorname{tr}(\varphi(A) - \varphi(B)) = \int_{\mathbb{R}} \varphi'(\lambda)\xi(\lambda) d\lambda. \qquad (2.17)$$

□

We will make extensive use of the spectral shift function in the forthcoming chapters.

IV   Trace class properties

In this chapter, we prove the trace class property for $e^{-tD^2} - e^{-t\tilde{D}'^2}$, $\tilde{D}$ a perturbation of D as defined in II 4, i.e. the defining data of $\tilde{D}$ are $\in$ gen comp (defining data of D). We introduced in II 4 a hierarchy of perturbations, admitting step by step larger perturbations, i.e. the goal of this chapter is to prove that

$$e^{-tD^2} - e^{-t\tilde{D}'^2}$$

is for $t > 0$ of trace class where $\tilde{D}'$ is an appropriate transform of D' into the Hilbert space of D. We decompose the perturbations into several steps, 1) $\nabla \longrightarrow \nabla'$, all other fixed, 2) $h, \nabla \longrightarrow h', \nabla', \cdot, g$ fixed, 3) $h, \nabla, \cdot \longrightarrow h', \nabla', \cdot', g$ fixed and finally 4) $h, \nabla, \cdot, g \longrightarrow h', \nabla', \cdot', g'$. The last step consists in even admitting compact topological perturbations.

1   Variation of the Clifford connection

The first and simplest case is settled by

Theorem 1.1 *Assume* $(E, \nabla) \longrightarrow (M^n, g)$ *with* $(I), (B_k)$, (E, ∇) *with* (B_k), $k \geq r > n + 2$, $n \geq 2$, $\nabla' \in \text{comp}(\nabla) \cap \mathcal{C}_E(B_k) \subset \mathcal{C}_E^{1,r}(B_k)$, $D = D(g, \nabla)$, $D' = D(g, \nabla')$ *generalized Dirac operators. Then*

$$e^{-tD^2} - e^{-tD'^2} \quad \text{and} \quad De^{-tD^2} - D'e^{-tD'^2}$$

are trace class operators for $t > 0$ and their trace norm is uniformly bounded on compact t–intervalls $[a_0, a_1]$, $a_0 > 0$.   □

Here $\nabla' \in \text{comp}^{1,r}(\nabla)$ means in particular $|\nabla - \nabla'|_{\nabla,1,r} < \infty$ and both connections satisfy $(B_k(E))$. Denote $\nabla - \nabla' = \eta$.

As we indicated in III 2, we have, writing $D^2 - D'^2 = D(D - D') + (D - D')D'$, to estimate

$$e^{-tD^2} - e^{-tD'^2} = -\int_0^t e^{-sD^2}(D^2 - D'^2)e^{-(t-s)D'^2}\, ds$$

$$= -\int_0^t e^{-sD^2} D(D - D')e^{-(t-s)D'^2}\, ds$$

$$-\int_0^t e^{-sD^2}(D - D')D' e^{-(t-s)D'^2}\, ds$$

$$= \int_0^t e^{-sD^2} D\eta\, e^{-(t-s)D'^2}\, ds$$

$$+ \int_0^t e^{-sD^2} \eta D' e^{-(t-s)D'^2}\, ds,$$

where $\eta = \eta^{op}$ is defined by $\eta^{op}(\Psi)|_x = \sum_{i=1}^n e_i \eta_{e_i}(\Psi)$ and $|\eta^{op}|_{op,x} \leq C \cdot |\eta|_x$, C independent of x. We split $\int_0^t = \int_0^{t/2} + \int_{t/2}^t$,

$$e^{-tD^2} - e^{-tD'^2} = \int_0^{t/2} e^{-sD^2} D\eta\, e^{-(t-s)D'^2}\, ds \quad (I_1)$$

$$+ \int_0^{t/2} e^{-sD^2} \eta D' e^{-(t-s)D'^2}\, ds \quad (I_2)$$

$$+ \int_{t/2}^t e^{-sD^2} D\eta\, e^{-(t-s)D'^2}\, ds \quad (I_3)$$

$$+ \int_{\frac{t}{2}}^{t} e^{-sD^2} \eta D' e^{-(t-s)D'^2} \, ds. \qquad (I_4)$$

We want to show that each integral $(I_1) - (I_4)$ is a product of Hilbert–Schmidt operators and to estimate their Hilbert–Schmidt norm. Consider the integrand of (I_4),

$$(e^{-sD^2}\eta)(D'e^{-(t-s)D'^2}).$$

There holds

$$|e^{-(t-s)D'^2}|_{L_2 \to H^1} \leq C \cdot (t-s)^{-\frac{1}{2}}$$
$$|D'e^{-(t-s)D'^2}|_{L_2 \to L_2} \leq |D'|_{H^1 \to L_2} \cdot |e^{-(t-s)D'^2}|_{L_2 \to H^1}$$
$$\leq C \cdot (t-s)^{-\frac{1}{2}}.$$

Write

$$(e^{-sD^2}\eta)(D'e^{-(t-s)D'^2}) = (e^{-\frac{s}{2}D^2}f)(f^{-1}e^{-\frac{s}{2}D^2}\eta)(D'e^{-(t-s)D'^2}).$$

Here f shall be a scalar function which acts by multiplikation. The main point is the right choice of f. $e^{-\frac{s}{2}D^2}f$ has the integral kernel

$$W(\frac{s}{2}, m, p)f(p) \qquad (1.1)$$

and $f^{-1}e^{-\frac{s}{2}D^2}\eta$ has the kernel

$$f^{-1}(m)W(\frac{s}{2}, m, p)\eta(p). \qquad (1.2)$$

We have to make a choice such that (1.1), (1.2) are square integrable over $M \times M$ and that their L_2–norm is uniformly bounded on compact t–intervals.

We decompose the L_2–norm of (1.1) as

$$\int_M \int_M |W(\frac{s}{2},m,p)|^2 |f(m)|^2 \, dm \, dp \qquad (1.3)$$

$$= \int_M \int_{dist(m,p)\geq c} |W(\frac{s}{2},m,p)|^2 |f(m)|^2 \, dp \, dm \qquad (1.4)$$

$$+ \int_M \int_{dist(m,p)<c} |W(\frac{s}{2},m,p)|^2 |f(m)|^2 \, dp \, dm \qquad (1.5)$$

We use the fact that for any $T > 0$ and sufficiently small $\varepsilon > 0$ there exists $C > 0$ such that

$$|W(t,m,p)| \leq e^{-(t-\varepsilon)\inf \sigma(D^2)} \cdot C \cdot C(m) \cdot C(p)$$

for all $t \in]T, \infty[$ and obtain for $s \in]\frac{t}{2}, t[$

$$(1.5) \leq \int_M C_1 |f(m)|^2 \mathrm{vol} B_c(m) \, dm \leq C_2 \int_M |f(m)|^2 \, dm$$

Moreover, for any $\varepsilon > 0$, $T > 0$, $\delta > 0$ there exists $C > 0$ such that for $r > 0$, $m \in M$, $T > t > 0$ holds

$$\int_{M \setminus B_r(m)} |W(t,m,p)|^2 dp \leq C \cdot C(m) \cdot e^{-\frac{(r-\varepsilon)^2}{(4+\delta)t}},$$

which yields

$$\int_M \int_{dist(m,p)\geq c} |W(\frac{s}{2},m,p)|^2 |f(m)|^2 \, dp \, dm$$

$$\leq \int_M C_3 e^{-\frac{(c-\varepsilon)^2}{4+\delta}\frac{s}{2}} |f(m)|^2 \, dm$$

$$\leq C_3 \cdot e^{-\frac{(c-\varepsilon)^2}{4+\delta}\frac{s}{2}} \int_M |f(m)|^2 \, dm, \quad c > \varepsilon. \qquad (1.6)$$

Hence the estimate of $\int_M \int_M |W(\frac{s}{2}, m, p)|^2 |f(m)|^2 dp\, dm$ for $s \in [\frac{t}{2}, t]$ is done if

$$\int_M |f(m)|^2 \, dm < \infty$$

and then $|e^{-\frac{s}{2}D^2} f|_2 \leq C_4 \cdot |f|_{L_2}$, where $C_4 = C_4(t)$ contains a factor e^{-at}, $a > 0$, if $\inf \sigma(D^2) > 0$.

For (1.2) we have to estimate

$$\int_M \int_M |f(m)|^{-2} |(W(\frac{s}{2}, m, p), \eta^{op}(p) \cdot)_p|^2 \, dp\, dm \quad (1.7)$$

We recall a simple fact about Hilbert spaces. Let X be a Hilbert space, $x \in X, x \neq 0$. Then $|x| = \sup_{|y|=1} |\langle x, y \rangle|$,

$$|x|^2 = \left(\sup_{|y|=1} |\langle x, y \rangle| \right)^2. \quad (1.8)$$

This follows from $|\langle x, y \rangle| \leq |x| \cdot |y|$ and equality for $y = \frac{x}{|x|}$. We apply this to $E \to M$, $X = L_2(M, E, dp)$, $x = x(m) = W(t, m, p), \eta^{op}(p) \cdot)_p = W(t, m, p) \circ \eta^{op}(p)$ and have to estimate

$$\sup_{\substack{\Phi \in C_c^\infty(E) \\ |\Phi|_{L_2} = 1}} N(\Phi) = \sup_{\substack{\Phi \in C_c^\infty(E) \\ |\Phi|_{L_2} = 1}} |\langle \delta(m), e^{-tD^2} \eta^{op} \Phi \rangle|_{L_2} \quad (1.9)$$

The heat kernel is of Sobolev class,

$$W(t, m, \cdot) \in H^{\frac{r}{2}}(E), \quad |W(t, m, \cdot)|_{H^{\frac{r}{2}}} \leq C_5(t). \quad (1.10)$$

Hence we have can restrict in (1.9) to

$$\sup_{\substack{\Phi \in C_c^\infty(E) \\ |\Phi|_{L_2} = 1 \\ |\Phi|_{H^{\frac{r}{2}}} \leq C_5}} N(\Phi) \quad (1.11)$$

In the sequel we estimate (1.11). For doing this, we recall some simple facts concerning the wave equation

$$\frac{\partial \Phi_s}{\partial s} = iD\Phi_s, \quad \Phi_0 = \Phi, \quad \Phi \ C^1 \text{ with compact support.} \quad (1.12)$$

It is well known that (1.12) has a unique solution Φ_s which is given by
$$\Phi_s = e^{isD}\Phi \qquad (1.13)$$
and
$$\operatorname{supp} \Phi_s \subset U_{|s|}(\operatorname{supp} \Phi) \qquad (1.14)$$
$U_{|s|} = |s|$ – neighbourhood. Moreover,
$$|\Phi_s|_{L_2} = |\Phi|_{L_2}, \quad |\Phi_s|_{H^{\frac{r}{2}}} = |\Phi|_{H^{\frac{r}{2}}}. \qquad (1.15)$$

We fix a uniformly locally finite cover $\mathcal{U} = \{U_\nu\}_\nu = \{B_d(x_\nu)\}_\nu$ by normal charts of radius $d < r_{inj}(M, g)$ and associated decomposition of unity $\{\varphi_\nu\}_\nu$ satisfying
$$|\nabla^i \varphi_\nu| \leq C \text{ for all } \nu, \ 0 \leq i \leq k+2 \qquad (1.16)$$

Write
$$\begin{aligned}
N(\Phi) &= |\langle \delta(m), e^{-tD^2} \eta^{op}\Phi \rangle| \\
&= \frac{1}{\sqrt{4\pi t}} \left|\langle \delta(m), \int_{-\infty}^{+\infty} e^{\frac{-s^2}{4t}} e^{isD}(\eta^{op}\Phi) \, ds \rangle\right|_{L_2(dp)} = \\
&= \frac{1}{\sqrt{4\pi t}} \left|\int_{-\infty}^{+\infty} e^{\frac{-s^2}{4t}} (e^{isD}\eta^{op}\Phi)(m) \, ds\right|_{L_2(dp)}. \qquad (1.17)
\end{aligned}$$

We decompose
$$\eta^{op}(\Phi) = \sum_\nu \varphi_\nu \eta^{op}\Phi. \qquad (1.18)$$

(1.18) is a locally finite sum, (1.12) linear. Hence
$$(\eta^{op}(\Phi))_s = \sum_\nu (\varphi_\nu \eta^{op}\Phi)_s. \qquad (1.19)$$

Denote as above
$$|\ |_{p,i} \equiv |\ |_{W^{p,i}},$$
in particular
$$|\ |_{2,i} \equiv |\ |_{W^{2,i}} \sim |\ |_{H^i}, \quad i \leq k. \qquad (1.20)$$

Then we obtain from (1.15), (1.16) and an Sobolev embedding theorem

$$|(\varphi_\nu \eta^{op}\Phi)_s|_{H^{\frac{r}{2}}} = |\varphi_\nu \eta^{op}\Phi|_{H^{\frac{r}{2}}} \leq C_6 |\varphi_\nu \eta^{op}\Phi|_{2,\frac{r}{2}}$$

$$\leq C_7 |\eta^{op}\Phi|_{2,\frac{r}{2},U_\nu} \leq C_8 |\eta|_{2,\frac{r}{2},U_\nu} \leq C_9 |\eta|_{1,r-1,U_\nu} \quad (1.21)$$

since $r - 1 - \frac{n}{i} \geq \frac{r}{2} - \frac{n}{2}, r - 1 \geq \frac{r}{2}, 2 \geq i$ for $r > n + 2$ and $|\Phi|_{H^{\frac{r}{2}}} \leq C_5$. This yields together with the Sobolev embedding the estimate

$$|(\eta^{op}\Phi)_s(m)| \leq C_{10} \cdot \sum_{\substack{\nu \\ m \in U_s(U_\nu)}} |(\varphi_\nu \eta^{op}\Phi)_s|_{2,\frac{r}{2}} \leq$$

$$\leq C_{11} \cdot \sum_{\substack{\nu \\ m \in U_s(U_\nu)}} |\eta|_{1,r-1,U_\nu} \leq C_{12} \cdot |\eta|_{1,r-1,B_{2d+|s|}(m)} =$$

$$= C_{12} \cdot \text{vol}(B_{2d+|s|}(m)) \cdot \left(\frac{1}{\text{vol}B_{2d+|s|}(m)} \cdot |\eta|_{1,r-1,B_{2d+|s|}(m)} \right).$$
(1.22)

There exist constants A and B, independent of m s. t.

$$\text{vol}(B_{2d+|s|}(m)) \leq A \cdot e^{B|s|}.$$

Write

$$e^{-\frac{s^2}{4t}} \cdot \text{vol}(B_{2d+|s|}(m)) \leq C_{13} \cdot e^{-\frac{9}{10}\frac{s^2}{4t}}, \quad C_{13} = A \cdot e^{10B^2 t}, \quad (1.23)$$

thus obtaining

$$N(\Phi) \leq C_{14} \int_0^\infty e^{-\frac{9}{10}\frac{s^2}{4t}} \left(\frac{1}{\text{vol}B_{2d+|s|}(m)} \cdot |\eta|_{1,r-1,B_{2d+|s|}(m)} \right) ds,$$

$$C_{14} = C_{12} \cdot C_{13} = C_{12} \cdot A \cdot e^{10B^2 t}.$$

Now we apply Buser/Hebey's inequality in chapter II, proposition 1.9,

$$\int_M |u - \overline{u}_c| \, \text{dvol}_x(g) \leq C \cdot c \int_M |\nabla u| \, \text{dvol}_x(g)$$

for $u \in W^{1,1}(M) \cap C^1(M)$, $c \in]0, R[$, Ric$(g) \geq k$, $C = C(n, k, R)$ and
$$\bar{u}_c(x) := \frac{1}{\text{vol} B_c(x)} \int_{B_c(x)} u(y) \, \text{dvol}_y$$

with $R = 3d + s$ and infer

$$\int_M \frac{1}{\text{vol} B_{2d+|s|}(m)} \cdot |\eta|_{1,r-1,B_{2d+|s|}(m)} \, dm$$
$$\leq |\eta|_{1,r-1} + C(3d+s) \cdot (2d+s)|\nabla \eta|_{1,r-1}$$
$$\leq |\eta|_{1,r-1} + C(3d+s) \cdot (2d+s)|\eta|_{1,r-1}. \quad (1.24)$$

$C(3d+s)$ depends on $3d+s$ at most linearly exponentially, i. e.
$$C(3d+s) \cdot (2d+s) \leq A_1 e^{B_1 s}.$$

This implies

$$\int_0^\infty e^{-\frac{9}{10}\frac{s^2}{4t}} \int_M \frac{1}{\text{vol} B_{2d+|s|}(m)} \cdot |\eta|_{1,r-1,B_{2d+|s|}(m)} \, dm \, ds \quad (1.25)$$
$$\leq = \int_0^\infty e^{-\frac{9}{10}\frac{s^2}{4t}} (|\eta|_{1,r-1} + C(3d+s) \cdot (2d+s)|\eta|_{1,r-1}) \, ds$$
$$\leq \int_0^\infty e^{-\frac{8}{10}\frac{s^2}{4t}} \, ds (|\eta|_{1,r-1} + A_1 e^{10 B_1^2 t} |\eta|_{1,r}$$
$$= \sqrt{t} \cdot \frac{1}{2}\sqrt{5\pi}(|\eta|_{1,r-1} + A_1 e^{10 B_1^2 t} |\eta|_{1,r}) < \infty.$$

The function $\mathbb{R}_+ \times M \to \mathbb{R}$,
$$(s, m) \to e^{-\frac{9}{10}\frac{s^2}{4t}} \left(\frac{1}{\text{vol} B_{2d+|s|}(m)} \cdot |\eta|_{1,r-1,B_{2d+|s|}(m)} \right)$$

is measurable, nonnegative, the integrals (1.24), (1.25) exist, hence according to the principle of Tonelli, this function is 1–summable, the Fubini theorem is applicable and

$$\tilde{\eta} := C_{10} \cdot \int_0^\infty e^{-\frac{9}{10}\frac{s^2}{4t}} \left(\frac{1}{\text{vol} B_{2d+|s|}(m)} \cdot |\eta|_{1,r-1,B_{2d+|s|}(m)} \right) ds$$

is (for $\eta \not\equiv 0$) everywhere $\neq 0$ and 1–summable. We proved

$$\int |(W(t,m,p), \eta^{op}\cdot)_p|^2 \leq \tilde{\eta}(m)^2. \qquad (1.26)$$

Now we set
$$f(m) = (\tilde{\eta}(m))^{\frac{1}{2}} \qquad (1.27)$$
and infer $f(m) \neq 0$ everywhere, $f \in L_2$ and

$$|f^{-1}e^{-\frac{s}{2}D^2} \circ \eta|^2_{L_2}$$
$$= \int_M \int_M f(m)^{-2} |((W(\frac{s}{2}, m, p), \eta^{op})_p|^2 \, dp \, dm$$
$$\leq \int_M \frac{1}{\tilde{\eta}(m)} \tilde{\eta}(m)^2 \, dm = \int_M \tilde{\eta}(m) \, dm$$
$$\leq C_{12} \cdot A \cdot e^{10B^2 s} \sqrt{s} \cdot \frac{1}{2} \sqrt{5\pi} (|\eta|_{1,r-1} + A_1 e^{10B_1^2 s} |\eta|_{1,r})$$
$$\leq C_{15} \sqrt{s} e^{10B^2 s} |\eta|_{1,r}, \qquad (1.28)$$

i. e.
$$|f^{-1}e^{-\frac{s}{2}D^2} \circ \eta|_2 \leq C_{15}^{\frac{1}{2}} \cdot s^{\frac{1}{4}} \cdot e^{5B^2 s} \cdot |\eta|_{1,r}^{\frac{1}{2}}. \qquad (1.29)$$

Here according to the term $A_1 e^{10B_1^2 s}$, C_{15} still depends on s. We obtain

$$|e^{-\frac{s}{2}D^2} \circ f|_{L_2} \cdot |f^{-1} \circ e^{-\frac{s}{2}D^2} \circ \eta|$$
$$\leq C_4 |f|_{L_2} \cdot C_{15}^{\frac{1}{2}} \cdot s^{\frac{1}{4}} \cdot e^{5B^2 s} \cdot |\eta|_{1,r}^{\frac{1}{2}}$$
$$\leq C_4 \cdot C_{15} \sqrt{s} e^{10B^2 s} |\eta|_{1,r} = C_{16} \cdot \sqrt{s} \cdot e^{10B^2 s} |\eta|_{1,r}. \quad (1.30)$$

This yields $e^{-sD^2} \circ \eta$ is of trace class,

$$|e^{-sD^2}\eta|_1 \leq |e^{-\frac{s}{2}D^2} \circ f|_2 \cdot |f^{-1}e^{-\frac{s}{2}D^2}\eta|_2 \leq C_{16}\sqrt{s}e^{10B^2 s}|\eta|_{1,r}, \qquad (1.31)$$

$e^{-sD^2} \circ \eta \circ D' \circ e^{-(t-s)D'^2}$ is of trace class,

$$|e^{-sD^2} \circ \eta \circ D' \circ e^{-(t-s)D'^2}|_1 \leq |e^{-sD^2}\eta|_1 \cdot |D'e^{-(t-s)D'^2}|_{op}$$
$$\leq C_{16}\sqrt{s}e^{10B^2 s}|\eta|_{1,r} \cdot C' \cdot \frac{1}{\sqrt{t-s}}, \qquad (1.32)$$

$$\left| \int_{\frac{t}{2}}^{t} (e^{-sD^2} \circ \eta \circ D' \circ e^{-(t-s)D'^2} \, ds \right|_1$$

$$\leq \int_{\frac{t}{2}}^{t} |e^{-sD^2} \eta \circ D' e^{-(t-s)D'^2}|_1 \, ds$$

$$\leq C_{16} \cdot C' \cdot e^{10B^2 t} |\eta|_{1,r} \cdot \int_{\frac{t}{2}}^{t} \left(\frac{s}{t-s} \right)^{\frac{1}{2}} ds, \quad (1.33)$$

$$\int_{\frac{t}{2}}^{t} \left(\frac{s}{t-s} \right)^{\frac{1}{2}} ds = [\sqrt{s(t-s)} + \frac{t}{2} \arcsin \frac{2s-t}{t}]_{\frac{t}{2}}^{t}$$

$$= -\frac{t}{2} + \frac{t}{2} \frac{\pi}{2} = \frac{t}{2}(\frac{\pi}{2} - 1),$$

$$\left| \int_{\frac{t}{2}}^{t} (e^{-sD^2} \circ \eta \circ D' \circ e^{-(t-s)D'^2} \, ds \right|_1$$

$$\leq C_{16} \cdot C' \cdot e^{10B^2 t} \cdot (\frac{\pi}{2} - 1) \cdot \frac{t}{2} |\eta|_{1,r}$$

$$= C_{17} e^{10B^2 t} \cdot t \cdot |\eta|_{1,r}. \quad (1.34)$$

Here $C_{17} = C_{17}(t)$ and $C_{17}(t)$ can grow exponentially in t if the volume grows exponentially. (1.34) expresses the fact that (I_4) is of trace class and its trace norm is uniformly bounded on any t-intervall $[a_0, a_1]$, $a_0 > 0$. The treatment of $(I_1) - (I_3)$ is quite parallel to that of (I_4). Write the integrand of (I_3), (I_2) or (I_1) as

$$(De^{-\frac{s}{2}D^2})[(e^{-\frac{s}{4}D^2} f)(f^{-1} e^{-\frac{s}{4}D^2} \eta)] e^{-(t-s)D'^2} \quad (1.35)$$

or

$$(e^{-sD^2})[(\eta e^{-\frac{(t-s)}{4}D'^2} f^{-1})(f e^{-\frac{(t-s)}{4}D'^2})] D' e^{-\frac{t-s}{2}D'^2} \quad (1.36)$$

or

$$(e^{-sD^2}D)[(\eta e^{-\frac{(t-s)}{2}D'^2}f^{-1})(fe^{-\frac{(t-s)}{2}D'^2})], \quad (1.37)$$

respectively. Then in the considered intervals the expressions [...] are of trace class which can literally be proved as for (I_4). The main point in (I_4) was the estimate of $f^{-1}e^{-\tau D^2}\eta$. In (1.36), (1.37) we have to estimate expressions $\eta e^{-\tau D'^2}f^{-1}$. Here we use the fact that $\eta = \eta^{op}$ is symmetric with respect to the fibre metric h: the endomorphism $\eta_{e_i}(\cdot)$ is skew symmetric as the Clifford multiplication $e_i\cdot$ which yields together that η^{op} is symmetric. Then the L_2-estimate of $(\eta^{op}\cdot W'(\tau,m,p),\cdot)$ is the same as that of $W'(\tau,m,p),\eta^{op}(p)\cdot)$ and we can perform the same procedure as that starting with (1.6). The only distinction are other constants. Here essentially enters the equivalence of the D– and D'–Sobolev spaces i.e. the symmetry of our uniform structure. The factors outside [...] produce $\frac{1}{\sqrt{s}}$ on $[\frac{t}{2},t]$, $\frac{1}{\sqrt{t-s}}$ and $\frac{1}{\sqrt{s}}$ on $[0,\frac{t}{2}]$ (up to constants). Hence $(I_1)-(I_3)$ are of trace class with uniformly bounded trace norm on any t–intervall $[a_0,a_1]$, $a_0 > 0$. This finishes the proof of the first part of theorem 1.1.

We must still prove the trace class property of

$$e^{-tD^2}D - e^{-tD'^2}D'. \quad (1.38)$$

Consider the decomposition

$$\begin{aligned}e^{-tD^2}D - e^{-tD'^2}D' &= e^{-\frac{t}{2}D^2}D(e^{-\frac{t}{2}D^2} - e^{-\frac{t}{2}D'^2}) &(1.39)\\ &+ (e^{-\frac{t}{2}D^2}D - e^{-\frac{t}{2}D'^2}D')e^{-\frac{t}{2}D'^2} &(1.40)\end{aligned}$$

According to the first part, $e^{-\frac{t}{2}D^2} - e^{-\frac{t}{2}D'^2}$ is for $t > 0$ of trace class. Moreover, $e^{-\frac{t}{2}D^2}D = De^{-\frac{t}{2}D^2}$ is for $t > 0$ bounded, its operator norm is $\leq \frac{C}{\sqrt{t}}$. Hence their product is for $t > 0$ of trace class and has bounded trace norm for $t \in [a_0,a_1]$, $a_0 > 0$. (1.39) is done. We can write (1.40) as

$$(e^{-\frac{t}{2}D^2}D - e^{-\frac{t}{2}D'^2}D')e^{-\frac{t}{2}D'^2}$$
$$= [e^{-\frac{t}{2}D^2}(D-D') + (e^{-\frac{t}{2}D^2} - e^{-\frac{t}{2}D'^2})D'] \cdot e^{-\frac{t}{2}D'^2}$$
$$= [-e^{\frac{t}{2}D^2}\eta]e^{-\frac{t}{2}D'^2} + [\int_0^{\frac{t}{2}} e^{-sD^2}D\eta e^{-(\frac{t}{2}-s)D'^s}\,ds$$
$$+ \int_0^{\frac{t}{2}} e^{-sD^2}\eta D'e^{-(\frac{t}{2}-s)D'^2}\,ds](D'e^{-\frac{t}{2}D'^2}). \qquad (1.41)$$

Now

$$[e^{-\frac{t}{2}D^2}\eta] \cdot e^{-\frac{t}{2}D'^2} = [(e^{-\frac{t}{4}D^2}f)(f^{-1}e^{-\frac{t}{4}D^2}\eta)]e^{-\frac{t}{2}D'^2}. \qquad (1.42)$$

(1.42) is of trace class and its trace norm is uniformly bounded on any $[a_0, a_1]$, $a_0 > 0$, according the proof of the first part. If we decompose $\int_0^{\frac{t}{2}} = \int_0^{\frac{t}{4}} + \int_{\frac{t}{4}}^{\frac{t}{2}}$ then we obtain back from the integrals in (1.41) the integrals $(I_1) - (I_4)$, replacing $t \to \frac{t}{2}$. These are done. $D'e^{-\frac{t}{2}D'^2}$ generates $C/\sqrt{t}$ in the estimate of the trace norm. Hence we are done. $\square$

2 Variation of the Clifford structure

Our procedure is to admit much more general perturbations than those of $\nabla = \nabla^h$ only. Nevertheless, the discussion of more general perturbations is modelled by the case of ∇-perturbation. In this next step, we admit perturbations of $g, \nabla^h, \cdot$, fixing h, the topology and vector bundle structure of $E \longrightarrow M$. The next main result shall be formulated as follows.

Theorem 2.1 *Let $E = (E, h, \nabla = \nabla^h, \cdot) \longrightarrow (M^n, g)$ be a Clifford bundle with (I), $(B_k(M, g))$, $(B_k(E, \nabla))$, $k \geq r+1 > n+3$, $E' = (E, h, \nabla' = \nabla'^h, \cdot') \longrightarrow (M^n, g') \in \text{gen comp}_{L,diff,F}^{1,r+1}(E) \cap$*

$CL\mathcal{B}^{N,n}(I, B_k)$, $D = D(g, h, \nabla = \nabla^h, \cdot)$, $D' = D(g', h, \nabla' = \nabla'^h, \cdot')$ the associated generalized Dirac operators. Then for $t > 0$

$$e^{-tD^2} - e^{-tD'^2_{L_2}} \tag{2.1}$$

is of trace class and the trace norm is uniformly bounded on compact t-intervalls $[a_0, a_1]$, $a_0 > 0$.

Here $D'^2_{L_2}$ is the unitary transformation of D'^2 to $L_2 = L_2((M, E), g, h)$. 2.1 needs some explanations. D acts in $L_2 = L_2((M, E), g, h)$, D' in $L'_2 = L_2((M, E), g', h)$. L_2 and L'_2 are quasi isometric Hilbert spaces. As vector spaces they coincide, their scalar products can be quite different but must be mutually bounded at the diagonal after multiplication by constants. D is self adjoint on $\mathcal{D}_D$ in L_2, D' is self adjoint on $\mathcal{D}_{D'}$ in L'_2 but not necessarily in L_2. Hence $e^{-tD'^2}$ and $e^{-tD^2} - e^{-tD'^2}$ are not defined in L_2. One has to graft D^2 or D'^2. Write $\text{dvol}_q(g) \equiv dq(g) = \alpha(q) \cdot dq(g') \equiv \text{dvol}_q(g')$. Then

$$0 < c_1 \leq \alpha(q) \leq c_2, \alpha, \alpha^{-1} \text{ are } (g, \nabla^g) -$$
$$\text{and } (g', \nabla^{g'}) - \text{boundedup to order 3,}$$
$$|\alpha - 1|_{g,1,r+1}, |\alpha - 1|_{g',1,r+1} < \infty, \tag{2.2}$$

since $g' \in \text{comp}^{1,r+1}(g)$. Define $U : L_2 \longrightarrow L'_2$, $U\Phi = \alpha^{\frac{1}{2}}\Phi$. Then U is a unitary equivalence between L_2 and L'_2, $U^* = U^{-1}$. $D'_{L_2} := U^* D' U$ acts in L_2, is self adjoint on $U^{-1}(\mathcal{D}_{D'})$, since U is a unitary equivalence. The same holds for $D'^2_{L_2} = U^* D'^2 U = (U^* D' U)^2$. It follows from the definition of the spectral measure, the spectral integral and the spectral representations $D'^2 = \int \lambda^2 \, dE'_\lambda$, $e^{-tD'^2} = \int e^{-t\lambda^2} \, dE'_\lambda$ that $D'^2_{L_2} = U^* D'^2 U = U^* \int \lambda^2 \, dE'_\lambda U = \int \lambda^2 \, d(U^* E'_\lambda U)$ and

$$e^{-tD'^2_{L_2}} = \int e^{-t\lambda^2} \, d(U^* E'_\lambda U) = U^* \left(\int e^{-t\lambda^2} \, dE'_\lambda \right) U = U^* e^{-tD'^2} U. \tag{2.3}$$

In (2.1) $e^{-tD'^2_{L_2}}$ means $e^{-tD'^2_{L_2}} = e^{-t(U^* D' U)^2} = U^* e^{-tD'^2} U$. We obtain from $g' \in \text{comp}^{1,r+1}(g)$, $\nabla'^h \in \text{comp}^{1,r+1}(\nabla^h g)$, $\cdot' \in \text{comp}^{1,r+1}(\cdot)$, $D - \alpha^{-\frac{1}{2}} D' \alpha^{\frac{1}{2}} = D - D' - \frac{\text{grad}' \alpha \cdot'}{2\alpha}$ and (2.2) the following lemma concerning the equivalence of Sobolev spaces.

Lemma 2.2 $W^{1,i}(E, g, h, \nabla^h) = W^{1,i}(E, g', h, \nabla'^h)$ as equivalent Banach spaces, $0 \leq i \leq r+1$.

$\square$

Corollary 2.3 $W^{2,i}(E, g, h, \nabla^h) = W^{2,i}(E, g', h, \nabla'^h)$ as equivalent Hilbert spaces, $0 \leq j \leq \frac{r+1}{2}$. $\square$

Corollary 2.4 $H^j(E, D) \cong H^j K(E, D')$, $0 \leq j \leq \frac{r+1}{2}$. $\square$

2.2 has a parallel version for the endomorphism bundle $\mathrm{End}E$.

Lemma 2.5 $\Omega^{1,1,i}(EndE, g, h, \nabla^h) \cong \Omega^{1,1,i}(EndE, g', h, \nabla'^h)$
$0 \leq i \leq r+1$.
$\square$

Lemma 2.6 $\Omega^{1,2,j}(EndE, g, h, \nabla^h) \cong \Omega^{1,2,j}(EndE, g', h, \nabla'^h)$
$0 \leq j \leq \frac{r+1}{2}$.
$\square$

$e^{-tD'^2_{L_2}} : L_2 \longrightarrow L_2$ has evidently the heat kernel

$$W'_{L_2}(t, m, p) = \alpha^{-\frac{1}{2}}(m) W'(t, m, p) \alpha^{\frac{1}{2}}(p)$$

$W' \equiv W_{L'_2}$. Our next task is to obtain an explicit expression for $e^{-tD^2} - e^{-tD'^2_{L_2}}$. For this we apply again Duhamel's principle. The steps 1) – 4) in the proof of lemma 2.1 in chapter III remain. Then we set $\Phi(t, q) = W(t, m, q)$, $\Psi(t, q) = W'_{L_2}(t, m, q)$ and obtain

$$-\int_\alpha^\beta \int_M h_q(W(\tau, m, q), (D^2 + \frac{\partial}{\partial t}) W'_{L_2}(t - \tau, q, p))\, dq(g)\, d\tau$$

$$= \int_M [h_q(W(\beta, m, q), W'_{L_2}(t - \beta, q, p)$$

$$- h_q(W(\alpha, m, q), W'_{L_2}(t - \alpha, q, p)]\, dq(g).$$

Performing $\alpha \longrightarrow 0^+, \beta \longrightarrow t$ and using $dq(g) = \alpha(q)dq(g')$ yields

$$-\int_0^t \int_M h_q(W(s,m,q), (D^2 + \frac{\partial}{\partial t})W'(t-s,q,p))dq(g)ds$$

$$= -\int_0^t \int_M [h_q(W(s,m,q), (D^2 - D'^2_{L_2})W'_{L_2}(t-s,q,p)dq(g)ds$$

$$= W(t,m,p)\alpha(p) - W'_{L_2}(t,m,p). \tag{2.4}$$

(2.4) expresses the operator equation

$$e^{-tD^2}\alpha - e^{-tD'^2_{L_2}} = -\int_0^t e^{-sD^2}(D^2 - D'^2_{L_2})e^{-(t-s)D'^2_{L_2}} ds.$$

$$e^{-tD^2}\alpha - e^{-tD'^2_{L_2}} = e^{-tD^2}(\alpha - 1) + e^{-tD^2} - e^{-tD'^2_{L_2}}, \text{ hence}$$

$$e^{-tD^2} - e^{-tD'^2_{L_2}} = -e^{-tD^2}(\alpha - 1)$$

$$-\int_0^t e^{-sD^2}(D^2 - D'^2_{L_2})e^{-(t-s)D'^2_{L_2}} ds.$$

$$\tag{2.5}$$

As we mentioned in (2.2), $(\alpha - 1) = \frac{dq(g)}{dq(g')} - 1 = \frac{\sqrt{\det g}}{\sqrt{\det g'}} - 1 \in \Omega^{0,1,r+1}$ since $g \in \text{comp}^{1,r+1}(g)$. We write $e^{-tD^2}(\alpha - 1) = (e^{-\frac{t}{2}D^2}f)(f^{-1}e^{-\frac{t}{2}D^2}(\alpha-1))$, determine f as in the proof of theorem 1.1 from $\eta_\alpha = \alpha - 1$ and obtain $e^{-tD^2}(\alpha - 1)$ is of trace class with trace norm uniformly bounded on any t–interval $[a_0, a_1]$, $a_0 > 0$. Decompose $D^2 - D'^2_{L_2} = D(D - D'_{L_2}) + (D - D'_{L_2})D'_{L_2}$. We need explicit analytic expressions for this. $D(D - D'_{L_2}) = D(D - \alpha^{-\frac{1}{2}}D'\alpha^{\frac{1}{2}}) = D(D - D') - D\frac{\text{grad }'\alpha \cdot'}{2\alpha}$, $(D - D'_{L_2})D'_{L_2} = ((D - D') - \frac{\text{grad }'\alpha \cdot'}{2\alpha})\alpha^{-\frac{1}{2}}D'\alpha^{\frac{1}{2}}$. If we set again $D - D' = -\eta$ then we have to consider as before with $\frac{\text{grad }'\alpha}{2\alpha} = \frac{\text{grad }'\alpha \cdot'}{2\alpha}$ where

$\operatorname{grad}' \equiv \operatorname{grad}_{g'}$

$$\int_0^{\frac{t}{2}} e^{-sD^2} D(\eta - \frac{\operatorname{grad}'\alpha}{2\alpha}) e^{-(t-s)D'^2_{L_2}} \, ds$$

$$+ \int_0^{\frac{t}{2}} e^{-sD^2} (\eta - \frac{\operatorname{grad}'\alpha}{2\alpha}) D'_{L_2} e^{-(t-s)D'^2_{L_2}} \, ds$$

$$+ \int_{\frac{t}{2}}^t e^{-sD^2} D(\eta - \frac{\operatorname{grad}'\alpha}{2\alpha}) e^{-(t-s)D'^2_{L_2}} \, ds$$

$$+ \int_{\frac{t}{2}}^t e^{-sD^2} (\eta - \frac{\operatorname{grad}'\alpha}{2\alpha}) D'_{L_2} e^{-(t-s)D'^2_{L_2}} \, ds.$$

It follows immediately from $g' \in \operatorname{comp}^{1,r+1}(g)$ that the vector field $\frac{\operatorname{grad}'\alpha}{\alpha} \in \Omega^{0,1,r}(TM)$. If we write $\eta_0^{op} = -\frac{\operatorname{grad}'\alpha\cdot'}{\alpha}$ then η_0^{op} is a zero order operator, $|\eta_0|_r < \infty$ and we literally repeat the procedure for $(I_1) - (I_4)$ as before, inserting $\eta_0 = -\frac{\operatorname{grad}'\alpha\cdot'}{\alpha}$ for η there. Hence there remains to discuss the integrals

$$\int_0^t e^{-sD^2} D\eta\, e^{-(t-s)D'^2_{L_2}} \, ds + \int_0^t e^{-sD^2} \eta D'_{L_2} e^{-(t-s)D'^2_{L_2}} \, ds. \quad (2.6)$$

The next main step is to insert explicit expressions for $D - D'$. Let $m_0 \in M$, $U = U(m_0)$ a manifold and bundle coordinate neighbourhood with coordinates $x^1, \ldots, x^n$ and local bundle basis $\Phi_1, \ldots, \Phi_n : U \longrightarrow E|_U$. Setting $\nabla_{\frac{\partial}{\partial x_i}} \Phi_\alpha \equiv \nabla_i \Phi_\alpha = \Gamma_{i\alpha}^\beta \Phi_\beta$, $\nabla \Phi_\alpha = dx^i \otimes \Gamma_{i\alpha}^\beta \Phi_\beta$, we can write $D\Phi_\alpha = \Gamma_{i\alpha}^\beta g^{ik} \frac{\partial}{\partial x^k} \cdot \Phi_\beta$, $D'\Phi_\alpha = \Gamma'^\beta_{i\alpha} g'^{ik} \frac{\partial}{\partial x^k} \cdot' \Phi_\beta$, or for a local section Φ

$$D\Phi = g^{ik} \frac{\partial}{\partial x^k} \cdot \nabla_i \Phi, \quad D'\Phi = g'^{ik} \frac{\partial}{\partial x^k} \cdot' \nabla'_i \Phi. \quad (2.7)$$

This yields

$$\begin{aligned}
-(D-D')\Phi &= g^{ik}\frac{\partial}{\partial x^k}\cdot \nabla_i\Phi - g'^{ik}\frac{\partial}{\partial x^k}\cdot{}'\nabla'_i\Phi \\
&= [(g^{ik}-g'^{ik})\frac{\partial}{\partial x^k}\cdot\nabla_i + g'^{ik}\frac{\partial}{\partial x^k}\cdot(\nabla_i-\nabla'_i) \\
&\quad + g'^{ik}\frac{\partial}{\partial x^k}(\cdot - \cdot')\nabla'_i]\Phi,
\end{aligned} \qquad (2.8)$$

i. e. we can write

$$-(D-D')\Phi = (\eta_1^{op}+\eta_2^{op}+\eta_3^{op})\Phi, \qquad (2.9)$$

where locally

$$\eta_1^{op}\Phi = (g^{ik}-g'^{ik})\frac{\partial}{\partial x^k}\cdot\nabla_i\Phi, \qquad (2.10)$$

$$\eta_2^{op}\Phi = g'^{ik}\frac{\partial}{\partial x^k}\cdot(\nabla_i-\nabla'_i)\Phi, \qquad (2.11)$$

$$\eta_3^{op}\Phi = g'^{ik}\frac{\partial}{\partial x^k}(\cdot - \cdot')\nabla'_i\Phi. \qquad (2.12)$$

Here $(g'^{ik}) = (g'_{jl})^{-1}$. We simply write η_ν instead η_ν^{op}, hence

$$(2.6) = \int_0^t e^{-sD^2}D(\eta_1+\eta_2+\eta_3)e^{-(t-s)D'^2_{L_2}}\,ds \;+\; (2.13)$$

$$+ \int_0^t e^{-sD^2}(\eta_1+\eta_2+\eta_3)D'_{L_2}e^{-(t-s)D'^2_{L_2}}\,ds. \quad (2.14)$$

We have to estimate

$$\int_0^t e^{-sD^2}D\eta_\nu e^{-(t-s)D'^2_{L_2}}\,ds \qquad (2.15)$$

and

$$\int_0^t e^{-sD^2}\eta_\nu D'_{L_2}e^{-(t-s)D'^2_{L_2}}\,ds. \qquad (2.16)$$

Decompose $\int_0^t = \int_0^{\frac{t}{2}} + \int_{\frac{t}{2}}^t$ which yields

$$\int_0^{\frac{t}{2}} e^{-sD^2} D\eta_\nu e^{-(t-s)D'^2_{L_2}} \, ds, \qquad (I_{\nu,1})$$

$$\int_0^{\frac{t}{2}} e^{-sD^2} \eta_\nu D'_{L_2} e^{-(t-s)D'^2_{L_2}} \, ds, \qquad (I_{\nu,2})$$

$$\int_{\frac{t}{2}}^t e^{-sD^2} D\eta_\nu e^{-(t-s)D'^2_{L_2}} \, ds, \qquad (I_{\nu,3})$$

$$\int_{\frac{t}{2}}^t e^{-sD^2} \eta_\nu D'_{L_2} e^{-(t-s)D'^2_{L_2}} \, ds. \qquad (I_{\nu,4})$$

$(I_{\nu,1}) - (I_{\nu,4})$ look as $(I_1) - (I_4)$ as before. But in distinction to that, not all $\eta_\nu = \eta_\nu^{op}$ are operators of order zero. Only η_2 is a zero order operator, generated by an End$E$ valued 1-form η_2. η_1 and η_3 are first order operators. We start with $\nu = 2$, $\eta_2 \cdot |\eta_2|_{1,r} < \infty$ is a consequence of $E' \in \text{comp}^{1,r+1}_{L,diff}(E)$ and we are from an analytical point of view exactly in the situation as before. $(I_{2,1}) - (I_{2,4})$ can be estimated quite parallel to $(I_1) - (I_4)$ and we are done. There remains to estimate $(I_{\nu,j})$, $\nu \neq 2$, $j = 1, \ldots, 4$. We start with $\nu = 1$, $j = 3$ and write

$$e^{-sD^2} D\eta_1 e^{-(t-s)D'^2} = (De^{-\frac{s}{2}D^2})(e^{-\frac{s}{4}D^2} \cdot f)(f^{-1} e^{-\frac{s}{4}D^2} \eta_1)(e^{-(t-s)D'^2}).$$

$$(2.17)$$

$De^{-\frac{s}{2}D^2}$ and $e^{-(t-s)D'^2}$ are bounded in $[\frac{t}{2}, t]$ and we perform their estimate as in section 1. $e^{-\frac{s}{4}D^2} \cdot f$ is Hilbert–Schmidt if $f \in L_2$. There remains to show that for appropriate f

$$f^{-1} e^{-\frac{s}{4}D^2} \eta_1$$

is Hilbert–Schmidt. Recall $r+1 > n+3$, $n \geq 2$, which implies $\frac{r}{2} > \frac{n}{2} + 1$, $r - 1 - n \geq \frac{r}{2} - \frac{n}{2}$, $r - 1 \geq \frac{r}{2}$, $2 \geq i$. If we write in the sequel pointwise or Sobolev norms we should always write $|\Psi|_{g',h,m'}$, $|\Psi|_{H^\nu(E,D')}$, $|\Psi|_{g',h,\nabla',2,\frac{r}{2}}$, $|g - g'|_{g',m}$, $|g - g'|_{g',1,r}$ etc. or the same with respect to g, h, ∇, D, depending on the situation. But we often omit the reference to $g', h, \nabla', D, m, g, h \ldots$ in the notation. The justification for doing this in the Sobolev case is the symmetry of our uniform structure.
Now

$$(\eta_1 \Phi)(m) = ((g^{ik} - g'^{ik})\frac{\partial}{\partial x^k} \cdot \nabla_i \Phi)|_m, \qquad (2.18)$$

$|\eta_1 \Phi|_m = |\eta_1 \Phi|_{g,h,m}$

$$\leq C_1 \cdot |g - g'|_{g,m} \cdot \left(\sum_{k=1}^n \left|\frac{\partial}{\partial x^k}\right|^2_{g,m}\right)^{\frac{1}{2}} \cdot \left(\sum_{i=1}^n |\nabla_i \Phi|^2_{h,m}\right)^{\frac{1}{2}}.$$

To estimate $\sum_{k=1}^n \left|\frac{\partial}{\partial x^k}\right|^2_{g,m}$ more concretely we assume that $x^1, \ldots, x^n$ are normal coordinates with respect to g, i.e. we assume a (uniformly locally finite) cover of M by normal charts of fixed radius $\leq r_{inj}(M,g)$. Then $\left|\frac{\partial}{\partial x^k}\right|^2_{g,m} = g\left(\frac{\partial}{\partial x^k}, \frac{\partial}{\partial x^k}\right) = g_{kk}(m)$, and there is a constant $C_2 = C_2(R, r_{inj}(M,g))$ s. t. $\left(\sum_{i=1}^n |\nabla_i \Phi|^2_{h,m}\right)^{\frac{1}{2}} \leq C_2$. Using finally $|\nabla_X \Phi| \leq |X| \cdot |\nabla \Phi|$, we obtain

$$|\eta_1 \Phi|_m \leq C \cdot |g - g'|_g \cdot |\nabla \Phi|_{h,m}. \qquad (2.19)$$

(2.19) extends by the Leibniz rule to higher derivatives $|\nabla^k \eta_1 \Phi|_m$, where the polynomials on the right hand side are integrable by the module structure theorem (this is just the content of this theorem). (2.18), (2.19) also hold (with other constants) if we perform some of the replacements $g \longrightarrow g'$, $\nabla \longrightarrow \nabla'$: We remark that the expressions $D(g, h, \nabla^h, \cdot)$, $D(g', h, \nabla^h, \cdot)$ are invariantly defined, hence

$$[D(g, h, \nabla^h, \cdot) - D(g, h, \nabla^h, \cdot)](\Phi|_U) = ((g^{ik} - g'^{ik})\partial_k) \cdot \nabla_i(\Phi|_U). \qquad (2.20)$$

We have to estimate the kernel of

$$h_p(W(t,m,p), \eta_1^{op} \cdot) \qquad (2.21)$$

in $L_2((M,E), g, h)$ and to show that this represents the product of two Hilbert–Schmidt operators in $L_2 = L_2((M,E), g, h)$. We cannot immediately apply the procedure as before since η_1^{op} is not of zero order but we would be done if we could write (2.21) as

$$(\eta_{1,1}^{op}(p) W(t,m,p), \eta_{1,0}^{op} \cdot), \qquad (2.22)$$

$\eta_{1,1}^{op}$ of first order, $\eta_{1,0}^{op}$ of zeroth order. Then we would replace W by $\eta_{1,1}^{op}(p) W(t,m,p)$, apply $k \geq r+1 > n+3$, and obtain

$$\eta_{1,1}^{op} W(t,m,\cdot) \in H^{\frac{r}{2}}(E), \quad |W(t,m,\cdot)|_{H^{\frac{r}{2}}} \leq C(t) \qquad (2.23)$$

and would then literally proceed as before.

Let $\Phi \in C_c^\infty(U)$. Then

$$\int (W(t,m,p), \eta_1^{op}(p) \Phi(p))_p \, dvol_p(g) =$$

$$\int (((g^{ik} - g'^{ik}) \partial_k) \cdot \nabla_i W, \Phi)_p \, dvol_p(g) -$$

$$- \int (W, (\nabla_i(g^{ik} - g'^{ik}) \partial_k) \cdot \Phi) \, dvol_p(g) =$$

$$- \int (\nabla_i W, (g^{ik} - g'^{ik}) \partial_k \cdot \Phi)_p \, dvol_p(g) -$$

$$- \int (W, (\nabla_i((g^{ik} - g'^{ik}) \partial_k)) \cdot \Phi)_p \, dvol_p(g).$$

This can easily be globalized by introducing a u. l. f. cover by normal charts $\{U_\alpha\}_\alpha$ of fixed radius, an associated decomposition of unity $\{\varphi_\alpha\}_\alpha$ as follows:

$$\int (W, \eta_1^{op}(\sum_\alpha \varphi_\alpha \Phi)) = \sum_\alpha \int (W, \eta_1^{op}(\varphi_\alpha \Phi))$$

$$= \sum_\alpha \int (\nabla_{\alpha,i} W, ((g_\alpha^{ik} - g'^{ik}_\alpha) \partial_{\alpha,k} \cdot \varphi_\alpha \Phi) -$$

$$- \sum_\alpha \int (W, (\nabla_{\alpha,i}((g_\alpha^{ik} - g'^{ik}_\alpha) \partial_k)) \cdot \varphi_\alpha \Phi)$$

$$= -\int (\sum_\alpha \nabla_{\alpha,i} W, \varphi_\alpha((g_\alpha^{ik} - g'^{ik}_\alpha)\partial_{\alpha,k}) \cdot \Phi) - \quad (2.24)$$

$$-\int (W, \sum_\alpha \varphi_\alpha(\nabla_{\alpha,i}((g_\alpha^{ik} - g'^{ik}_\alpha)\partial_k)) \cdot \Phi) \quad (2.25)$$

Using (2.24), (2.25), we write

$$\begin{aligned} N(\Phi) &= |\langle \delta(m), e^{tD^2}\eta_1^{op}\Phi\rangle|_{L_2(M,E,dp)} \\ &= |(W(t,m,p), \eta_1^{op}\Phi)_p|_{L_2(M,E,dp)} \\ &= |(\eta_{1,1}^{op}(p)W(t,m,p), \eta_{1,0}^{op}\Phi)_p + \\ &+ (W(t,m,p,\eta_{1,0,0}^{op}\Phi))_p|_{L_2(M,E,dp)}. \end{aligned} \quad (2.26)$$

Now we use $|\nabla_X \chi| \le |X| \cdot |\nabla \chi|$, that the cover is u.l.f. and $|\nabla W| \le C_1 \cdot (|DW| + W)$ (since we have bounded geometry) and obtain

$$\begin{aligned} N(\Phi) &\le C \cdot (|(DW(t,m,p), \eta_{1,0}^{op}\Phi|_{L_2(M,E,dp)} \\ &+ |W(t,m,p,\eta_{1,0,0}^{op}\Phi|_{L_2(dp)} \\ &\equiv C \cdot (N_1(\Phi) + N_2(\Phi)). \end{aligned} \quad (2.27)$$

Hence we have to estimate

$$\sup_{\substack{\Phi \in C_c^\infty(E) \\ |\Phi|_{L_2}=1}} N_1(\Phi) = \sup_{\substack{\Phi \in C_c^\infty(E) \\ |\Phi|_{L_2}=1}} |\langle \delta(m), (De^{-tD^2})\eta_{1,0}^{op}\Phi\rangle|_{L_2(dp)}$$

$$(2.28)$$

and

$$\sup_{\substack{\Phi \in C_c^\infty(E) \\ |\Phi|_{L_2}=1}} N_2(\Phi) = \sup_{\substack{\Phi \in C_c^\infty(E) \\ |\Phi|_{L_2}=1}} |\langle \delta(m), (e^{-tD^2})\eta_{1,0,0}^{op}\Phi\rangle|_{L_2(dp)}.$$

$$(2.29)$$

According to $k > r + 1 > n + 3$,

$$D(W(t,m,\cdot), W(t,m,\cdot) \in H^{\frac{r}{2}}(E),$$
$$|(D(W(t,m,\cdot)|_{H^{\frac{r}{2}}}, |W(t,m,\cdot)|_{H^{\frac{r}{2}}} \le C_1(t) \quad (2.30)$$

and we can restrict in (2.28), (2.29) to

$$\sup_{\substack{\Phi \in C_c^\infty(E) \\ |\Phi|_{L_2}=1 \\ |\Phi|_{H^{\frac{r}{2}}} \le C_1(t)}} N_i(\Phi). \quad (2.31)$$

$\eta_{1,0}^{op}$, $\eta_{1,0,0}^{op}$ are of order zero and we estimate them by

$$C \cdot |g-g'|_{g,2,\frac{r}{2}} \leq C'|g-g'|_{g,1,r-1} \quad (2.32)$$

and $\quad D \cdot |\nabla(g-g')|_{g,2,\frac{r}{2}} \leq D'|\nabla(g-g')|_{g,1,r-1}$
$$\leq D''|g-g'|_{g,1,r} \quad (2.33)$$

respectively. As we have seen already, into the estimate (2.33) enters $|\nabla \eta|_{1,r-1}$, i. e. in our case $|\nabla^2(g-g')|_{r-1} \sim |g-g'|_{r+1}$. For this reason we assumed $E' \in \mathrm{comp}_{L,diff,F}^{1,r+1}(E)$. In the expression for $N_1(\Phi)$ there is now a slight deviation,

$$N_1(\Phi) = \frac{1}{\sqrt{4\pi t}} \frac{1}{2t} \left| \int_{-\infty}^{+\infty} s \cdot e^{-\frac{s^2}{4t}} e^{isD} \eta_{1,0}^{op} \Phi(m) \, ds \right|. \quad (2.34)$$

We estimate in (2.34) $s \cdot e^{-\frac{1}{18}\frac{s^2}{4t}}$ by a constant and write

$$e^{-\frac{17}{18}\frac{s^2}{4t}} \cdot vol(B_{2d+s}(m)) \leq C \cdot e^{-\frac{9}{10}\frac{s^2}{4t}}$$

and proceed now for $N_1(\Phi)$, $N_2(\Phi)$ literally as before. Hence (2.17) is of trace class, its trace norm in uniformly bounded on any t–intervall $[a_0, a_1]$, $a_0 > 0$. $(I_{1,3})$ is done. $(I_{1,4})$ is absolutely parallel to $(I_{1,3})$, even better, since the left hand factor D is missing. $|D'_{L_2} e^{-(t-s)D'^2_{L_2}}|_{op}$ now produces the factor $\frac{1}{\sqrt{t-s}}$ which is integrable over $[\frac{t}{2}, t]$. Write the integrand of $(I_{1,1})$ as

$$(De^{-sD^2})(\eta_1 e^{-\frac{(t-s)}{2}D'^2_{L_2}} f^{-1})(fe^{-\frac{(t-s)}{2}D'^2_{L_2}}). \quad (2.35)$$

We proceed with (2.35) as before. Here η_1 already stands at the right place, we must not perform partial integration. Into the estimate enters again the first derivative of W'. De^{-sD^2} generates the factor $\frac{1}{\sqrt{s}}$ which is intgrable on $[0, \frac{t}{2}]$. We write $(I_{1,2})$ as

$$\int_0^{\frac{t}{2}} e^{-sD^2}[(\eta_1 e^{-\frac{(t-s)}{4}D'^2_{L_2}} f^{-1})(fe^{-\frac{(t-s)}{4}D'^2_{L_2}})]e^{-\frac{(t-s)}{2}D'^2_{L_2}} D'^2_{L_2} \, ds$$

$$(2.36)$$

and proceed as before.

Consider finally the case $\nu = 3$, locally

$$\eta_3^{op}\Phi = g'^{ik}\frac{\partial}{\partial x^k}(\cdot - \cdot')\nabla_i'\Phi.$$

The first step in this procedure is quite similar as in the case $\nu = 1$ to shift the derivation to the left of W and to shift all zero order terms to the right.

Let X be a tangent vector field and Φ a section.

Lemma 2.7 $X(\cdot - \cdot')\nabla_i'\Phi = \nabla_i'(X(\cdot - \cdot')\Phi)$ + zero order terms.

Proof. $X(\cdot - \cdot')\nabla_i'\Phi = [X(\cdot - \cdot')\nabla_i'\Phi - \nabla_i'(X(\cdot - \cdot')\Phi)] + \nabla_i'(X(\cdot - \cdot')\Phi)$. We are done if $[\ldots]$ on the right hand side contains no derivatives of Φ. But an easy calculation yields

$$[X(\cdot - \cdot')\nabla_i'\Phi - \nabla_i'(X(\cdot - \cdot')\Phi)] =$$
$$= X \cdot (\nabla_i' - \nabla_i)\Phi - (\nabla_i' - \nabla_i)(X \cdot \Phi)$$
$$+ (\nabla_i' - \nabla_i)X \cdot' \Phi + (\nabla_i X)(\cdot' - \cdot)\Phi. \tag{2.37}$$

$\square$

Hence for $\Phi, \Psi \in C_c^\infty(U)$

$$\int h(\Psi, g'^{ik}\frac{\partial}{\partial x^k}(\cdot - \cdot')\nabla_i'\Phi)_p dp(g) =$$
$$= \int h(\Psi, \nabla_i'(g'^{ik}\frac{\partial}{\partial x^k}(\cdot - \cdot')\Phi)_p\, dp(g) + \tag{2.38}$$
$$+ \int h(\Psi, g'^{ik}\frac{\partial}{\partial x^k}(\nabla_i' - \nabla_i)\Phi - (\nabla_i' - \nabla_i)g'^{ik}\frac{\partial}{\partial x^k} \cdot \Phi)_p +$$
$$+ (\nabla_i' - \nabla_i)X \cdot' \Phi + \left(\nabla_i g'^{ik}\frac{\partial}{\partial x^k}\right)(\cdot' - \cdot)\Phi)_p\, dp(g). \tag{2.39}$$

(2.38) equals to

$$\int h(\nabla_i'^*\Psi, g'^{ik}\frac{\partial}{\partial x^k}(\cdot - \cdot')\Phi)_p\, dp(g). \tag{2.40}$$

If Φ is Sobolev and $\Psi = W$ then we obtain again by a u.l.f. cover by normal charts $\{U_\alpha\}_\alpha$ and an associated decomposition of unity $\{\varphi_\alpha\}_\alpha$

$$\int h(W, \eta_3^{op} \Phi)_p \, dp(g) =$$

$$= \int h(W, \sum_\alpha g'^{ik}_\alpha \frac{\partial}{\partial x^k}(\cdot - \cdot')\nabla'_{\alpha,i}(\varphi_\alpha \Phi))_p \, dp(g) =$$

$$= \int h(\nabla'^*_{\alpha,i} W, \sum_\alpha \varphi_\alpha g'^{ik}_\alpha \frac{\partial}{\partial x^k_\alpha}(\cdot - \cdot')\Phi)_p \, dp(g) + \quad (2.41)$$

$$+ \int h(W, \eta_{3,0}^{op} \Phi)_p \, dp(g), \quad (2.42)$$

where $\eta_{3,0}^{op} \Phi$ is the right component in $h(\cdot, \cdot)$ under the integral (2.41), multiplied with φ_α and summed up over α.

Now we proceed literally as before. Start with

$$(I_{3,3}) = \int_{\frac{t}{2}}^{t} e^{-sD^2} D \eta_3^{op} e^{-(t-s)D'^2_{L_2}} \, ds =$$

$$= \int_{\frac{t}{2}}^{t} (De^{-\frac{s}{2}D^2})[(e^{-\frac{s}{4}D^2}f)(f^{-1}e^{-\frac{s}{4}D^2}\eta_3^{op})]e^{-(t-s)D'^2_{L_2}} \, ds.$$

$$(2.43)$$

We want that for suitable $f \in L_2$, $f^{-1}e^{\frac{s}{4}D^2}\eta_3^{op}$ is Hilbert–Schmidt. For this we have to estimate $h(W(t,m,p), \eta_3^{op} \cdot)_p$ and to show it defines an integral operator with finite $L_2((M,E), dp)$–norm. We estimate

$$N(\Phi) = |\langle \delta(m), e^{-tD^2}\eta_3^{op}\Phi\rangle|_{L_2((M,E),dp)} = \quad (2.44)$$
$$= |h(W(t,m,p), \eta_3^{op}\Phi)_p|_{L_2((M,E),dp)}. \quad (2.45)$$

Using (2.41) and (2.42), we write

$$N(\Phi) = |h(W(t,m,p), \eta_3^{op}\Phi)_p|_{L_2(dp)} =$$
$$= |h(\eta_{3,1}^{op} W(t,m,p), \eta_{3,0}^{op}\Phi)_p$$
$$+ h(W(t,m,p), \eta_{3,0,0}^{op}\Phi)_p|_{L_2(dp)}. \quad (2.46)$$

215

Now we use $|\nabla'_X{}^* \chi| \leq C_1 |\nabla'_X \chi| \leq C_2 |X| \cdot |\nabla' \chi| \leq C_3 |X|(|\nabla \chi| + |\chi|)$, that the cover is u.l.f. and $|\nabla W| \leq C_4(|DW| + |W|)$ and obtain

$$\begin{aligned} N(\Phi) &\leq C(|hDW(t,m,p), \eta^{op}_{3,0}\Phi)_p|_{L_2(dp)} \\ &\quad + |h(W(t,m,p), \eta^{op}_{3,0,0}\Phi)_p|_{L_2(dp)} = \\ &= C(N_1(\Phi) + N_2(\Phi)). \end{aligned}$$

Here we again essentially use the bounded geometry. Hence we have to estimate

$$\sup_{\substack{\Phi \in C_c^\infty(E) \\ |\Phi|_{L_2} = 1}} N_1(\Phi) = \sup_{\substack{\Phi \in C_c^\infty(E) \\ |\Phi|_{L_2} = 1}} |\langle \delta(m), (De^{-tD^2})\eta^{op}_{3,0}\Phi\rangle|_{L_2(dp)} \quad (2.47)$$

and

$$\sup_{\substack{\Phi \in C_c^\infty(E) \\ |\Phi|_{L_2} = 1}} N_2(\Phi) = \sup_{\substack{\Phi \in C_c^\infty(E) \\ |\Phi|_{L_2} = 1}} |\langle \delta(m), e^{-tD^2}\eta^{op}_{3,0,0}\Phi\rangle|_{L_2(dp)}. \quad (2.48)$$

According to $k > r + 1 > n + 3$,

$$DW(t,m,\cdot), W(t,m,\cdot) \in H^{\frac{r}{2}}(E),$$
$$|DW(t,m,\cdot)|_{H^{\frac{r}{2}}}, |W(t,m,\cdot)|_{H^{\frac{r}{2}}} \leq C_1(t) \quad (2.49)$$

and we can restrict on (2.48), (2.49) to

$$\sup_{\substack{\Phi \in C_c^\infty(E) \\ |\Phi|_{L_2} = 1 \\ |\Phi|_{H^{\frac{r}{2}}} \leq C_1(t)}} \quad (2.50)$$

$\eta^{op}_{3,0}, \eta^{op}_{3,0,0}$ are of order zero and can be estimated by

$$C_0 |\cdot - \cdot'|_{2, \frac{r}{2}} \leq C_1 |\cdot - \cdot'|_{1, r-1} \quad (2.51)$$

and

$$D_0 \cdot (|\nabla - \nabla'|_{2, \frac{r}{2}} + |\cdot - \cdot'|_{2, \frac{r}{2}}) \leq D_1 \cdot (|\nabla - \nabla'|_{2, r-1} + |\cdot - \cdot'|_{1, r-1}) \quad (2.52)$$

respectively.

Now we proceed literally as for $(I_{1,3})$, replacing (2.34) by

$$N_1(\Phi) = \frac{1}{\sqrt{4\pi t}} \frac{1}{2t} \left| \int_{-\infty}^{+\infty} s e^{-\frac{s^2}{4t}} e^{isD} \eta_{3,0}^{op} \Phi(m) \, ds \right|. \qquad (2.53)$$

$(I_{3,3})$ is done, $(I_{3,4})$, $(I_{3,1})$, $(I_{3,2})$ are absolutely parallel to the case $\nu = 1$.

This finishes the proof of theorem 2.1. □

Theorem 2.8 *Suppose the hypotheses of 2.1. Then*

$$De^{-tD^2} - D'_{L_2} e^{-tD'^2_{L_2}}$$

is of trace class and the trace norm is uniformly bounded on compact t-intervalls $[a_0, a_1]$, $a_0 > 0$.

Proof. The proof is a simple combination of the proofs of 1.1 and 2.1. □

Now we additionally admit perturbation of the fibre metric h. Before the formulation of the theorem we must give some explanations. Consider the Hilbert spaces $L_2(g, h) = L_2((M, E), g, h)$, $L_2(g', h) = L_2((M, E), g', h)$, $L_2(g', h') = L_2((M, E), g', h') \equiv L'_2$ and the maps

$$i_{(g',h),(g',h')} : L_2(g', h) \longrightarrow L_2(g', h'), \quad i_{(g',h),(g',h')} \Phi = \Phi$$

$$U_{(g,h),(g',h)} : L_2(g, h) \longrightarrow L_2(g', h), \quad U_{(g,h),(g',h)} \Phi = \alpha^{\frac{1}{2}} \Phi$$

where $dp(g) = \alpha(p) dp(g')$. Then we set

$$\begin{aligned} D'_{L_2(g,h)} &= D'_{L_2} \\ &:= U^*_{(g,h),(g',h)} i^*_{(g',h),(g',h')} D' i_{(g',h),(g',h')} U_{(g,h),(g',h)} \\ &\equiv U^* i^* D' i U. \end{aligned} \qquad (2.54)$$

Here i^* is even locally defined (since g' is fixed) and $i^*_p = \text{dual}_h^{-1} \circ i' \circ \text{dual}_{h'}$, where $\text{dual}_h(\Phi(p)) = h_p(\cdot, \Phi(p))$. In a local basis field $\Phi_1, \ldots, \Phi_N$, $\Phi(p) = \xi^i(p) \Phi_i(p)$,

$$i^*_p \Phi(p) = h^{kl} h'_{ik} \xi^i \Phi_l(p). \qquad (2.55)$$

It follows from (2.55) that for $h' \in \text{comp}^{1,r+1}(h)$ i^*, i^{*-1} are bounded up to order k,

$$i^* - 1, i^{*-1} - 1 \in \Omega^{0,1,r+1}(Hom((E, h', \nabla^{h'}) \longrightarrow$$
$$\longrightarrow (M, g'), (E, h, \nabla^h) \longrightarrow (M, g'))) \qquad (2.56)$$

and

$$i^* - 1, i^{*-1} - 1 \in \Omega^{0,2,\frac{r+1}{2}}(Hom((E, h', \nabla^{h'}) \longrightarrow$$
$$\longrightarrow (M, g'), (E, h, \nabla^h) \longrightarrow (M, g'))). \qquad (2.57)$$

$D' \equiv \overline{D'}$ is self adjoint on $D_{\overline{D'}} = \overline{C_c^\infty(E)}^{|\cdot|_{D'}}$, where $|\Phi|^2_{D'} = |\Phi|^2_{L'_2} + |D'\Phi|^2_{L'_2}$. $i : L_2(g', h) \longrightarrow L_2(g', h') \equiv L'_2$ and $i^* : L_2(g', h') \longrightarrow L_2(g', h)$ are for $h' \in \text{comp}^{1,r+1}(h)$ quasi isometries with bounded derivatives, they map $C_c^\infty(E)$ 1–1 onto $C_c^\infty(E)$ and $i^*D'i$ is self adjoint on $\overline{C_c^\infty(E)}^{|\cdot|_{i^*D'i}} = D_{i^*D'i} \subset L_2((M, E), g', h) \equiv L_2(g', h)$. We obtain as a consequence that $e^{-t(i^*D'i)^2}$ is defined and selfadjoint in $L_2((M, E), g', h) = L_2(g', h)$, maps for $t > 0$ and $i, j \in \mathbb{Z}$ $H^i(E, i^*D'i)$ continuously into $H^j(E, i^*D'i)$ and has the heat kernel $W'_{g',h}(t, m, p) = \langle \delta(m), e^{-t(i^*D'i)^2}\delta(p)\rangle$, $W'(t, m, p)$ satisfies the same general estimates as $W(t, m, p)$. By exactly the same arguments we obtain that $e^{-tU^*(i^*D'i)^2U} = e^{-t(U^*i^*D'iU)^2} = U^*e^{-t(i^*D'i)^2}U$ is defined in $L_2 = L_2((M, E), g, h)$, self adjoint and has the heat kernel $W'_{L'_2}(t, m, p) = W'_{g,h}(t, m, p) = \alpha^{-\frac{1}{2}}(m)W'_{g',h}(t, m, p)\alpha(p)^{\frac{1}{2}}$. Here we assume $g' \in \text{comp}^{1,r+1}(g)$. Now we are able to formulate our main theorem.

Theorem 2.9 Let $E = ((E, h, \nabla = \nabla^h, \cdot) \longrightarrow (M^n, g))$ be a Clifford bundle with (I), $(B_k(M, g))$, $(B_k(E, \nabla))$, $k \geq r + 1 > n + 3$, $E' = ((E, h, \nabla' = \nabla^{h'}, \cdot') \longrightarrow (M^n, g)) \in \text{gen comp}^{1,r+1}_{L,diff}(E) \cap CL\mathcal{B}^{N,n}(I, B_k)$, $D = D(g, h, \nabla = \nabla^h, \cdot)$, $D' = D(g', h, \nabla' = \nabla^{h'}, \cdot')$ the associated generalized Dirac operators, $dp(g) = \alpha(p)dp(g')$, $U = \alpha^{\frac{1}{2}}$. Then for $t > 0$

$$e^{-tD^2} - U^*e^{-t(i^*D'i)^2}U \qquad (2.58)$$

is of trace class and the trace norm is uniformly bounded on compact t-intervalls $[a_0, a_1]$, $a_0 > 0$.

Proof. We are done if we could prove the assertions for
$$e^{-t(UD'U^*)^2} - e^{-t(i^*D'i)^2} = Ue^{-tD^2}U^* - e^{-t(i^*D'i)^2} \quad (2.59)$$
since $U^*(2.59)U = (2.58)$. To get a better explicit expression for (2.59), we apply again Duhamel's principle. This holds since Greens formula for UD^2U^* holds,
$$\int h_q(UD^2U^*\Phi, \Psi) - h(\Phi, UD^2U^*\Psi) \, dq(g') = 0.$$
We obtain
$$-\int_0^t \int_M h_q(\alpha^{\frac{1}{2}}(m)W(s,m,q)\alpha^{-\frac{1}{2}}(q),$$
$$\left(UD^2U^* + \frac{\partial}{\partial t}\right)W'_{g',h}(t-s,q,p))\, dq(g')\, ds$$
$$= -\int_0^t \int_M h_q(\alpha^{\frac{1}{2}}(m)W(s,m,q)\alpha^{-\frac{1}{2}}(q),$$
$$(UD^2U^* - (i^*D'i)^2)W'_{g',h}(t-s,q,p)\, dq(g'))\, ds$$
$$= \alpha^{\frac{1}{2}}(m)W(t,m,q)\alpha^{-\frac{1}{2}}(q) - W'_{g',h'}(t,m,p)$$
$$= W_{g',h}(t,m,p) - W'_{g',h}(t,m,p). \quad (2.60)$$

(2.60) expresses the operator equation
$$e^{-t(UDU^*)^2} - e^{-t(i^*D'i)^2}$$
$$= -\int_0^t e^{-s(U^*DU)^2}((UDU^*)^2 - (i^*D'i)^2)e^{-(t-s)(i^*D'i)^2}\, ds$$
$$= -\int_0^t e^{-s(UDU^*)^2}UDU^*(UDU^* - i^*D'i)e^{-(t-s)(i^*D'i)^2}\, ds$$
$$\quad (2.61)$$
$$-\int_0^t e^{-s(UDU^*)^2}(UDU^* - i^*D'i)(i^*D'i)e^{-(t-s)(i^*D'i)^2}\, ds.$$
$$\quad (2.62)$$

We write (2.62) as

$$-\int_0^t \alpha^{\frac{1}{2}}e^{-sD^2}D\alpha^{-\frac{1}{2}}(\alpha^{\frac{1}{2}}D\alpha^{-\frac{1}{2}} - i^*D'i)e^{-(t-s)(i^*D'i)^2}\,ds$$

$$= -\int_0^t \alpha^{\frac{1}{2}}e^{-sD^2}D\alpha^{-\frac{1}{2}}(D - i^*D'i - \frac{\operatorname{grad}\alpha\cdot}{2\alpha})e^{-(t-s)(i^*D'i)^2}\,ds$$

$$= -\int_0^t \alpha^{\frac{1}{2}}e^{-sD^2}D\alpha^{-\frac{1}{2}}i^*((i^{*-1}-1)D + (D-D')$$

$$-i^{*-1}\frac{\operatorname{grad}\alpha\cdot}{2\alpha})e^{-(t-s)(i^*D'i)^2}\,ds$$

$$= \int_0^t \alpha^{\frac{1}{2}}e^{-sD^2}D(\eta_0 + \eta_1 + \eta_2 + \eta_3 + \eta_4)e^{-(t-s)(i^*D'i)^2}\,ds,$$

$\eta_0 = \frac{\operatorname{grad}\alpha\cdot}{2\alpha^{\frac{3}{2}}}$, $\eta_i = -\alpha^{-\frac{1}{2}}i^*\eta_i(2)$, $i = 1, 2, 3$, $\eta_1(2) = (2.10)$, $\eta_2(2) = (2.11)$, $\eta_3(2) = (2.12)$, $\eta_4 = \alpha^{-\frac{1}{2}}i^{*-1}(i^*-1)D$. Here η_0 and η_2 are of zeroth order. η_1 and η_3 can be discussed as in (2.18)–(2.53). η_4 can be discussed analogous to η_1, η_3 as before, i.e. η_4 will be shifted via partial integration to the left (up to zero order terms) and $\alpha^{-\frac{1}{2}}i^*(i^*-1)$ thereafter again to the right. In the estimates one has to replace W by DW and nothing essentially changes as we exhibited in (2.35). We perform in (2.62) the same decomposition and have to estimate 20 integrals,

$$\int_0^{\frac{t}{2}} \alpha^{\frac{1}{2}}e^{-sD^2}D\eta_\nu e^{-(t-s)(i^*D'i)^2}\,ds, \qquad (I_{\nu,1})$$

$$\int_0^{\frac{t}{2}} \alpha^{\frac{1}{2}}e^{-sD^2}\eta_\nu(i^*D'i)e^{-(t-s)(i^*D'i)^2}\,ds, \qquad (I_{\nu,2})$$

$$\int_{\frac{t}{2}}^t \alpha^{\frac{1}{2}}e^{-sD^2}D\eta_\nu e^{-(t-s)(i^*D'i)^2}\,ds, \qquad (I_{\nu,3})$$

$$\int_{\frac{t}{2}}^{t} \alpha^{\frac{1}{2}} e^{-sD^2} \eta_\nu(i^*D'i) e^{-(t-s)(i^*D'i)^2} \, ds, \qquad (I_{\nu,4})$$

$\nu = 0, \ldots, 4$ and to show that these are products of Hilbert–Schmidt operators and have uniformly bounded trace norm on compact t–intervals. This has been completely modelled in the proof of 2.1. □

Finally we obtain

Theorem 2.10 *Assume the hypotheses of 2.9. Then for $t > 0$*

$$e^{tD^2}D - U^* e^{-t(i^*D'i)^2}(i^*D'i) U$$

is of trace class and its trace norm is uniformly bounded on compact t–intervalls $[a_0, a_1]$, $a_0 > 0$.

□

The operators $i^*D'^2 i$ and $(i^*D'i)^2$ are different in general. We should still compare $e^{-ti^*D'^2 i}$ and $e^{-t(i^*D'i)^2}$.

Theorem 2.11 *Assume the hypotheses of 2.9. Then for $t > 0$*

$$e^{-t(i^*D'^2 i)} - e^{-t(i^*D'i)^2}$$

is of trace class and the trace norm is uniformly bounded on compact t–intervalls $[a_0, a_1]$, $a_0 > 0$.

Proof. We obtain again immediately from Duhamel's principle

$$e^{-ti^*D'^2i} - e^{-t(i^*D'i)^2} =$$

$$= -\int_0^t e^{-s(i^*D'^2i)}(i^*D'^2i - (i^*D'i)^2)e^{-(t-s)(i^*D'i)^2}\, ds =$$

$$= -\int_0^t e^{-s(i^*D'^2i)} i^*D'(1 - ii^*)D'ie^{-(t-s)(i^*D'i)^2}\, ds =$$

$$= -\int_0^t e^{-s(i^*D'^2i)}(i^*D'i)i^{-1}(1 - ii^*)i^{*-1}(i^*D'i)e^{-(t-s)(i^*D'i)^2}\, ds.$$

(2.63)

In $[\frac{t}{2}, t]$ we shift $i^*D'i$ again to the left of the kernel $W'_{e^{-s(i^*D^2i)}}$ via partial integration and estimate

$$(i^*D'ie^{-\frac{s}{2}(i^*D'^2i)})[(e^{-\frac{s}{4}(i^*D'^2i)})f)$$
$$(f^{-1}e^{-\frac{s}{4}(i^*D'^2i)}i^{-1}(1-ii^*)i^{*-1})]$$
$$((i^*D'i)e^{-(t-s)(i^*D'i)^2})$$

as before. In $[0, \frac{t}{2}]$ we write the integrand of (2.63) as

$$(e^{-s(i^*D'^2i)}i^*D'i)[((i^*i)^{-1}e^{-\frac{t-s}{4}(i^*D'i)^2}f^{-1})(fe^{-\frac{t-s}{4}(i^*D'i)^2})]$$
$$(e^{-\frac{t-s}{2}(i^*D'i)^2}(i^*D'i))$$

and proceed as in the corresponding cases. □

Theorem 2.12 *Assume the hypotheses of 2.9. Then for $t > 0$*

$$e^{-tD^2} - e^{-t(U^*i^*D^2iU)} \equiv e^{-tD^2} - U^*e^{-t(i^*D^2i)}U$$

is of trace class and the trace norm is uniformly bounded on any t-intervall $[a_0, a_1]$, $a_0 > 0$.

Proof. This immediately follows from 2.9 and 2.11. □

3 Additional topological perturbations

Finally the last class of admitted perturbations are compact topological perturbations which will be studied now.
Let $E = ((E, h, \nabla^h) \longrightarrow (M^n, g)) \in \mathrm{CL}\mathcal{B}^{N,n}(I, B_k)$ be a Clifford bundle, $k \geq r+1 > n+3$, $E' = ((E, h', \nabla^{h'}) \longrightarrow (M'^n, g')) \in \mathrm{comp}^{1,r+1}_{L,diff,rel}(E) \cap \mathrm{CL}\mathcal{B}^{N,n}(I, B_k)$. Then there exist $K \subset M$, $K' \subset M'$ and a vector bundle isomorphism (not necessarily an isometry) $f = (f_E, f_M) \in \tilde{\mathcal{D}}^{1,r+2}(E|_{M \setminus K}, E'|_{M' \setminus K'})$ s. t.

$$g|_{M \setminus K} \text{ and } f_M^* g'|_{M \setminus K} \text{ are quasi isometric,} \tag{3.1}$$

$$h|_{E|_{M \setminus K}} \text{ and } f_E^* h'|_{E|_{M \setminus K}} \text{ are quasi isometric,} \tag{3.2}$$

$$|g|_{M \setminus K} - f_M^* g'|_{M \setminus K}|_{g,1,r+1} < \infty, \tag{3.3}$$

$$|h|_{E_{M \setminus K}} - f_E^* h'|_{E|_{M \setminus K}}|_{g,h,\nabla^h,1,r+1} < \infty, \tag{3.4}$$

$$|\nabla^h|_{E|_{M \setminus K}} - f_E^* \nabla^{h'}|_{E|_{M \setminus K}}|_{g,h,\nabla^h,1,r+1} < \infty, \tag{3.5}$$

$$|\cdot|_{M \setminus K} - f_E^* \cdot'|_{M \setminus K}|_{g,h,\nabla^h,1,r+1} < \infty. \tag{3.6}$$

(3.1) – (3.6) also hold if we replace f by f^{-1}, $M \setminus K$ by $M' \setminus K'$ and $g, h, \nabla^h, \cdot$ by $g', h', \nabla^{h'}, \cdot'$. If we consider the complete pull back $f_E^*(E'|_{M' \setminus K'})$, i.e. the pull back together with all Clifford datas, then we have on $M \setminus K$ two Clifford bundles, $E|_{M \setminus K}$, $f_E^*(E'|_{M' \setminus K'})$ which are as vector bundles isomorphic and we denote $f_E^*(E'|_{M' \setminus K'})$ again by E' on $M \setminus K$, i.e. $g'_{new} = (f_M|_{M \setminus K})^* g'_{old}$ etc.. (3.1) – (3.6) and the symmetry of our uniform structure $\mathfrak{U}^{1,r+1}_{L,diff,rel}$ imply

$$W^{1,i}(E|_{M \setminus K}) \cong W^{1,i}(E'|_{M \setminus K}), \quad 0 \leq i \leq r+1,$$

$$W^{2,j}(E|_{M \setminus K}) \cong W^{2,j}(E'|_{M \setminus K}), \quad 0 \leq j \leq \frac{r+1}{2},$$

$$H^j(E|_{M \setminus K}, D) \cong H^j(E'|_{M \setminus K}, D'), \quad 0 \leq j \leq \frac{r+1}{2}, \tag{3.7}$$

$$\Omega^{1,1,i}(End(E|_{M \setminus K})) \cong \Omega^{1,1,i}(End(E'|_{M \setminus K})), \quad 0 \leq i \leq r+1,$$

$$\Omega^{1,2,j}(End(E|_{M \setminus K})) \cong \Omega^{1,2,j}(End(E'|_{M \setminus K})), \quad 0 \leq j \leq \frac{r+1}{2}.$$

Here the Sobolev spaces are defined by restriction of corresponding Sobolev sections.

We now fix our set up for compact topological perturbations. Set $\mathcal{H} = L_2((K, E|_K), g, h) \oplus L_2((K', E'|_{K'}), g', h') \oplus L_2((M \setminus K, E), g, h)$ and consider the following maps

$$i_{L_2, K'} : L_2((K', E'|_{K'}), g', h') \longrightarrow \mathcal{H},$$
$$i_{L_2, K'}(\Phi) = \Phi,$$
$$i^{-1} : L_2((M' \setminus K', E'|_{M' \setminus K'}), g', h')$$
$$\longrightarrow L_2((M' \setminus K', E'|_{M' \setminus K'}), g', h),$$
$$i^{-1}\Phi = \Phi,$$
$$U^* : L_2((M' \setminus K', E'|_{M' \setminus K'}), g', h)$$
$$\longrightarrow L_2((M' \setminus K', E'|_{M' \setminus K'}), g, h),$$
$$U^*\Phi = \alpha^{-\frac{1}{2}},$$

where $dq(g) = \alpha(q) dq(g')$. We identify $M \setminus K$ and $M' \setminus K'$ as manifolds and $E'|_{M' \setminus K'}$ and $E|_{M \setminus K}$ as vector bundles. Then we have natural embeddings

$$i_{L_2, M} : L_2((M, E), g, h) \longrightarrow \mathcal{H},$$
$$i_{L_2, K'} \oplus U^* i^{-1} : L_2((M', E'), g', h') \longrightarrow \mathcal{H},$$
$$(i_{L_2, K'} \oplus U^* i^{-1})\Phi = i_{L_2, K'} \chi_{K'} \Phi + U^* i^{-1} \chi_{M' \setminus K'} \Phi.$$

The images of these two embeddings are closed subspaces of $\mathcal{H}$. Denote by P and P' the projection onto these closed subspaces. D is defined on $\mathcal{D}_D \subset \operatorname{im} P$. We extend it onto $(\operatorname{im} P)^\perp$ as zero operator. The definition of (the shifted) D' is a little more complicated. For the sake of simplicity of notation we write $U^* i^{-1} \equiv i_{L_2, K'} \oplus U^* i^{-1} = id \oplus U^* i^{-1}$, keeping in mind that $i_{L_2, K'}$ fixes χ_K, Φ and the scalar product. Moreover we set also $i U \chi_{K'} \Phi = U^* i^* \chi_{K'} \Phi = \chi_{K'} \Phi$. Let $\Phi \in \mathcal{D}_{D'}$, $\chi_{K'} \Phi + U^* i^{-1} \chi_{M' \setminus K'} \Phi$ its image in $\mathcal{H}$. Then $(U^* i^* D' i U)(\chi_{K'} \Phi + U^* i^{-1} \chi_{M' \setminus K'}, \Phi) = U^* i^* D' \Phi = \chi_{K'} D' \Phi + U^* i^* \chi_{M' \setminus K'} D' \Phi$. Now we set as $\mathcal{D}_{U^* i^* D' i U} \subset H$

$$\mathcal{D}_{U^* i^* D' i U} = \{\chi_{K'} \Phi + U^* i^{-1} \chi_{M' \setminus K'} \Phi | \Phi \in \mathcal{D}_{D'}\} \oplus (\operatorname{im} P')^\perp. \quad (3.8)$$

It follows very easy from the selfadjointness of D' on $\mathcal{D}_{D'}$ and (3.7) that $U^*i^*D'iU$ is self adjoint on $\mathcal{D}_{U^*i^*D'iU}$, if we additionally set $U^*i^*D'iU = 0$ on $(\text{im } P')^\perp$.

Remark 3.1 If g and h do not vary then we can spare the whole $i - U$–procedure, $i = U = \text{id}$. Nevertheless this case still includes interesting perturbations. Namely perturbations of $\nabla, \cdot$ and compact topological perturbations. $\square$

We set for the sake of simplicity $\tilde{D}' = U^*i^*D'iU$. The first main result of this section is the following

Theorem 3.2 Let $E = ((E, h, \nabla^h) \longrightarrow (M^n, g)) \in \text{CL}\mathcal{B}^{N,n}(I, B_k)$, $k \geq r+1 > n+3$, $E' \in \text{gen comp}_{L,diff,rel}^{1,r+1}(E) \cap \text{CL}\mathcal{B}^{N,n}(I, B_k)$. Then for $t > 0$

$$e^{-tD^2}P - e^{-t\tilde{D}'^2}P' \quad (3.9)$$

and

$$e^{-tD^2}D - e^{-t\tilde{D}'^2}\tilde{D}' \quad (3.10)$$

are of trace class and their trace norms are uniformly bounded on any t–intervall $[a_0, a_1]$, $a_0 > 0$.

For the proof we make the following construction. Let $V \subset M \setminus K$ be open, $\overline{M \setminus K} \setminus V$ compact, $\text{dist}(V, \overline{M \setminus K} \setminus (M \setminus K)) \geq 1$ and denote by $B \in L(\mathcal{H})$ the multiplication operator $B = \chi_V$. The proof of 3.2 consists of two steps. First we prove 3.2 for the restriction of (3.9), (3.10) to V, i.e. for $B(3.9)B$, thereafter for $(1-B)(3.9)B$, $B(3.9)(1-B)$ and the same for (3.10).

Theorem 3.3 *Assume the hypotheses of 3.2. Then*

$$B(e^{-tD^2}P - e^{-t\tilde{D}'^2}P')B, \qquad (3.11)$$

$$B(e^{-tD^2}D - e^{-t\tilde{D}'^2}\tilde{D}')B, \qquad (3.12)$$

$$B(e^{-tD^2}P - e^{-t\tilde{D}'^2}P')(1-B), \qquad (3.13)$$

$$(1-B)(e^{-tD^2}P - e^{-t\tilde{D}'^2}P')B, \qquad (3.14)$$

$$B(e^{-tD^2}D - e^{-t\tilde{D}'^2}\tilde{D}')(1-B), \qquad (3.15)$$

$$(1-B)(e^{-tD^2}D - e^{-t\tilde{D}'^2}\tilde{D}')B, \qquad (3.16)$$

$$(1-B)(e^{-tD^2}P - e^{-t\tilde{D}'^2}P')(1-B), \qquad (3.17)$$

$$(1-B)(e^{-tD^2}D - e^{-t\tilde{D}'^2}\tilde{D}')(1-B) \qquad (3.18)$$

are of trace class and their trace norms are uniformly bounded on any t-intervall $[a_0, a_1]$, $a_0 > 0$.

3.2 immediately follows from 3.3. We start with the assertion for (3.11). Introduce functions $\varphi, \psi, \gamma \in C^\infty(M, [0,1])$ with the following properties.
1) supp $\varphi \subset M \setminus K$, $(1-\varphi) \in C_c^\infty(M \setminus K)$, $\varphi|_V = 1$.
2) ψ with the same properties as φ and additionally $\psi = 1$ on supp φ, i.e. supp $(1-\psi) \cap$ supp $\varphi = 0$.
3) $\gamma \in C_c^\infty(M)$, $\gamma = 1$ on supp $(1-\varphi)$, $\gamma|_V = 0$.
Define now an approximate heat kernel $E(t, m, p)$ on M by

$$E(t, m, p) := \gamma(m)W(t, m, p)(1 - \varphi(p)) + \psi(m)\tilde{W}'(t, m, p)\varphi(p).$$

Applying Duhamel's principle yields

$$-\int_\alpha^\beta \int_M h_q(W(s, m, q), \left(\frac{\partial}{\partial t} + D^2\right) E(t-s, q, p))\chi_\nu(p) \, dq(g) \, ds$$

$$= \int_M [h_q(W(\beta, m, q), E(t-\beta, q, p)) - h_q(W(\alpha, m, q),$$

$$E(t-\alpha, q, p))]\chi_\nu(p) \, dq(g). \qquad (3.19)$$

Performing $\alpha \longrightarrow 0^+$, $\beta \longrightarrow t^-$ in (3.19), we obtain

$$-\int_\alpha^\beta \int_M h_q(W(s,m,q), \left(D^2 + \frac{\partial}{\partial t}\right) E(t-s,q,p))\chi_\nu(p) \, dq(g) \, ds$$

$$= \lim_{\beta \to t^-} \int_M [h_q(W(\beta,m,q), E(t-\beta,q,p))\chi_\nu(p) \, dq(g)$$

$$-E(t,m,p)\chi_\nu(p). \tag{3.20}$$

Now we use

$$\chi_V(p)(1 - \varphi(p)) = 0 \tag{3.21}$$

and obtain

$$\lim_{\beta \to t^-} \int_M [h_q(W(\beta,m,q), E(t-\beta,q,p))\chi_\nu(p) \, dq(g)$$

$$= \lim_{\beta \to t^-} \int_M [h_q(W(\beta,m,q), \psi(q)\tilde{W}'(t-\beta,q,p))\varphi(p)\chi_\nu(p) \, dq(g)$$

$$= W(t,m,p)$$

since $\tilde{W}'(\tau,q,p)$ is the heat kernel of $e^{-\tau\tilde{D}'^2}$. This yields

$$-\int_0^t \int_M h_q(W(s,m,q), \left(D^2 + \frac{\partial}{\partial t}\right) E(t-s,q,p))\chi_V(p) \, dq(g) \, ds$$

$$= -\int_0^t \int_M h_q(W(s,m,q), (D^2\psi(q) - \psi(q)\tilde{D}'^2)$$

$$\tilde{W}'(t-s,q,p))\chi_V(p) \, dq(g) \, ds$$

$$= [W(t,m,p) - \tilde{W}'(t,m,p)] \cdot \chi_V(p). \tag{3.22}$$

(3.22) expresses the operator equation

$$(e^{-tD^2}P - e^{-t\tilde{D}'^2}P')B = -\int_0^t e^{-sD^2}(D^2\psi - \psi\tilde{D}'^2)e^{-(t-s)\tilde{D}'^2}B \, ds$$

$$\tag{3.23}$$

in $\mathcal{H}$ at kernel level.

We rewrite (3.23) as in the foregoing cases.

$$\begin{aligned}
(3.23) \quad &= -\int_0^t e^{-sD^2}(D(D-\tilde{D}')\psi + (D-\tilde{D}')\tilde{D}'\psi \\
&\quad + \tilde{D}'^2\psi - \psi\tilde{D}'^2)e^{-(t-s)\tilde{D}'^2}\,ds \\
&= -\Bigg[\int_0^{\frac{t}{2}} e^{-sD^2}(D(D-\tilde{D}')\psi e^{-(t-s)\tilde{D}'^2}\,ds \quad (3.24)\\
&\quad + \int_0^{\frac{t}{2}} e^{-sD^2}(D-\tilde{D}')\tilde{D}'\psi e^{-(t-s)\tilde{D}'^2}\,ds \quad (3.25)\\
&\quad + \int_0^{\frac{t}{2}} e^{-sD^2}(\tilde{D}'^2\psi - \psi\tilde{D}'^2)\,ds \quad (3.26)\\
&\quad + \int_{\frac{t}{2}}^t e^{-sD^2}D(D-\tilde{D}')\psi e^{-(t-s)\tilde{D}'^2}\,ds \quad (3.27)\\
&\quad + \int_{\frac{t}{2}}^t e^{-sD^2}(D-\tilde{D}')\tilde{D}'\psi e^{-(t-s)\tilde{D}'^2}\,ds \quad (3.28)\\
&\quad + \int_{\frac{t}{2}}^t e^{-sD^2}(\tilde{D}'^2\psi - \psi\tilde{D}'^2)e^{-(t-s)\tilde{D}'^2}\,ds\Bigg]. \quad (3.29)
\end{aligned}$$

Write the integrand of (3.27) as

$$(e^{-\frac{s}{2}D^2}D)[(e^{-\frac{s}{4}D^2}f)(f^{-1}e^{-\frac{s}{4}D^2}(D-\tilde{D}')\psi)]e^{-(t-s)\tilde{D}'^2},$$

$|e^{-\frac{s}{2}D^2}D|_{op} \leq \frac{C}{\sqrt{s}}$, $|e^{-(t-s)\tilde{D}'^2}|_{op} \leq C'$ and $[\ldots]$ is the product of two Hilbert–Schmidt operators if f can be chosen $\in L_2$ and such that $f^{-1}e^{-\frac{s}{4}D^2}(D-\tilde{D}')\psi$ is Hilbert–Schmidt. We know from the preceding considerations, sufficient for this is that $(D-\tilde{D}')\psi$ has Sobolev coefficients of order $r+1$ (and $p=1$).

$$(D - \tilde{D}')\psi = (D - \alpha^{-\frac{1}{2}}i^*D'i\alpha^{\frac{1}{2}})\psi$$
$$= \left(D - i^*\frac{\operatorname{grad}'\alpha}{2\alpha}\cdot' - i^*D'\right)\psi$$
$$= i^*\left((i^{*-1} - 1)D + (D - D') - \frac{\operatorname{grad}'\alpha\cdot'}{2\alpha}\right)\psi$$
$$= i^*\bigg[i^{*-1}(\operatorname{grad}\psi\cdot + \psi D) + \operatorname{grad}\Psi(\cdot - \cdot')$$
$$+ (\operatorname{grad}\psi - \operatorname{grad}'\psi)\cdot' + \psi(D - D') - \frac{\operatorname{grad}'\alpha\cdot'}{2\alpha}\psi\bigg].$$

i^* is bounded up to order k, $i^{*-1} - 1$ is $(r+1)$-Sobolev, $\operatorname{grad}\psi$, $\operatorname{grad}'\psi$ have compact support, $0 \leq \psi \leq 1$, $\frac{\operatorname{grad}\alpha\cdot'}{2\alpha}$ is $(r+1)$-Sobolev and $\psi(D-D')$ is completely discussed in (2.9) – (2.53). Hence (3.27) is completely done.

Write the integrand of (3.28) as

$$[(e^{\frac{s}{2}D^2}f)(f^{-1}e^{\frac{s}{2}D^2}(D-D'))](D'e^{-(t-s)\tilde{D}'^2}).$$

[...] is the product of two Hilbert–Schmidt operators with bounded trace norm on t-intervalls $[a_0, a_1]$, $a_0 > 0$. An easy calculation yields

$$\tilde{D}'\psi = \alpha^{-\frac{1}{2}}i^*D'i\alpha^{\frac{1}{2}}\psi = i^*\operatorname{grad}\psi\cdot' + \psi\tilde{D}',$$

hence

$$|\tilde{D}'\psi e^{-(t-s)\tilde{D}'^2}|_{op} = |(i^*\operatorname{grad}'\psi\cdot' + \psi\tilde{D}')e^{-(t-s)\tilde{D}'^2}|_{op}$$
$$\leq C + \frac{C'}{\sqrt{t-s}},$$

(3.28) is done.

Rewrite finally the integrands of (3.24), (3.25) as

$$(e^{-sD^2}D)[((D-\tilde{D}')\psi e^{-\frac{t-s}{2}\tilde{D}'^2}f^{-1})(fe^{-\frac{t-s}{2}\tilde{D}'^2})]$$
$$= e^{-sD^2}Di^*[(((i^{*-1}-1)(\text{ grad }\psi\cdot+\psi D)+$$
$$+\text{ grad }\psi(\cdot-\cdot')+(\text{ grad }\psi-\text{ grad }'\psi)\cdot'+$$
$$+\psi(D-D')-\frac{\text{grad }\alpha}{2\alpha}\cdot'\psi)e^{-\frac{t-s}{2}\tilde{D}'^2}f^{-1})(fe^{-\frac{t-s}{2}\tilde{D}'^2})]$$

and

$$e^{-sD^2}i^*[((D-\tilde{D}')\psi e^{-\frac{t-s}{4}\tilde{D}'^2}f^{-1})(fe^{-\frac{t-s}{4}\tilde{D}'^2})](\tilde{D}'e^{-\frac{t-s}{2}\tilde{D}'^2})$$
$$= e^{-sD^2}i^*[(((i^{*-1}-1)(\text{ grad }\psi\cdot+\psi D)+\text{ grad }\psi(\cdot-\cdot')$$
$$+(\text{ grad }\psi-\text{ grad }'\psi)\cdot'+\psi(D-D')-\frac{\text{grad }\alpha}{2\alpha}\cdot'\psi)$$
$$e^{-\frac{t-s}{4}\tilde{D}'^2}(f^{-1})(fe^{-\frac{t-s}{4}\tilde{D}'^2})](\tilde{D}'e^{-\frac{t-s}{2}\tilde{D}'^2}),$$

respectively, and (3.24), (3.25) are done. The remaining integrals are (3.26) and (3.29). We have to find an appropriate expression for $\tilde{D}'^2\psi - \psi\tilde{D}'^2$.

$$\begin{aligned}\tilde{D}'^2 &= (\alpha^{-\frac{1}{2}}i^*D'i\alpha^{\frac{1}{2}})(\alpha^{-\frac{1}{2}}i^*D'i\alpha^{\frac{1}{2}}) \quad (3.30)\\ &= \alpha^{-\frac{1}{2}}i^*D'i^*\left(\frac{\text{grad }'\alpha}{2\alpha^{\frac{1}{2}}}\cdot'+\alpha^{\frac{1}{2}}D'\right)\\ &= i^*\left(D'\alpha^{-\frac{1}{2}}+\frac{\text{grad }'\alpha}{2\alpha^{\frac{3}{2}}}\cdot'\right)i^*\left(\frac{\text{grad }'\alpha}{2\alpha^{\frac{1}{2}}}\cdot'+\alpha^{\frac{1}{2}}D'\right)\\ &= i^*D'i^*D'+i^*D'i^*\frac{\text{grad }'\alpha}{2\alpha}\cdot'\\ &+i^*\frac{\text{grad }'\alpha}{2\alpha}\cdot'i^*\frac{\text{grad }'\alpha}{2\alpha}\cdot'++i^*\frac{\text{grad }'\alpha}{2\alpha}\cdot'i^*D'.\end{aligned}$$
(3.31)

Hence

$$\tilde{D}'^2\psi - \psi\tilde{D}'^2 = i^*D'i^*D'\psi - \psi i^*D'i^*D'$$
$$+i^*D'i^*\frac{\text{grad }'\alpha}{2\alpha}\cdot'\psi - \psi i^*D'i^*\frac{\text{grad }'\alpha}{2\alpha}\cdot' \qquad (3.32)$$
$$+i^*\frac{\text{grad }'\alpha}{2\alpha}\cdot'i^*\frac{\text{grad }'\alpha}{2\alpha}\cdot'\psi - \psi i^*\frac{\text{grad }'\alpha}{2\alpha}\cdot'i^*\frac{\text{grad }'\alpha}{2\alpha}\cdot'$$
$$+i^*\frac{\text{grad }'\alpha}{2\alpha}\cdot'i^*D'\psi - \psi i^*\frac{\text{grad }'\alpha}{2\alpha}\cdot'i^*D'$$
$$= i^*D'i^*\text{ grad }'\psi\cdot' + i^*\text{ grad }'\psi\cdot'i^*D' + \psi i^*D'i^*D'$$
$$-\psi i^*D'i^*D' + i^*(\text{ grad }'\psi\cdot' + \psi D')i^*\frac{\text{grad }'\alpha}{2\alpha}\cdot' \qquad (3.33)$$
$$-\psi i^*D'i^*\frac{\text{grad }'\alpha}{2\alpha}\cdot' + i^*\frac{\text{grad }'\alpha}{2\alpha}\cdot'i^*(\text{ grad }'\psi\cdot' + \psi D')$$
$$-\psi i^*\frac{\text{grad }'\alpha}{2\alpha}\cdot'i^*D'$$
$$= i^*D'i^*\text{ grad }'\psi\cdot' + i^*\text{ grad }'\psi\cdot'i^*D' \qquad (3.34)$$
$$+i^*\text{ grad }'\psi\cdot'i^*\frac{\text{grad }'\alpha}{2\alpha}\cdot' \qquad (3.35)$$
$$+i^*\frac{\text{grad }'\alpha}{2\alpha}\cdot'i^*\text{ grad }'\psi\cdot'. \qquad (3.36)$$

The terms in (3.34) are first order operators but grad $'\psi$ has compact support and we are done. The terms in (3.35), (3.36) are zero order operators and we are also done since grad $'\psi$ has compact support.

Hence $(e^{-tD^2}P - e^{-t\tilde{D}'^2}P')B$, $B(e^{-tD^2}P - e^{-t\tilde{D}'^2}P')B$ are of trace class and the trace norm in uniformly bounded on any compact t-interval $[a_0, a_1]$, $a_0 > 0$. The assertions for (3.11) are done.
Next we study the operator

$$(e^{\frac{t}{2}D^2}P - e^{-\frac{t}{2}\tilde{D}'^2}P')(1-B). \qquad (3.37)$$

Denote by M_ε the multiplication operator with $\exp(-\varepsilon\text{dist}(m, K)^2)$. We state that for ε small enough

$$M_\varepsilon e^{-tD^2}B, \quad M_\varepsilon e^{-tD'^2}B \qquad (3.38)$$

and
$$M_\varepsilon^{-1} e^{-tD^2} \chi_G, \quad M_\varepsilon^{-1} e^{-tD'^2} \chi_G \qquad (3.39)$$
are Hilbert–Schmidt for every compact $G \subset M$ or $G' \subset M'$. Write

$$(e^{\frac{t}{2}D^2} P - e^{-\frac{t}{2}\tilde{D}'^2} P')(1-B) =$$
$$= [e^{-\frac{t}{2}D^2} P M_\varepsilon] \cdot [M_\varepsilon^{-1}(e^{-\frac{t}{2}D^2} P - e^{-\frac{t}{2}\tilde{D}'^2} P')(1-B)] \quad (3.40)$$
$$+ [e^{-\frac{t}{2}D^2} P - e^{-\frac{t}{2}\tilde{D}'^2} P') M_\varepsilon] \cdot [M_\varepsilon^{-1} e^{-\frac{t}{2}\tilde{D}'^2} P'(1-B)]. \quad (3.41)$$

According to (1.1) – (1.5) and (3.38), each of the factors $[\cdots]$ in (3.40), (3.41) is Hilbert–Schmidt and we obtain that (3.34) is of trace class and has uniformly bounded trace norm in any t–interval $[a_0, a_1]$, $a_0 > 0$. The same holds for

$$B(e^{\frac{t}{2}D^2} P - e^{-\frac{t}{2}\tilde{D}'^2} P')(1-B) \qquad (3.42)$$
$$(1-B)(e^{\frac{t}{2}D^2} P - e^{-\frac{t}{2}\tilde{D}'^2} P') B \qquad (3.43)$$
$$(1-B)(e^{\frac{t}{2}D^2} P - e^{-\frac{t}{2}\tilde{D}'^2} P')(1-B) \qquad (3.44)$$

by multiplication of (3.37) from the left by B etc., i.e. the assertions for (3.13), (3.14), (3.17) are done. Write now

$$(e^{-\frac{t}{2}D^2} D - e^{-\frac{t}{2}\tilde{D}'^2} D') B =$$
$$(e^{-\frac{t}{2}D^2} D)(e^{\frac{t}{2}D^2} P - e^{-\frac{t}{2}\tilde{D}'^2} P') B + \qquad (3.45)$$
$$(e^{-\frac{t}{2}D^2} D - e^{-\frac{t}{2}\tilde{D}'^2} D')(e^{-\frac{t}{2}D'^2} P) B. \qquad (3.46)$$

(3.45) is done already by (3.23) and $|e^{-\frac{t}{2}D^2} D|_{op} \leq \frac{C}{\sqrt{t}}$. Decompose (3.46) as the sum of

$$e^{-\frac{t}{2}D^2} P(D - \tilde{D}') \cdot (e^{-\frac{t}{2}\tilde{D}'^2} \tilde{D}') = [e^{-\frac{t}{2}D^2} P(-\eta)] \cdot (e^{-\frac{t}{2}\tilde{D}'^2} \tilde{D}') B \qquad (3.47)$$

and
$$(e^{-\frac{t}{2}D^2} P - e^{-\frac{t}{2}\tilde{D}'^2} P')(e^{-\frac{t}{2}\tilde{D}'^2} \tilde{D}') B \qquad (3.48)$$

$[\ldots]$ in (3.47) is done. Rewrite $e^{-\frac{t}{2}D^2} P - e^{-\frac{t}{2}\tilde{D}'^2} P'$ as

$$(e^{-\frac{t}{2}D^2} P - e^{-\frac{t}{2}\tilde{D}'^2} P) B + \qquad (3.49)$$
$$+ (e^{\frac{t}{2}D^2} P - e^{-\frac{t}{2}\tilde{D}'^2} P')(1-B). \qquad (3.50)$$

(3.49), (3.50) are done already, hence (3.48) and hence $(e^{\frac{t}{2}D^2}P - e^{-\frac{t}{2}\tilde{D}'^2}P')B$, (3.12), (3.15), (3.16), (3.18). This finishes the proof of 3.3. □

The proof of theorem 3.2 now follows from 3.3 by adding up the four terms containing $e^{\frac{t}{2}D^2}P - e^{-\frac{t}{2}\tilde{D}'^2}P'$ or $e^{\frac{t}{2}D^2}D - e^{-\frac{t}{2}\tilde{D}'^2}\tilde{D}'$, respectively. □

Remark 3.4 We could perform the proof of 3.2, 3.3 also along the lines of (2.59) – (2.62), performing first a unitary transformation, proving the trace class porperty and performing the back transformation, as we indicate in (2.59). This procedure is completely equivalent to the proof of 3.2, 3.3 presented above. □

The operators $U^*i^*D'^2iU$ and $(U^*i^*D'^2iU)^2$ are distinct in general and we have still to compare $e^{-t(U^*i^*D'^2iU)}P'$ and $e^{-t(U^*i^*D'^2iU)^2}P'$. According to our remark above, it is sufficient to prove the trace class property of

$$e^{-t(i^*D'^2i)}P - e^{-t(i^*D'i)^2}P' \qquad (3.51)$$

in

$$\mathcal{H}' = L_2((K,E),g,h) \oplus L_2((K',E'),g',h') \oplus L_2((M\setminus K,E),g',h).$$

Here we have an embedding

$$i_{L_2,K'} \oplus i^{-1} : L_2((M',E'),g',h') \longrightarrow \mathcal{H}'$$
$$(i_{L_2,K'} \oplus i^{-1})\Phi = i_{L_2,K'}\chi_{K'}\Phi + i^{-1}\chi_{M'\setminus K'}\Phi, \qquad (3.52)$$

where

$$i^{-1} : L_2((M' \setminus K', E'|_{M'\setminus K'}), g', h')$$
$$\longrightarrow L_2((M' \setminus K', E'|_{M'\setminus K'}), g'h),$$
$$i^{-1}\Phi = \Phi,$$

and

$$i^*D'i(\chi_K,\Phi + i^{-1}\chi_{M'\setminus K'}\Phi) := i^*D'\Phi = \chi_{K'}D'\Phi + i^*\chi_{M'\setminus K'}D'\Phi,$$

$i^*D'^2i$ similar, all with the canonical domains of definition analogous to (3.8). P' is here the projection onto im $(i_{L_2,K'} \oplus i^{-1})$. We define $i^*D'^2i$, $(i^*D'i)^2$ to be zero on im $P'^\perp$.

Remark 3.5 Quite similar we could embed $L_2((M,E),g,h)$ into $\mathcal{H}'$, define P, UDU^* and the assertion 3.2 would be equivalent to the assertion for

$$e^{-t(UDU^*)^2}P - e^{-t(i^*D'i)^2}P'. \qquad (3.53)$$

Applying the (extended) U^* from the right, U from the left, yields just the expression (3.9).

□

Theorem 3.6 *Assume the hypotheses of 3.2. Then*

$$e^{-t(i^*D'^2i)}P' - e^{-t(i^*D'i)^2}P' \qquad (3.54)$$

is of trace class and its trace norm is uniformly bounded on compact t–intervals $[a_0, a_1]$, $a_0 > 0$.

Proof. We prove this by establishing the assertion for the four cases arising from multiplication by B, $1-B$. Start with (3.54). B. Duhamel's principle again yields

$$(e^{-t(i^*D'^2i)}P' - e^{-t(i^*D'i)^2}P')B$$
$$= -\int_0^t e^{-s(i^*D'^2i)}((i^*D'^2i) - (i^*D'i)^2)e^{-(t-s)(i^*D'i)^2} \, ds. \qquad (3.55)$$

An easy calculation yields

$$(i^*D'^2i)\psi - \psi(i^*D'i)^2 = i^*D'^2\psi - \psi i^*D'i^*D'$$
$$= i^*D' \operatorname{grad}'\psi \cdot' + i^* \operatorname{grad}'\psi \cdot' D + \psi i^*D'^2$$
$$-(\psi i^*D'^2 + \psi i^*D'(i^* - 1)D')$$
$$= i^*D' \operatorname{grad}'\psi \cdot' + i^* \operatorname{grad}'\psi \cdot' D \qquad (3.56)$$
$$-\psi i^*D'(i^* - 1)D'. \qquad (3.57)$$

The first order operators in (3.56) contain the compact support factor grad $'\psi$ and we are done. Here i^*D' (coming from the first term or from grad $'\psi \cdot' D' =$ grad $'\psi \cdot' i^{*-1} i^* D$) will be connected with $e^{-s(i^*D'^2 i)}$ or $e^{-(t-s)(i^*D'i)^2}$, depending on the interval $[\frac{t}{2}, t]$ or $[0, \frac{t}{2}]$. The (D')'s of the second order operator (3.57) can be distributed analogous to the proof of 3.9. The remaining main point is $0 \leq \psi \leq 1$ and $i^* - 1$ Sobolev of order $r+1$, i.e. $i^* - 1 \in \Omega^{0,1,r+1}(Hom((E'|_{M'\setminus K'}, g', h'), (E|_{M\setminus K}, g', h)))$.

The assertion for (3.54)·B is done. Quite analogously (and parallel to the proofs of (3.11), (3.13), (3.14), (3.17)) one discusses the other 3 cases. □

We obtain as a corollary from 3.2 and 3.6

Theorem 3.7 *Assume the hypothesis of 3.2. Then for $t > 0$*

$$e^{-tD^2}P - e^{-t(U^*i^*D'^2 iU)}P' \qquad (3.58)$$

is of trace class in $\mathcal{H}'$ and the trace norm is uniformly bounded on compact t-intervals $[a_0, a_1]$, $a_0 > 0$. □

The simplest standard example for a Clifford bundle is $E = (\Lambda^*T^*M \otimes \mathbb{C}, g_{\Lambda^*}, \nabla^{g_{\Lambda^*}}) \longrightarrow (M^n, g)$ with Clifford multiplication

$$X \otimes \omega \in T_m M \otimes \Lambda^*T^*M \otimes \mathbb{C} \longrightarrow X \cdot \omega = \omega_X \wedge \omega - i_X \omega,$$

where $\omega_X := g(, X)$. In this case E as a vector bundle remains fixed but the Clifford module structure varies smoothly with $g, g' \in \text{comp}(g)$. It is well known that in this case $D = d + d^*$, $D^2 = (d + d^*)^2$ Laplace operator Δ.
Theorem 2.12 then yields

Theorem 3.8 *Assume (M^n, g) with (I), (B_k), $k \geq r + 1 > n + 3$, $g' \in \mathcal{M}(I, B_k)$, $g' \in \text{comp}^{1,r+1}(g) \subset \mathcal{M}(I, B_k)$. Denote by $\Delta'_{L_2(g)} = U^*i^*\Delta(g')iU$ the transformation of $\Delta' = \Delta(g')$ from $L_2(g', g') \equiv L_2((M, \Lambda^*T^*M \otimes \mathbb{C}), g', g'_{\Lambda^*})$ to $L_2(g) \equiv L_2(g, g) = L_2((M, \Lambda^*T^*M \otimes \mathbb{C}, g, g_{\Lambda^*})$, where $i : L_2(g, g') =$*

$L_2((M, \Lambda^*T^*M \otimes \mathbb{C}, g, g_{\Lambda^*}) \longrightarrow L_2(g', g')$ and $U : L_2(g, g) \longrightarrow L_2(g, g')$, $U = \alpha^{\frac{1}{2}}$, $dq(g) = \alpha(q) \cdot dq(g')$, are the canonical maps, i^*, U^* their adjoints. Then for $t > 0$

$$e^{-t\Delta(g)} - e^{-t\Delta'_{L_2}(g)}$$

is of trace class and the trace norm is uniformly bounded on compact t-intervalls $[a_0, a_1]$, $a_0 > 0$. □

Applying 3.7 to the case $E = \Lambda^*T^*M \otimes \mathbb{C}$, $D^2 = \Delta$, we obtain

Theorem 3.9 Let $(\Lambda^*T^*M \otimes \mathbb{C}, g_{\Lambda^*} \in \mathrm{CL}\mathcal{B}^{2^n}(I, B_k)$, $k \geq r + 2 > n + 3$, $(\Lambda^*T^*M' \otimes \mathbb{C}, g'_{\Lambda^*}) \in \mathrm{gen\,comp}^{1,r+1}_{L,diff,rel}(\Lambda^*T^*M \otimes \mathbb{C}, g_{\Lambda^*}) \cap \mathrm{CL}\mathcal{B}^{2^n,n}(I, B_k)$, $\Delta = \Delta(g)$, $\Delta' = \Delta(g')$ the graded Laplace operator. Then for $t > 0$

$$e^{-t\Delta}P - e^{-t(U^*i^*\Delta'iU}P' \tag{3.59}$$

is of trace class in $\mathcal{H}'$ and the trace norm is uniformly bounded on compact t-intervals $[a_0, a_1]$, $a_0 > 0$. □

Roughly or more concretely speaking, as one prefers, the means the following. Given an open manifold (M^n, g) satisfying (I) and (B_k), $k \geq r + 2 > n + 3$. Cut out a compact submanifold K and glue the compact submanifold K' along $\partial M = \partial K$, getting thus M', endow M' with a metric g' satisfying (I) and (B_k) and

$$|g - g'|_{M \setminus K, g, 1, r+1} < \infty.$$

Then for $t > 0$

$$e^{-t\Delta}P - e^{-tU^*i^*\Delta'iU}P'$$

has the asserted properties.

We can apply a slight modification of 3.8. also to Schrödinger operators.

Lemma 3.10 *Let (M^n, g) be open with (I) and (B_0), $r > n$, $V \in \Omega^{1,r}(M^n, g)$ a real–valued function, $0 \leq q \leq n$. Then the operator*

$$\Delta_q + (V \cdot)$$

is essentially self-adjoint on $C_c^\infty(\Lambda^q)$.

Proof. According to the Sobolev embedding theorem,

$$|V(x)| \leq b.$$

Hence

$$|V \cdot \varphi|_{L_2} \leq o \cdot |\Delta \varphi|_{L_2} + b |\varphi|_{L_2} \tag{3.60}$$

$\square$

Proposition 3.11 *Suppose $g \in \mathcal{M}(I, B_k)$, $k \geq r > n + 2$, $V \in \Omega^{1,r}(M^n, g)$ a real valued function. Then for $t > 0$*

$$e^{-t\Delta_q} - e^{-t(\Delta_q + (V \cdot))} \tag{3.61}$$

is of trace class and the trace norm is uniformly bounded on compact t-intervals $[a_0, a_1]$, $a_0 > 0$.

Proof. We infer from Duhamel's formula

$$e^{-t\Delta_q} - e^{-t(\Delta_q + (V \cdot))} = \int_0^t e^{-s\Delta_q} V e^{-(t-s)(\Delta_q + (V \cdot))} ds, \tag{3.62}$$

decompose $\int_0^t = \int_0^{\frac{t}{2}} + \int_{\frac{t}{2}}^t$ and estimate these integrals as in (1.1) –
(1.42), replacing η by V. $\square$

Theorem 3.12 *Assume the hypotheses of 3.8 and $V \in \Omega^{1,r}(M^n, g)$, $V' \in \Omega^{1,r}(M'^n g)$. Then*

$$e^{-t(\Delta_q + (V \cdot))} P - e^{-tU^* i^*(\Delta'_q + (V \cdot)) iU} P' \tag{3.63}$$

is of trace class in $\mathcal{H}'$ and the trace norm is uniformly bounded on compact t-intervals $[a_0, a_1]$, $a_0 > 0$.

Proof. We write

$$e^{-t(\Delta_q+(V\cdot))}P_e^{-tU^*i^*(\Delta'_q+(V'\cdot))iU}P'$$
$$= (e^{-t(\Delta_q+(V\cdot))} - e^{-t\Delta_q})P \qquad (3.64)$$
$$+e^{-t\Delta_q}P - e^{-tU^*i^*\Delta'_qiU})P' \qquad (3.65)$$
$$+(e^{-tU^*i^*\Delta'_qiU} - e^{-tU^*i^*(\Delta'_q+(V'\cdot))})P'. \qquad (3.66)$$

(3.64) and (3.66) are of trace class according to proposition 3.10, (3.65) is of trace class according to theorem 3.8. □

We proved that after fixing $E \in \text{CL}\mathcal{B}^{N,n}(I, B_k)$, $k \geq r+1 > n+3$, we can attach to any $E' \in \text{gen comp}^{1,r+1}_{L,diff,rel}(E)$ two number valued invariants, namely

$$E' \longrightarrow \text{tr}(e^{-tD^2}P - e^{-t(U^*i^*D'iU)^2}P') \qquad (3.67)$$

and

$$E' \longrightarrow \text{tr}(e^{-tD^2}P - e^{-tU^*i^*D'^2iU}P'). \qquad (3.68)$$

This is a contribution to the classification inside a component but still unsatisfactory insofar as it

1) could depend on t.

2) will depend on the $K \subset M$, $K' \subset M'$ in question,

3) is not yet clear the meaning of this invariant.

The answers to these open question will be the content of chapters V, VI and VII.

V Relative index theory

1 Relative index theorems, the spectral shift function and the scattering index

In many applications the Clifford bundles under consideration are endowed with an involution $\tau : E \longrightarrow E$, s.t.

$$\tau^2 = 1, \quad \tau^* = \tau \tag{1.1}$$
$$[\tau, X]_+ = 0 \text{ for } X \in TM \tag{1.2}$$
$$[\nabla, \tau] = 0 \tag{1.3}$$

Then $L_2((M, E), g, h) = L_2(M, E^+) \oplus L_2(M, E^-)$

$$D = \begin{pmatrix} 0 & D^- \\ D^+ & 0 \end{pmatrix}$$

and $D^- = (D^+)^*$. If M^n is compact then as usual

$$\operatorname{ind} D := \operatorname{ind} D^+ := \dim \ker D^+ - \dim \ker D^- \equiv \operatorname{tr}(\tau e^{-tD^2}), \tag{1.4}$$

where we understand τ as

$$\tau = \begin{pmatrix} I & 0 \\ 0 & -I \end{pmatrix}.$$

For open M^n $\operatorname{ind} D$ in general is not defined since τe^{-tD^2} is not of trace class. The appropriate approach on open manifolds is relative index theory for pairs of operators D, D'. If D, D' are selfadjoint in the same Hilbert space and $e^{tD^2} - e^{-tD'^2}$ would be of trace class then

$$\operatorname{ind}(D, D') := \operatorname{tr}(\tau(e^{-tD^2} - e^{-tD'^2})) \tag{1.5}$$

makes sense, but at the first glance (1.5) should depend on t.

If we restrict to Clifford bundles $E \in \text{CL}\mathcal{B}^{N,n}(I, B_k)$ with involution τ then we assume that the maps entering in the definition of $\text{comp}^{1,r+1}_{L,diff,F}(E)$ or $\text{gen comp}^{1,r+1}_{L,diff,rel}(E)$ are τ–compatible, i.e. after identification of $E|_{M\setminus K}$ and $f_E^* E'|_{M'\setminus K}$ holds

$$[f_E^* \nabla^{h'}, \tau] = 0, \quad [f^* \cdot', \tau]_+ = 0. \tag{1.6}$$

We call $E|_{M\setminus K}$ and $E'|_{M'\setminus K'}$ τ–compatible. Then, according to the preceding theorems,

$$\text{tr}(\tau(e^{-tD^2} P - e^{-t(U^* i^* D' iU)^2} P')) \tag{1.7}$$

makes sense.

Theorem 1.1 *Let* $((E, h, \nabla^h) \longrightarrow (M^n, g), \tau) \in \text{CL}\mathcal{B}^{N,n}(I, B_k)$ *be a graded Clifford bundle, $k \geq r > n+2$.*
a) *If* $\nabla'^h \in \text{comp}^{1,r}(\nabla) \subset C_E^{1,r}(B_k)$, ∇' τ*–compatible, i.e.* $[\nabla', \tau] = 0$ *then*
$$\text{tr}(\tau(e^{-tD^2} - e^{-tD'^2}))$$
is independent of t.
b) *If* $E' \in \text{gen comp}^{1,r+1}_{L,diff,rel}(E)$ *is* τ*–compatible with E, i.e.* $[\tau, X \cdot']_+ = 0$ *for $X \in TM$ and $[\nabla', \tau] = 0$, then*
$$\text{tr}(\tau(e^{-tD^2} P - e^{-t(U^* i^* D' iU)^2} P'))$$
is independent of t.

Proof. a) follows from our IV 1.1. b) follows from our 1.2 below. □

Proposition 1.2 *If* $E' \in \text{gen comp}^{1,r+1}_{L,diff,rel}(E)$ *and*
$$\tau(e^{-tD^2} P - e^{-t(U^* i^* D' iU)^2} P')$$
$$\tau(e^{-tD^2} D - e^{-t(U^* i^* D' iU)^2}(U^* i^* D' iU))$$
are for $t > 0$ of trace class and the trace norm of

$\tau(e^{-tD^2}D - e^{-t(U^*i^*D'iU)^2}(U^*i^*D'iU))$ *is uniformly bounded on compact t–intervals* $[a_0, a_1]$, $a_0 > 0$, *then*

$$\mathrm{tr}(\tau(e^{-tD^2}p - e^{-t(U^*i^*D'iU)^2}(U^*i^*D'iU)))$$

is independent of t.

Proof. Let $(\varphi_i)_i$ be a sequence of smooth functions $\in C_c^\infty(\overline{M \setminus K})$, satisfying $\sup|d\varphi_i| \xrightarrow[i\to\infty]{} 0$, $0 \le \varphi_i \le \varphi_{i+1}$ and $\varphi_i \xrightarrow[i\to\infty]{} 1$. Denote by M_i the multiplication operator with φ_i on $L_2((\overline{M \setminus K}, E|_{M\setminus K}), g, h)$. We extend M_i by 1 to the complement of $L_2((M \setminus K, E), g, h)$ in H. We have to show

$$\frac{d}{dt}\mathrm{tr}\tau(e^{-tD^2}P - e^{-t(U^*i^*D'iU)^2}P') = 0.$$

$e^{-tD^2}P - e^{-t(U^*i^*D'iU)^2}P'$ is of trace class, hence

$$\mathrm{tr}\tau(e^{-tD^2}P - e^{-t(U^*i^*D'iU)^2}P')$$
$$= \lim_{j\to\infty}\mathrm{tr}\tau M_j(e^{-tD^2}P - e^{-t(U^*i^*D'iU)^2}P')M_j.$$

M_j restricts to compact sets and we can differentiate under the trace and we obtain

$$\frac{d}{dt}\mathrm{tr}\tau M_j(e^{-tD^2}P - e^{-t(U^*i^*D'iU)^2}P')M_j$$
$$= \frac{d}{dt}\mathrm{tr}\tau(M_j U^*(e^{-t(UDU^*)^2}P - e^{-t(i^*D'i)^2}P')UM_j$$
$$= -\mathrm{tr}\tau(U^* M_j(e^{-t(UDU^*)^2}(UDU^*)^2 - e^{-t(i^*D'i)^2}(i^*D'i)^2)M_j U)$$

Consider

$$\mathrm{tr}\tau(U^* M_j(e^{-t(UDU^*)^2}(UDU^*)^2 M_j U) = \mathrm{tr}\tau M_j e^{-tD^2}D^2 M_j.$$

There holds

$$\mathrm{tr}\tau(M_j e^{-tD^2}D^2 M_j) = \mathrm{tr} M_j \,\mathrm{grad}\, \varphi_i \cdot \tau D e^{-tD^2}.$$

Quite similar

$$\begin{aligned}
&\mathrm{tr}\tau(M_j(e^{-t(i^*D'i)^2}(i^*D'i)^2)M_j)\\
&= \mathrm{tr}\tau\varphi_j e^{-\frac{t}{2}(i^*D'i)^2}(i^*Di)(i^*D'i)e^{-\frac{t}{2}(i^*D'i)^2}\varphi_j\\
&= \mathrm{tr}(i^*Di)e^{-\frac{t}{2}(i^*D'i)^2}\varphi_j\tau\varphi_j e^{-\frac{t}{2}(i^*D'i)^2}(i^*D'i)\\
&= \mathrm{tr}e^{-\frac{t}{2}(i^*D'i)^2}(i^*D'i)\varphi_j^2\tau e^{-\frac{t}{2}(i^*D'i)^2}(i^*D'i)\\
&= \mathrm{tr}e^{-\frac{t}{2}(i^*D'i)^2}i^*(2\varphi_j \text{ grad }'\varphi_j\cdot' +\varphi_j^2 D')i\tau e^{-\frac{t}{2}(i^*D'i)^2}(i^*D'i)\\
&= \mathrm{tr}2i^*M_j \text{ grad }'\varphi_j\cdot' \, i\tau(i^*D'i)e^{-t(i^*D'i)^2}\\
&\quad -\mathrm{tr}\tau M_j e^{-t(i^*D'i)^2}(i^*D'i)^2 M_j,
\end{aligned}$$

hence

$$\begin{aligned}
&\mathrm{tr}\tau(M_j e^{-t(i^*D'i)^2}(i^*D'i)^2 M_j)\\
&= \mathrm{tr}M_j i^* \text{ grad }'\varphi_j\cdot' \, i\tau(i^*D'i)e^{-t(i^*D'i)^2}
\end{aligned}$$

and finally

$$\begin{aligned}
&\frac{d}{dt}\mathrm{tr}\tau M_j(e^{-tD^2}P - e^{-t(U^*i^*D'iU)^2}P')M_j\\
&= \mathrm{tr}\tau M_j[\text{ grad } \varphi_j \cdot e^{-tD^2}D\\
&\quad - \text{ grad }'\varphi_j\cdot' \, e^{-t(U^*i^*D'iU)^2}(U^*i^*D'iU)]\\
&= \mathrm{tr}\tau M_j[(\text{ grad }\varphi_j - \text{ grad }'\varphi_j)\cdot e^{-tD^2}\\
&\quad + \text{ grad }'\varphi_j(\cdot - \cdot')e^{-tD^2}\\
&\quad + \text{ grad }'\varphi_j\cdot' \, (e^{-tD^2} - e^{-t(U^*i^*D'iU)^2}(U^*i^*D'iU))].
\end{aligned}$$

But this tends to zero uniformly for t in compact intervals since grad φ_j, grad $'\varphi_j$ do so. $\square$

We denote $Q^\pm = D^\pm$

$$Q = \begin{pmatrix} 0 & Q^+ \\ Q^- & 0 \end{pmatrix},$$

$$H = \begin{pmatrix} H^+ & 0 \\ 0 & H^- \end{pmatrix} = \begin{pmatrix} Q^-Q^+ & 0 \\ 0 & Q^+Q^- \end{pmatrix} = Q^2, \quad (1.8)$$

$Q'^{\pm} = U^*i^*D'^{\pm}iU = (U^*i^*D'iU)^{\pm}$, Q', H' analogous, assuming (1.1) – (1.3) as before and $\cdot'$, ∇' τ–compatible. H, H' form by definition a supersymmetric scattering system if the wave operators

$$W^{\mp}(H,H') := \lim_{t\to\mp\infty} e^{itH}e^{-itH'} \cdot P_{ac}(H') \text{ exist and are complete}$$
(1.9)

and

$$QW^{\mp}(H,H') = W^{\mp}(H,H')H' \text{ on } \mathcal{D}_{H'} \cap \mathcal{H}'_{ac}(H'). \quad (1.10)$$

Here $P_{ac}(H')$ denotes the projection on the absolutely continuous subspace $\mathcal{H}'_{ac}(H') \subset \mathcal{H}$ of H'.

A well known sufficient criterion for forming a supersymmetric scattering system is given by

Proposition 1.3 *Assume for the graded operators Q, Q' (= supercharges)*

$$e^{-tH} - e^{-tH'}$$

and

$$e^{-tH}Q - e^{-tH'}Q$$

are for $t > 0$ of trace class. Then they form a supersymmetric scattering system. □

Corollary 1.4 *Assume the hypotheses of 1.1. Then D, D' or $D, U^*i^*D'iU$ form a supersymmetric scattering system, respectively. In particular, the restriction of D, D' or $D, U^*i^*D'iU$ to their absolutely continuous spectral subspaces are unitarily equivalent, respectively.* □

Until now we have seen that under the hypotheses of 1.1

$$\text{ind}(D,\tilde{D}') = \text{tr}\tau(e^{-tD^2}P - e^{-t\tilde{D}'^2}P'), \quad (1.11)$$

$\tilde{D}' = D'$ or $\tilde{D}' = U^*i^*D'iU$, is a well defined number, independent of $t > 0$ and hence yields an invariant of the pair (E, E'),

still depending on K, K'. Hence we should sometimes better write
$$\operatorname{ind}(D, \tilde{D}', K, K'). \tag{1.12}$$
We want to express in some good cases $\operatorname{ind}(D, \tilde{D}', K, K')$ by other relevant numbers. Consider the abstract setting (1.8). If $\inf \sigma_e(H) > 0$ then $\operatorname{ind} D := \operatorname{ind} D^+$ is well defined.

Lemma 1.5 *If $e^{-tH}p - e^{-tH'}P'$ is of trace class for all $t > 0$ and $\inf \sigma_e(H)$,*
$\inf \sigma_e(H') > 0$ *then*
$$\lim_{t \to \infty} \operatorname{tr}\tau(e^{-tH}P - e^{-tH'}P') = \operatorname{ind} Q^+ - \operatorname{ind} Q^-. \tag{1.13}$$
□

We infer from this

Theorem 1.6 *Assume the hypotheses of 1.1 and $\inf \sigma_e(D^2) > 0$. Then $\inf \sigma_e(D'^2), \inf \sigma_e(U^*i^*D'iU)^2 > 0$ and for each $t > 0$*
$$\operatorname{tr}\tau(e^{-tD^2} - e^{-t\tilde{D}'^2}) = \operatorname{ind} D^+ - \operatorname{ind} D'^+. \tag{1.14}$$

Proof. In the case 1.1 a), $\inf \sigma_e(D'^2) > 0$ follows from a standard fact and (1.14) then follows from 1.5. Consider the case 1.1 b). We can replace the comparison of $\sigma_e(D^2)$ and $\sigma_e((U^*i^*D'iU)^2)$ by that of $\sigma_e(UD^2U^*)$ and $\sigma_e((i^*D'i)^2)$. Moreover, for self adjoint A, $0 \notin \sigma_e(A)$ if and only if $\inf \sigma_e(A^2) > 0$. Assume $0 \notin \sigma_e(UDU^*)$ and $0 \in \sigma_e(i^*D'i)$. We must derive a contradiction. Let $(\Phi_\nu)_\nu$ be a Weyl sequence for $0 \in \sigma_e(i^*D'i)$ satisfying additionally $|\Phi_\nu|_{L_2} = 1$, $\operatorname{supp} \Phi_\nu \subseteq M \setminus K = M' \setminus K'$ and for any compact $L \subset M \setminus K = M' \setminus K'$
$$|\Phi_\nu|_{L_2(M \setminus L)} \xrightarrow[\nu \to \infty]{} 0. \tag{1.15}$$

We have $\lim_{\nu \to \infty} i^*D'i\Phi_\nu = 0$. Then also $\lim_{\nu \to \infty} D'\Phi_\nu = 0$. We use in the sequel the following simple fact. If β is an L_2–function, in particular if β is even Sobolev, then
$$|\beta \cdot \Phi_\nu|_{L_2} \longrightarrow 0. \tag{1.16}$$

Now $(UDU^*)\Phi_\nu = (UDU^* - D')\Phi_\nu + D'\Phi_\nu$. Here $D'\Phi_\nu \xrightarrow[\nu\to\infty]{} 0$.
Consider $(UDU^* - D')\Phi_\nu = (\alpha^{\frac{1}{2}}D\alpha^{-\frac{1}{2}} - D')\Phi_\nu = \left(-\frac{\text{grad }\alpha}{2\alpha}\cdot + D - D'\right)\Phi_\nu$. Assume $\alpha \not\equiv 1$. Then $\beta = \left|\frac{\text{grad }\alpha}{2\alpha}\right| \in \Omega^{0,2,\frac{1}{2}}(T(M\setminus K))$ satisfies the assumptions above and

$$\lim_{\nu\to\infty}\left|-\frac{\text{grad }\alpha}{2\alpha}\cdot\Phi_\nu\right|_{L_2} = 0. \tag{1.17}$$

If $\alpha \equiv 1$ this term does not appear. Write, according to IV (2.9) – (2.12),

$$(D - D')\Phi_\nu = \eta_1^{op}\Phi_\nu + \eta_2^{op}\Phi_\nu + \eta_3^{op}\Phi_\nu. \tag{1.18}$$

$\left|g'^{ik}\frac{\partial}{\partial x^k}\cdot\right|$ is bounded (we use a uniformly locally finite cover by normal charts, an associated bounded decomposition of unity etc.). $\beta = |\nabla - \nabla'|$ is Sobolev hence L_2 and by (1.16)

$$|\eta_2^{op}\Phi_\nu|_{L_2} \xrightarrow[\nu\to\infty]{} 0. \tag{1.19}$$

Now $|\nabla\Phi_\nu|_{L_2} \le C_1(|\Phi_\nu|_{L_2} + |D\Phi_\nu|_{L_2}) \le C_2(|\Phi_\nu|_{L_2} + |D'\Phi_\nu|_{L_2})$. $g - g'$ is Sobolev, hence, according to (1.16) with $\beta = |g - g'|$, $\||g - g'|\cdot\Phi_\nu|_{L_2} \xrightarrow[\nu\to\infty]{} 0$ and finally $\||g - g'|\cdot D'\Phi_\nu|_{L_2} \xrightarrow[\nu\to\infty]{} 0$. This yields

$$|\eta_1^{op}\Phi_\nu|_{L_2} \xrightarrow[\nu\to\infty]{} 0. \tag{1.20}$$

We conclude in the same manner from $\cdot - \cdot'$ Sobolev and $|\nabla'\Phi_\nu|_{L_2} \le C_3(|\Phi_\nu|_{L_2} + |D'\Phi_\nu|_{L_2})$ that

$$|\eta_3^{op}\Phi_\nu|_{L_2} \xrightarrow[\nu\to\infty]{} 0. \tag{1.21}$$

(1.17) – (1.21) yield $(UDU^*)\Phi_\nu \longrightarrow 0$, $0 \in \sigma_e(UDU^*)$, $\inf\sigma_e(D^2) = 0$, a contradiction, hence $\inf\sigma_e((U^*i^*D'iU)^2) > 0$, $\inf\sigma_e((i^*D'i)^2) > 0$, $0 \notin \sigma_e(i^*D'i)$, $0 \notin \sigma_e(D')$, $\inf\sigma_e(D'^2) > 0$. We infer from 1.2 and 1.5 that for $t > 0$

$$\text{tr}\tau e^{-tD^2} - e^{-t(U^*i^*D'iU)^2} = \text{ind}D^+ - \text{ind}(U^*i^*D'iU)^+. \tag{1.22}$$

We are done if we can show

$$\text{ind}(U^*i^*D'iU)^+ = \text{ind}D'^+. \tag{1.23}$$

$\Phi \in \ker(U^* i^* D' i U)^+$ means

$$\begin{aligned}(U^* i^* D' i U)^+ \Phi &= (U^* i^* D' i U)(\chi_{K'} \Phi + U^* i^{-1} \chi_{M' \setminus K'} \Phi) \\ &= \chi_{K'} D'^+ \Phi + U^* i^* \chi_{M' \setminus K'} D'^+ \Phi \\ &= 0.\end{aligned}$$

But this is equivalent to $D'^+ \Phi = 0$. Similar for D'^-. (1.23) holds and hence (1.14). □

It would be desirable to express $\text{ind}(D, \tilde{D}', K, K')$ by geometric/topological terms. In particular, this would be nice in the case $\inf \sigma_e(D^2) > 0$. In the compact case, one sets $\text{ind}_a D := \text{ind}_a D^+ = \dim \ker D^+ - \dim \ker (D^+)^* = \dim \ker D^+ - \dim \ker D^- = \lim_{t \to \infty} \text{tr} \tau e^{-tD^2}$. On the other hand, for $t \to 0^+$ there exists the well known asymptotic expansion for the kernel of τe^{-tD^2}. Its integral at the diagonal yields the trace. If $\text{tr} \tau e^{-tD^2}$ is independent of t (as in the compact case), we get the index theorem where the integrand appearing in the L_2–trace consists only of the t–free term of the asymptotic expansion. Here one would like to express things in the asymptotic expansion of the heat kernel of $e^{-tD'^2}$ instead of $e^{-t(U^* i^* D' i U)^2}$. For this reason we restrict in the definition of the topological index to the case $E' \in \text{comp}^{1,r+1}_{L,diff,F}(E)$ or $E' \in \text{comp}^{1,r+1}_{L,diff,F,rel}(E)$, i.e. we admit Sobolev perturbation of $g, \nabla^h, \cdot$ but the fibre metric h should remain fixed. Then for $D' = D(g', h, \nabla'^h, \cdot')$ in $L_2((M, E), g, h)$ the heat kernel of $e^{-t(U^* D' U)^2} = U^* e^{-tD'^2} U$ equals to $\alpha(q)^{-\frac{1}{2}} W'(t, q, p) \alpha(p)^{\frac{1}{2}}$. At the diagonal this equals to $W'(t, m, m)$, i.e. the asymptotic expansion at the diagonal of the original $e^{-tD'^2}$ and the transformed to $L_2((M, E), g, h)$ coincide. Consider

$$\text{tr} \tau W(t, m, m) \underset{t \to 0^+}{\sim} t^{-\frac{n}{2}} b_{-\frac{n}{2}}(D, m) + \cdots + b_0(D, m) + \cdots \quad (1.24)$$

and

$$\text{tr} \tau W'(t, m, m) \underset{t \to 0^+}{\sim} t^{-\frac{n}{2}} b_{-\frac{n}{2}}(D', m) + \cdots + b_0(D', m) + \cdots. \quad (1.25)$$

We prove in VI 1.1 and 1.2

Lemma 1.7

$$b_i(D,m) - b_i(D',m) \in L_1, \quad -\frac{n}{2} \leq i \leq 1. \tag{1.26}$$

□

Define for $E' \in \text{gen comp}_{L,diff,F}^{1,r+1}(E)$

$$\text{ind}_{top}(D,D') := \int_M b_0(D,m) - b_0(D',m). \tag{1.27}$$

According to (1.26), $\text{ind}_{top}(D,D')$ is well defined.

Theorem 1.8 *Assume* $E' \in \text{gen comp}_{L,diff,F,rel}^{1,r+1}(E)$
a) *Then*

$$\text{ind}(D,D',K,K') = \int_K b_0(D,m) - \int_{K'} b_0(D',m) + \tag{1.28}$$

$$+ \int_{M\setminus K = M'\setminus K'} b_0(D,m) - b_0(D',m).$$

$$\tag{1.29}$$

b) *If* $E' \in \text{gen comp}_{L,diff,F}^{1,r+1}(E)$ *then*

$$\text{ind}(D,D') = \text{ind}_{top}(D,D'). \tag{1.30}$$

c) *If* $E' \in \text{gen comp}_{L,diff,F}^{1,r+1}(E)$ *and* $\inf \sigma_e(D^2) > 0$ *then*

$$\text{ind}_{top}(D,D') = \text{ind}_a D - \text{ind}_a D'. \tag{1.31}$$

Proof. All this follows from 1.1, the asymptotic expansion, (1.26) and the fact that the L_2-trace of a trace class integral operator equals to the integral over the trace of the kernel. □

Remarks 1.9 1) If $E' \in \text{gen comp}_{L,diff,rel}^{1,r+1}(E)$, g and g', ∇^h and ∇'^h, $\cdot$ and $\cdot'$ coincide in $V = M \setminus L = M' \setminus L'$, $L \supseteq K$, $L' \supseteq K'$, then in IV (2.5) – (2.53) $\alpha - 1$ and the η's have compact support and we conclude from IV (3.38), (3.39) and the standard heat kernel estimates that

$$\int_V |W(t,m,m) - W'(t,m,m)|\, dm \leq C \cdot e^{-\frac{d}{t}} \qquad (1.32)$$

and obtain

$$\text{ind}(D, D', L, L') = \int_L b_o(D,m) - \int_{L'} b_0(D',m). \qquad (1.33)$$

This follows immediately from 1.8 a).

2) The point here is that we admit much more general perturbations than in preceding approaches to prove relative index theorems.

3. $\inf \sigma_e(D^2) > 0$ is an invariant of $\text{gen comp}_{L,diff,F}^{1,r+1}(E)$. If we fix E, D as reference point in $\text{gen comp}_{L,diff,F}^{1,r+1}(E)$ then 1.8 c) enables us to calculate the analytical index for all other D's in the component from $\text{ind}\,D$ and a pure integration.

4) $\inf \sigma_e(D^2) > 0$ is satisfied e.g. if in $D^2 = \nabla^*\nabla + \mathcal{R}$ the operator $\mathcal{R}$ satisfies outside a compact K the condition

$$\mathcal{R} \geq \kappa_0 \cdot id, \kappa_0 > 0. \qquad (1.34)$$

(1.34) is an invariant of $\text{gen comp}_{L,diff,F}^{1,r+1}(E)$ (with possibly different K, κ_0). □

It is possible that $\text{ind}\,D$, $\text{ind}\,D'$ are defined even if $0 \in \sigma_e$. For the corresponding relative index theorem we need the scattering index.

To define the scattering index and in the next section relative ζ-functions, we must now use spectral shift functions $\xi(\lambda)$ which we introduced in III section 2. According to theorem 2.8 of chapter III, $\xi(\lambda) \equiv \xi(\lambda; A, A')$ exists if A, A' are self-adjoint and $V = A - A'$ is of trace class. Then, with $R'(z) = (A' - z)^{-1}$,

$$\xi(\lambda) = \xi(\lambda, A, A') := \pi^{-1} \lim_{\varepsilon \to 0} \arg \det(1 + VR'(\lambda + i\varepsilon)) \quad (1.35)$$

exists for a.e. $\lambda \in \mathbb{R}$. $\xi(\lambda)$ is real valued, $\in L_1(\mathbb{R})$ and

$$\mathrm{tr}(A - A') = \int_I R\xi(\lambda)\, d\lambda, \quad |\xi|_{L_1} \leq |A - A'|_1. \quad (1.36)$$

If $I(A, A')$ is the smallest interval containing $\sigma(A) \cup \sigma(A')$ then $\xi(\lambda) = 0$ for $\lambda \notin I(A, A')$.
Let

$$\mathcal{G} = \{f : \mathbb{R} \longrightarrow \mathbb{R} \mid f \in L_1 \text{ and } \int_{\mathbb{R}} |\hat{f}(p)|(1 + |p|)\, dp < \infty\}.$$

Then for $\varphi \in \mathcal{G}$, $\varphi(A) - \varphi(A')$ is of trace class and

$$\mathrm{tr}(\varphi(A) - \varphi(A')) = \int_{\mathbb{R}} \varphi'(\lambda)\xi(\lambda)\, d\lambda. \quad (1.37)$$

We state without proof from [49]

Lemma 1.10 *Let $H, H' \geq 0$, selfadjoint in a Hilbert space $\mathcal{H}$, $e^{-tH} - e^{-tH'}$ for $t > 0$ of trace class. Then there exist a unique function $\xi = \xi(\lambda) = \xi(\lambda, H, H') \in L_{1,loc}(\mathbb{R})$ such that for $t > 0$, $e^{-t\lambda}\xi(\lambda) \in L_1(\mathbb{R})$ and the following holds.*

a) $\mathrm{tr}(e^{-tH} - e^{-tH'}) = -t \int_0^\infty e^{-t\lambda} \xi(\lambda)\, d\lambda.$

b) *For every $\varphi \in \mathcal{G}$, $\varphi(H) - \varphi(H')$ is of trace class and*

$$\mathrm{tr}(\varphi(H) - \varphi(H')) = \int_I R\varphi'(\lambda)\xi(\lambda)\, d\lambda.$$

c) $\xi(\lambda) = 0$ *for $\lambda < 0$.* □

We apply this to our case $E' \in \text{gen comp}_{L,diff,rel}^{1,r+1}(E)$. According to corollary 1.4, D and $U^*i^*D'iU$ form a supersymmetric scattering system, $H = D^2$, $H' = (U^*i^*D'iU)^2$. In this case

$$e^{2\pi i \xi(\lambda, H, H')} = \det S(\lambda),$$

where, according to II (2.5) and (2.6), $S = (W^+)^*W^- = \int S(\lambda) \, dE'(\lambda)$ and $H'_{ac} = \int \lambda \, dE'(\lambda)$.

Let $P_d(D)$, $P_d(U^*i^*D'iU)$ be the projector on the discrete subspace in $\mathcal{H}$, respectively and $P_c = 1 - P_d$ the projector onto the continuous subspace. Moreover we write

$$D^2 = \begin{pmatrix} H^+ & 0 \\ 0 & H^- \end{pmatrix}, \quad (U^*i^*D'iU)^2 = \begin{pmatrix} H'^+ & 0 \\ 0 & H'^- \end{pmatrix}. \tag{1.38}$$

We make the following additional assumption.

$$e^{-tD^2}P_d(D), e^{-t(U^*i^*D'iU)^2}P_d(U^*i^*D'iU)$$
are for $t > 0$ of trace class. \hfill (1.39)

Then for $t > 0$

$$e^{-tD^2}P_c(D) - e^{-t(U^*i^*D'iU)^2}P_c(U^*i^*D'iU)$$

is of trace class and we can in complete analogy to (1.35) define

$$\xi^c(\lambda, H^\pm, H'^\pm) := -\pi \lim_{\varepsilon \to 0^+} \arg \det[1 + (e^{-tH^\pm}P_c(H^\pm)$$
$$- e^{-tH'^\pm}P_c(H'^\pm))(e^{-tH'^\pm}P_c(H'^\pm) - e^{-\lambda t} - i\varepsilon)^{-1}] \quad (1.40)$$

According to (1.36),

$$\text{tr}(e^{-tH^\pm}P_c(H^\pm) - e^{-tH'^\pm}P_c(H'^\pm)) = -t \int_0^\infty \xi^c(\lambda, H^\pm, H'^\pm) e^{-t\lambda} \, d\lambda. \tag{1.41}$$

We denote as after (1.11) $\tilde{D}' = D'$ in the case $\nabla' \in \text{comp}^{1,r}(\nabla)$ and $\tilde{D}' = U^*i^*D'iU$ in the case $E' \in \text{gen comp}_{L,diff,rel}^{1,r+1}(E)$. The assumption (1.39) in particular implies that for the restriction of D and $\tilde{D}'$ to their discrete subspace the analytical index is well defined and we write $\operatorname{ind}_{a,d}(D, \tilde{D}') = \operatorname{ind}_{a,d}(D) - \operatorname{ind}_{a,d}(\tilde{D}')$ for it. Set

$$n^c(\lambda, D, \tilde{D}') := -\xi^c(\lambda, H^+, H'^+) + \xi^c(\lambda, H^-, H'^-). \quad (1.42)$$

Theorem 1.11 *Assume the hypotheses of 1.1 and (1.39). Then $n^c(\lambda, D, \tilde{D}') = n^c(D, \tilde{D}')$ is constant and*

$$\operatorname{ind}(D, \tilde{D}') - \operatorname{ind}_{a,d}(D, \tilde{D}') = n^c(D, \tilde{D}'). \quad (1.43)$$

Proof.
$$\begin{aligned}
\operatorname{ind}(D, \tilde{D}') &= \operatorname{tr}\tau(e^{-tD^2}P - e^{-t\tilde{D}'^2}P') = \\
&= \operatorname{tr}\tau e^{-tD^2}P_d(D)P - \operatorname{tr}\tau e^{-t\tilde{D}'^2}P_d(\tilde{D}')P' + \\
&\quad + \operatorname{tr}\tau(e^{-tD^2}P_c(D) - e^{-t\tilde{D}'^2}P_c(\tilde{D}')) = \\
&= \operatorname{ind}_{a,d}(D, \tilde{D}') + t\int_0^\infty e^{-t\lambda} n^c(\lambda, D, \tilde{D}')\, d\lambda.
\end{aligned}$$

According to 1.1, $\operatorname{ind}(D, \tilde{D}')$ is independent of t. The same holds for $\operatorname{ind}_{a,d}(D, \tilde{D}')$. Hence $t\int_0^\infty e^{-t\lambda} n^c(\lambda, D, \tilde{D}')\, d\lambda$ is independent of t. This is possible only if $\int_0^\infty e^{-t\lambda} n^c(\lambda, D, D')\, d\lambda = \frac{1}{t}$ or $n^c(\lambda, D, \tilde{D}')$ is independent of λ. □

Corollary 1.12 *Assume the hypotheses of 1.11 and additionally*
$\inf \sigma_e(D^2|_{(\ker D^2)^\perp}) > 0$. *Then $n^c(D, \tilde{D}') = 0$.* □

VI Relative ζ–functions, η–functions, determinants and torsion

In this chapter, we apply our preceding considerations and results to the construction of relative zeta functions and related invariants. We will attach to an appropriate pair of Clifford data a relative zeta function, which is essentially defined by the corresponding pair of asumptotic expansions of the heat kernel. Therefore we must first consider such a pair of expansions.

1 Pairs of asymptotic expansions

Assume $E' \in \text{gen comp}^{1,r+1}_{L,diff,F}(E)$. Then we have in $L_2((E,M), g, h)$ the asymptotic expansion

$$\operatorname{tr} W(t,m,m) \underset{t\to 0^+}{\sim} t^{-\frac{n}{2}} b_{-\frac{n}{2}}(m) + t^{-\frac{n}{2}+1} b_{-\frac{n}{2}+1} + \cdots \quad (1.1)$$

and analogously for

$$\operatorname{tr} \alpha^{-\frac{1}{2}}(m) W'(t,m,m) \alpha^{\frac{1}{2}}(m) = \operatorname{tr} W'(t,m,m)$$

with

$$b_{-\frac{n}{2}+l}(m) = b_{-\frac{n}{2}+l}(D(g,h,\nabla),m),$$
$$b'_{-\frac{n}{2}+l}(m) = b_{-\frac{n}{2}+l}(D(g',h,\nabla'),m).$$

Here we use that the odd coefficients vanish, i.e. terms with $t^{-\frac{n}{2}+\frac{1}{2}}$, $t^{-\frac{n}{2}+\frac{3}{2}}$ etc. do not appear. The heat kernel coefficients have for $l \geq 1$ a representation

$$b_{-\frac{n}{2}+l} = \sum_{k=1}^{l} \sum_{q=0}^{k} \sum_{i_1+i_2+\cdots+i_k=2(l-k)} \nabla^{i_1} R^g \ldots \nabla^{i_q} R^g$$
$$\operatorname{tr}(\nabla^{i_{q+1}} R^E \ldots \nabla^{i_k} R^E) C^{i_1,\ldots,i_k}, \quad (1.2)$$

where $C^{i_1,\ldots,i_k}$ stands for a contraction with respect to g, i.e. it is built up by linear combination of products of the g^{ij}, g_{ij}.

Lemma 1.1 $b_{-\frac{n}{2}+l} - b'_{-\frac{n}{2}+l} \in L_1(M,g)$, $0 \leq l \leq \frac{n+3}{2}$.

Proof. First we fix g. Forming the difference $b_{-\frac{n}{2}+l} - b'_{-\frac{n}{2}+l}$, we obtain a sum of terms of the kind

$$\nabla^{i_1} R^g \ldots \nabla^{i_q} R^g \operatorname{tr} [\nabla^{i_{q+1}} R^E \ldots \nabla^{i_k} R^E$$
$$- \nabla'^{i_{q+1}} R'^E \ldots \nabla'^{i_k} R'^E] C^{i_1,\ldots,i_k}. \tag{1.3}$$

The highest derivative of R^g with respect to ∇^g occurs if $q = k$, $i_1 = \cdots = i_{q-1} = 0$. Then we have

$$(\nabla^g)^{2l-2k} R^g. \tag{1.4}$$

By assumption, we have bounded geometry of order $\geq r > n+2$, i. e. of order $\geq n+3$. Hence $(\nabla^g)^i R^g$ is bounded for $i \leq n+1$. To obtain bounded $\nabla^j R^g$–coefficients of $[\ldots]$ in (1.3), we must assume

$$2l - 2 \leq n+1, \quad l \leq \frac{n+3}{2}. \tag{1.5}$$

Similarly we see that the highest occuring derivatives of R^E, R'^E in $[\ldots]$ are of order $2l - 2$. The corresponding expression is

$$R^E \nabla^{2l-2} R^E - R'^E \nabla'^{2l-2} R'^E$$
$$= (R^E - R'^E)(\nabla^{2l-2} R^E) + R'^E (\nabla^{2l-2} R^E - \nabla^{2l-2} R'^E). \tag{1.6}$$

We want to apply the module structure theorem. $\nabla - \nabla' \in \Omega^{1,1,r}(\mathcal{G}_E^{Cl}, \nabla) = \Omega^{1,1,r}(\mathcal{G}_E^{Cl}, \nabla')$ implies $R^E - R'^E \in \Omega^{2,1,r-1}$. We can apply the module structure theorem (and conclude that all norm products of derivatives of order $\leq 2l - 2$ are absolutely integrable) if $2l - 2 \leq r - 1$, $2l - 2 \leq n + 1$, $l \leq \frac{n+3}{2}$. Hence, (1.6) $\in L_1$ since R^E, R'^E, $C^{i_1\ldots i_k}$ bounded. It is now a very simple combinatorial matter to write $[\ldots]$ in (1.3) as a sum of terms each of them is a product of differences $(\nabla^i R^E - \nabla'^i R'^E)$ with bounded terms $\nabla^j R^E$, $\nabla'^{j'} R'^E$. We indicate this for an expression $\nabla^i R^E \nabla^j R^E - \nabla'^i R'^E \nabla'^j R'^E$,

$$\nabla^i R^E \nabla^j R^E - \nabla'^i R'^E \nabla'^j R'^E$$
$$= \nabla^i R^E (\nabla^j R^E - \nabla'^j R'^E) + \nabla^i R^E \nabla'^j R'^E - \nabla'^i R'^E \nabla'^j R'^E$$
$$= \nabla^i R^E (\nabla^j R^E - \nabla'^j R'^E) + (\nabla^i R^E - \nabla'^i R'^E) \nabla'^j R'^E. \tag{1.7}$$

The general case can be treated by simple induction. Remember $\nabla, \nabla' \in \mathcal{C}_E(B_k)$. Admit now change of g. We write $\nabla^{i_1} R^g \cdots \nabla^{i_q} R^g \text{tr}(\nabla^{i_{q+1}} R^E \cdots \nabla^{i_k} R^E) C^{i_1,\ldots,i_k}$ as $\mathcal{R}^1(g)\text{tr}\mathcal{R}^2(h,\nabla)C(g)$, similarly $\mathcal{R}^1(g')\text{tr}\mathcal{R}^2(h,\nabla')C(g')$. Then we have to consider expressions

$$\mathcal{R}^1(g)\text{tr}\mathcal{R}^2(h,\nabla)C(g) - \mathcal{R}^1(g')\text{tr}'\mathcal{R}^2(h,\nabla')C(g')$$
$$= [\mathcal{R}^1(g) - \mathcal{R}^1(g')]\text{tr}\mathcal{R}^2(h,\nabla)C(g)$$
$$+ \mathcal{R}^1(g')\text{tr}\mathcal{R}^2(h,\nabla)[C(g) - C(g')]$$
$$+ \mathcal{R}^1(g)\text{tr}[\mathcal{R}^2(h,\nabla) - \mathcal{R}^2(h,\nabla')]C(g'). \tag{1.8}$$

But each term of $[\mathcal{R}^1(g) - \mathcal{R}^1(g')]$ and $[\mathcal{R}^2(h,\nabla) - \mathcal{R}^2(h,\nabla')]$ can be written as a product of terms of type (1.7) composed with bounded terms (bounded morphisms). The terms $[C(g) - C(g')]$ are $\in L_1$ since $g' \in \text{comp}^{1,r+1}(g)$. Then the module structure theorem for $p = p_1 = p_2$ yields again that the whole expression (1.8) is $\in L_1$. This proves lemma 1.1. $\square$

Lemma 1.2 *There is an expansion*

$$\text{tr}\left(e^{-tD^2} - e^{-t(U^*D'U)^2}\right)$$
$$= t^{-\frac{n}{2}} a_{-\frac{n}{2}} + \cdots + t^{-\frac{n}{2} + [\frac{n+3}{2}]} a_{-\frac{n}{2} + [\frac{n+3}{2}]}$$
$$+ O(t^{-\frac{n}{2} + [\frac{n+3}{2}] + 1}). \tag{1.9}$$

Proof. Set

$$a_{-\frac{n}{2}+i} = \int \left(b_{-\frac{n}{2}+i}(m) - b'_{-\frac{n}{2}+i}(m)\right) dm \tag{1.10}$$

and use

$$\text{tr}\, W(t,m,m) = t^{-\frac{n}{2}} b_{-\frac{n}{2}} + \cdots + t^{-\frac{n}{2}+[\frac{n+3}{2}]} b_{-\frac{n}{2}+[\frac{n+3}{2}]}$$
$$+ O(m, t^{-\frac{n}{2}+[\frac{n+3}{2}]+1}), \tag{1.11}$$
$$\text{tr}\, W'(t,m,m) = t^{-\frac{n}{2}} b'_{-\frac{n}{2}} + \cdots + O'(m, t^{-\frac{n}{2}+[\frac{n+3}{2}]+1}) \tag{1.12}$$
$$\text{tr}\left(e^{-tD^2} - e^{-t(U^*D'U)^2}\right)$$
$$= \int \left(\text{tr}\, W(t,m,m) - \text{tr}\, W'(t,m,m)\right) dm.$$

Using lemma 1.1, the only critical point is

$$\int_M O(m, t^{-\frac{n}{2}+[\frac{n+3}{2}]+1}) - O'(m, t^{-\frac{n}{2}+[\frac{n+3}{2}]+1})\, dm = O(t^{-\frac{n}{2}+[\frac{n+3}{2}]+1}).$$

(1.13)

(1.11) requires a very careful investigation of the concrete representatives for $O(m, t^{-\frac{n}{2}+[\frac{n+3}{2}]})$. We did this step by step, following [38], p. 21/22, 66 – 69. Very roughly speaking, the m–dependence of $O(m, \cdot)$ is given by the parametrix construction, i. e. by differences of corresponding derivatives of the $\Gamma_{i\alpha}^\beta$, $\Gamma'^\beta_{i\alpha}$, which are integrable by assumption. □

If $E' \in \operatorname{gen\,compl}^{1,r+1}_{L,diff,F,rel}(E)$ then we immediately derive from IV, theorem 3.2

$$e^{-tD^2} P - e^{-t(u^* D'^2 u)} P'$$

is of trace class. The heat kernel and its asymptotic expansion split into its restriction to K and $M \setminus K = V$ or to K' and $M' \setminus K' = V$, respectively.

$$(e^{-tD^2} P)(t, m, m) = W_{g,h}(t, m, m)|_K + W_{g,h}(t, m, m)|_V,$$
$$(e^{-t(u^* D'^2 u)} P')(t, m, m) = W'_{g',h}(t, m, m)|_{K'} + W'_{g',h}(t, m, m)|_{V'},$$

hence

$$\operatorname{tr}(e^{-tD^2} P - e^{-t(u^* D'^2 u)} P')$$
$$= \int_K W(t, m, m) dm - \int_{K'} W'(t, m, m) dm$$
$$+ \int_V (W(t, m, m) - W'(t, m, m)) dm$$

$$= t^{-\frac{n}{2}} a_{-\frac{n}{2}}(K,g) + \cdots + t^{-\frac{n}{2}+[\frac{n+3}{2}]} a_{-\frac{n}{2}-[\frac{n+3}{2}]}(K,g)$$
$$+ O(t^{-\frac{n}{2}+[\frac{n+3}{2}]+1})$$
$$+ t^{-\frac{n}{2}} a_{-\frac{n}{2}}(K',g') + \cdots + t^{-\frac{n}{2}+[\frac{n+3}{2}]} a_{-\frac{n}{2}-[\frac{n+3}{2}]}(K',g')$$
$$+ O(t^{-\frac{n}{2}+[\frac{n+3}{2}]+1})$$
$$+ t^{-\frac{n}{2}} a_{-\frac{n}{2}}(V,g,g') + \cdots + t^{-\frac{n}{2}+[\frac{n+3}{2}]} a_{-\frac{n}{2}-[\frac{n+3}{2}]}(V,g,g')$$
$$+ O(t^{-\frac{n}{2}+[\frac{n+3}{2}]+1})$$
$$= t^{-\frac{n}{2}}(a_{-\frac{n}{2}}(K,g) - a_{-\frac{n}{2}}(K',g') + a_{-\frac{n}{2}}(V,g,g')) + \cdots$$
$$+ t^{-\frac{n}{2}+[\frac{n+3}{2}]} a_{-\frac{n}{2}-[\frac{n+3}{2}]}(K,g) - a_{-\frac{n}{2}-[\frac{n+3}{2}]}(K',g')$$
$$+ a_{-\frac{n}{2}-[\frac{n+3}{2}]}(V,g,g') + O(t^{-\frac{n}{2}+[\frac{n+3}{2}]+1}), \qquad (1.14)$$

where the $a_i(K,g)$ or $a_i(K',g')$ are the integral terms of the asymptotic expansion on K or K', respectively, and

$$a_i(V,g,g') = \int_V (b_i(m) - b'_i(m)) dm. \qquad (1.15)$$

The existence of the integrals (1.15) follows from the proof of the lemmas 1.1 and 1.2.

Hence we proved

Lemma 1.3 *Suppose $E' \in \text{gen comp}_{L,diff,F,rel}^{1,r+1}(E)$. Then there exists an asymptotic expansion*

$$\text{tr}(e^{-tD^2}P - e^{-t(u^*D'^2u)}P')$$
$$= t^{-\frac{n}{2}} c_{-\frac{n}{2}} + \cdots + t^{-\frac{n}{2}+[\frac{n+3}{2}]} c_{-\frac{n}{2}+[\frac{n+3}{2}]}$$
$$+ O(t^{-\frac{n}{2}+[\frac{n+3}{2}]+1}). \qquad (1.16)$$

□

2 Relative ζ–functions

For a closed oriented manifold (M^n, g), a Riemannian vector bundle $(E, h) \longrightarrow (M^n, g)$ and a non–negative self–adjoint elliptic differential operator $A : C^\infty(E) \longrightarrow C^\infty(A)$ there is a well–defined zeta function

$$\zeta(s, A) = \sum_{\substack{\lambda \in \sigma(A) \\ \lambda > 0}} \frac{1}{\lambda^s} \qquad (2.1)$$

which converges for $\operatorname{Re}(s) > \frac{n}{2}$ and which has a meromorphic extension to $\mathbb{C}$ with only simple poles. In particualar, $s = 0$ is not a pole. Hence

$$\zeta'(0) = \frac{d}{ds}\zeta(s, A)|_{s=0}$$

is well defined and one defines the ζ–determinant of A as

$$\det{}_\zeta A := e^{-\zeta'(0,A)}. \qquad (2.2)$$

This is the first step to define analytic torsion. On open manifolds, (2.1) does not make sense since $\sigma(A)$ is not nessecarily purely discrete. (2.1) can be rewritten as

$$\zeta(s, A) = \frac{1}{\Gamma(s)} \int_0^\infty t^{s-1}(\operatorname{tr} e^{-tA} - \dim \ker A)dt. \qquad (2.3)$$

But (2.3) has a meaningful extension to open manifolds as we will establish in this section.

Definition. Assume $E' \in \operatorname{gen}\operatorname{comp}_{L,diff,F}^{1,r+1}(E)$. Define

$$\zeta_1(s, D^2, (U^*D'U)^2) := \frac{1}{\Gamma(s)} \int_0^1 t^{s-1}\operatorname{tr}\left(e^{-tD^2} - e^{-t(U^*D'U)^2}\right) dt.$$
$$(2.4)$$

We insert the expansion (1.9) into the integrand of (2.4), thus obtaining

$$\int_0^1 t^{s-1} t^{-\frac{n}{2}} t^l \, dt = \frac{1}{s - \frac{n}{2} + l}, \qquad (2.5)$$

$$\int_0^1 t^{s-1} t^{-\frac{n}{2} + [\frac{n+3}{2}]} \, dt = \frac{1}{s + \frac{n}{2} + [\frac{n+3}{2}]},$$

$$\frac{1}{\Gamma(s)} \int_0^1 t^{s-1} O(t^{-\frac{n}{2} + [\frac{n+3}{2}] + 1}) \, dt \text{ holomorphic for}$$

$$\text{Re}(s) + (-\frac{n}{2}) + [\frac{n+3}{2}] + 1 > 0 \qquad (2.6)$$

and $[\frac{n+3}{2}] \geq \frac{n}{2} + 1$, we obtain a function meromorphic in $\text{Re}(s) > -1$, holomorphic in $s = 0$ with simple poles at $s = \frac{n}{2} - l$, $l \leq [\frac{n+3}{2}]$.

Much more troubles causes the integral $\int_1^\infty$. Here we must additionally assume

$$\mu_e(D) = \inf \sigma_e(D^2|_{(\ker D^2)^\perp}) > 0. \qquad (2.7)$$

(2.7) implies $\mu_e(D'^2) = \inf \sigma_e((U^*D'U)^2|_{(\ker(U^*D'U)^2)^\perp}) > 0$. Denote by $\mu_0(D^2)$, $\mu_0(D'^2) = \mu_0((U^*D'U)^2)$ the smallest positive eigenvalue of D^2, D'^2, respectively and set

$$\begin{aligned} \mu(D^2) &= \min\{\mu_e(D^2), \mu_0(D^2)\}, \\ \mu(D'^2) &= \min\{\mu_e(D'^2), \mu_0(D'^2)\}, \\ \mu(D^2, D'^2) &:= \min\{\mu(D^2), \mu(D'^2)\} > 0. \end{aligned} \qquad (2.8)$$

If there is no such eigenvalue for D^2 then set $\mu(D^2) = \mu_e(D^2)$, analogous for D'^2. D^2, D'^2, $(U^*D'U)^2$ have in $]0, \mu(D^2, D'^2)[$ no further spectral values.

We assert that the spectral function $\xi(\lambda) = \xi(\lambda, D^2, (U^*D'U)^2)$ is constant in the interval $[0, \mu(D^2, D'^2)/2[$.

Consider the function $\omega_\varepsilon(x) = \begin{cases} c_\varepsilon e^{-\frac{\varepsilon^2}{\varepsilon^2-x^2}} & |x| \le \varepsilon \\ 0 & |x| > \varepsilon \end{cases}$ and choose c_ε s. t. $\int \omega_\varepsilon(x)\, dx = 1$. Let $0 < 3\varepsilon < \frac{\delta}{2}$ and $\chi_{[-\delta-2\varepsilon,\delta+2\varepsilon]}$ the characteristic function of $[-\delta - 2\varepsilon, \delta + 2\varepsilon]$. Then $\varphi_{\delta,\varepsilon} := \chi_{[-\delta-2\varepsilon,\delta+2\varepsilon]} * \omega_\varepsilon$ satifies $0 \le \varphi_{\delta,\varepsilon} \le 1$, $\varphi_{\delta,\varepsilon}(x) = 1$ on $[-\delta - \varepsilon, \delta + \varepsilon]$, $\varphi_{\delta,\varepsilon}(x) = 0$ for $x \notin]-\delta - 3\varepsilon, \delta + 3\varepsilon[$, $\varphi'_{\delta,\varepsilon} \le K \cdot \varepsilon^{-1}$ and $\lim_{\varepsilon \to 0}\lim_{\delta \to 0} \varphi_{\delta,\varepsilon} = \delta-$ distribution. Assume $\delta + 3\varepsilon < \frac{\mu}{2}$. A regular distribution $f \in D'(]-\frac{\mu}{2}, \frac{\mu}{2}[)$ equals to zero if and only if $\langle f(\lambda), w_\varepsilon(\lambda - a)\sin(k(\lambda - a))\rangle = 0$, $\langle f(\lambda), w_\varepsilon(\lambda - a)\cos(k(\lambda - a))\rangle = 0$ for all sufficiently small ε and all a (s.t. $|a| + \varepsilon < \frac{\mu}{2}$) and for all k. ([71], p. 95). This is equivalent to $\langle f, w_\varepsilon(\lambda - a)\rangle = 0$ for all sufficiently small ε and a and the latter is equivalent to $\langle f(\lambda), \varphi_{\delta,\varepsilon} - \varphi_{\delta',\varepsilon}\rangle = 0$ for all δ, δ' and all ε (s.t. $\delta+3\varepsilon, \delta'+3\varepsilon < \frac{\mu}{2}$). We get for $\mu = \mu(D^2, D'^2)$, $0 < 3\varepsilon < \frac{\delta}{2}$, $\delta+3\varepsilon < \frac{\mu}{2}$ that $\varphi_{\delta,\varepsilon}(D^2) - \varphi_{\delta,\varepsilon}((U^*D'U)^2)$ is independent of δ, ε, $\operatorname{tr}(\varphi_{\delta,\varepsilon}(D^2) - \varphi_{\delta,\varepsilon}((U^*D'U)^2))$ is independent of δ, ε, $0 = \operatorname{tr}(\varphi_{\delta,\varepsilon}(D^2) - \varphi_{\delta,\varepsilon}((U^*D'U)^2)) - \operatorname{tr}(\varphi_{\delta',\varepsilon}(D^2) - \varphi_{\delta',\varepsilon}((U^*D'U)^2)) = \int_0^{\frac{\mu}{2}}(\varphi_{\delta,\varepsilon} - \varphi_{\delta',\varepsilon})'(\lambda)\xi(\lambda)\, d\lambda$, i.e. the distributional derivative of ξ equals to zero, $\xi(\lambda)$ is a constant regular distribution. We write $\xi(\lambda)|_{[0,\frac{\mu}{2}[} = \operatorname{tr}(\varphi_{\delta,\varepsilon}(D^2) - \varphi_{\delta,\varepsilon}((U^*D'U)^2)) \equiv -h$. Set quite parallel to [49] $\tilde{\xi}(\lambda) := \xi(\lambda) + h$ which yields $\tilde{\xi}(\lambda) = 0$ for $\lambda < \frac{\mu}{2}$ and $-t\int_0^\infty e^{-t\lambda}\xi(\lambda)\, d\lambda = h - \int_{\frac{\mu}{2}}^\infty e^{-t\lambda}\tilde{\xi}(\lambda)\, d\lambda$. The latter integral converges for $t > 0$ and can for $t \ge 1$ estimated by

$$e^{-t\frac{\mu}{4}} \int_{\frac{\mu}{2}}^\infty |\tilde{\xi}(\lambda)|e^{-t\frac{\lambda}{4}}\, d\lambda \le Ce^{-t\frac{\mu}{4}}.$$

Hence we proved

Proposition 2.1 *Assume* $E' \in \operatorname{comp}_{L,diff,F}^{1,r+1}(E)$, $\inf \sigma_e(D^2|_{(\ker D^2)^\perp}) > 0$ *and set* $h = \operatorname{tr}(\varphi_{\delta,\varepsilon}(D^2) - \varphi_{\delta,\varepsilon}((U^*D'U)^2))$ *as above. Then there exist* $c > 0$ *s.t.*

$$\operatorname{tr}(e^{-tD^2} - e^{-t(U^*D'U)^2}) = h + O(e^{-ct}). \tag{2.9}$$

Define for $\operatorname{Re}(s) < 0$

$$\zeta_2(s, D^2, D'^2) := \frac{1}{\Gamma(s)} \int_1^\infty t^{s-1}[\operatorname{tr}(e^{-tD^2} - e^{-t(U^*D'U)^2}) - h]\, ds.$$
(2.10)

Then $\zeta_2(s, D^2, D'^2)$ is holomorphic in $\operatorname{Re}(s) < 0$ and admits a meromorphic extension to $\mathbb{C}$ which is holomorphic in $s = 0$.
Define finally

$$\zeta(s, D^2, D'^2) := \frac{1}{\Gamma(s)} \int_0^\infty t^{s-1}[\operatorname{tr}(e^{-tD^2} - e^{-t(U^*D'U)^2}) - h]\, dt$$

$$= \zeta_1(s, D^2, D'^2) + \zeta_2(s, D^2, D'^2) - \frac{1}{\Gamma(s)} \int_0^1 t^{s-1} h\, dt$$

$$= \zeta_1(s, D^2, D'^2) + \zeta_2(s, D^2, D'^2) - \frac{h}{\Gamma(s+1)} \qquad (2.11)$$

We proved

Theorem 2.2 *Suppose $E' \in \operatorname{gen\,comp}_{L,diff,F,rel}^{1,r+1}(E)$, $\inf \sigma_e(D^2|_{(\ker D^2)^\perp}) > 0$ and set h as above. Then $\zeta(s, D^2, D'^2)$ is after meromorphic extension well defined in $\operatorname{Re}(s) > -1$ and holomorphic in $s = 0$.* □

We remark that $(U^*D'U)^2 = U^*D'^2U$, but $(U^*i^*D'iU)^2 \neq U^*i^*D'^2iU$.

The situation is more difficult if there is no spectral gap above zero. Then $\operatorname{tr}(e^{-tD^2} - e^{-t(u^*D'u)^2})$ must not – up to a constant – exponentially decrease to zero. But one can ask, whether there exists for $t \longrightarrow \infty$ an asymptotic expansion in negative powers of t. The answer is yes, if the spectral function $\xi(\lambda)$ has an asymptotic expansion in postive powers of λ near $\lambda = 0$ (cf. [49]). We perform our considerations at first for the general situation of a supersymmetric scattering system H, H' such that for $t > 0$ $e^{-tH} - e^{-tH'}$ is of trace class.

Proposition 2.3 *Suppose that for an interval $[0, \varepsilon]$ there exists a sequence*
$$0 \leq \gamma_0 < \gamma_1 < \gamma_2 < \cdots \longrightarrow \infty$$
such that for every $N \in \mathbb{N}$
$$\xi(\lambda) = \sum_{i=0}^{N} c_i \lambda^{\gamma_i} + O(\lambda^{\gamma_{N+1}}). \tag{2.12}$$

Then there exists for $t \longrightarrow 0$ an asymptotic expansion
$$\mathrm{tr}(e^{-tH} - e^{-tH'}) \sim \sum_{i=0}^{\infty} t^{-\gamma_i} b_{\gamma_i}.$$

Proof. According to II (2.17),
$$\mathrm{tr}(e^{-tH} - e^{-tH'}) = -t \int_0^\infty \xi(\lambda) e^{-t\lambda} d\lambda$$
$$= -t \left[\int_0^\varepsilon \xi(\lambda) e^{-t\lambda} d\lambda + \int_\varepsilon^\infty \xi(\lambda) e^{-t\lambda} d\lambda \right]$$

For $t \geq 1$
$$\left| \int_\varepsilon^\infty \xi(\lambda) e^{-t\lambda} d\lambda \right| \leq C \cdot e^{-t\varepsilon/2}$$

and
$$\int_0^\varepsilon \lambda^\gamma e^{-t\lambda} d\lambda = \int_0^\infty \lambda^\gamma e^{-t\lambda} d\lambda - \int_\varepsilon^\infty \lambda^\gamma e^{-t\lambda} d\lambda$$
$$= \frac{\gamma!}{t^{\gamma+1}} + O(e^{-t\varepsilon/2})$$
$$= \frac{b_\gamma}{t^{\gamma+1}} + O(e^{-t\varepsilon/2})$$

□

Looking at proposition 2.3, there arises the natural question, under which conditions an expansion (2.12) is available. We assume H and H' self-adjoint, ≥ 0 and $e^{-tH} - e^{-tH'}$ for $t > 0$ of trace class. Following [49], we can now establish

Proposition 2.4 *Suppose* $]0, \varepsilon[\subset \sigma_{ac}(H')$, $\xi(\lambda)$ *continuous in* $]0, \varepsilon[$ *and that there exists* $m \in \mathbb{N}$ *such that for* $\lambda \in]0, \varepsilon[$, $\det S(\lambda)$ *extends to a holomorphic function on* $\hat{D}_{\varepsilon,m}$, *where* $D_\varepsilon = \{z \in \mathbb{C} | |z| < \varepsilon\}$ *and* $\hat{D}_{\varepsilon,m} \longrightarrow D_\varepsilon$ *is the ramified covering given by* $z \longrightarrow z^m$. *Then there exist* ε_1, $0 < \varepsilon_1 < \varepsilon^{\frac{1}{m}}$, *and an expansion*

$$\xi(\lambda) = \sum_{k=0}^{\infty} c_k \lambda^{\frac{k}{m}}, \quad \lambda \in]0, \varepsilon_1[. \tag{2.13}$$

Proof. We have by assumption that $\det S(\lambda)$ is smooth on $]0, \varepsilon[$. $|\det S(\lambda)| = 1$ and $]0, \varepsilon[\subset \sigma_{ac}(H')$ imply that there exists $\varphi \in C^\infty(]0, \varepsilon[)$ such that $\det S(\lambda) = e^{-2\pi i \varphi(\lambda)}$ for $\lambda \in]0, \varepsilon[$. We know from II (2.8) $\det S(\lambda) = e^{-2\pi i \xi(\lambda)}$. Hence there exists $k \in \mathbb{Z}$ such that $\xi(\lambda) = \varphi(\lambda) + k$, $\lambda \in]0, \varepsilon[$. This implies $\xi(\lambda)$ smooth on $]0, \varepsilon[$. Differentiation of $\det S(\lambda) = e^{-2\pi i \xi(\lambda)}$ and the unitarity of $S(\lambda)$ yield

$$\frac{d\xi(\lambda)}{d\lambda} = -\frac{1}{2\pi i} \text{tr}\left(S^*(\lambda) \frac{dS(\lambda)}{d\lambda}\right). \tag{2.14}$$

We infer from the assumption $\hat{D}_{\varepsilon,m} \longrightarrow D_{\varepsilon,m}$ and ($\longrightarrow$)-extension of $\det S(\lambda)$ that $\lim_{\lambda \to 0^+} \varphi(\lambda)$ exists, hence $\lim_{\lambda \to 0^+} \xi(\lambda)$ too. Thus we obtain from (2.14)

$$\begin{aligned}\xi(\lambda) &= \xi(0^+) - \frac{1}{2\pi i} \int_0^\lambda \text{tr}\left(S^*(\sigma) \frac{dS(\sigma)}{d\sigma}\right) d\sigma \\ &= \xi(0^+) - \frac{1}{2\pi i} \int_0^{\frac{1}{m}} \text{tr}\left(S^*(\sigma^m) \frac{dS(\sigma^m)}{d\sigma}\right) d\sigma. \end{aligned} \tag{2.15}$$

But $\mathrm{tr}\left(S^*(z^m)\frac{dS(z^m)}{dz}\right)$ is holomorphic on $|z| < \varepsilon_1 \leq \varepsilon^{\frac{1}{m}}$. We get the expansion (2.13), inserting the power series of $\mathrm{tr}\left(S^*(z^m)\frac{dS(z^m)}{dz}\right)$ into (2.15). □

Corollary 2.5 a) *Under the hypothesis of 2.4 with $m = 1$, there exists for $t \to \infty$ an asymptotic expansion*

$$\mathrm{tr}(e^{-tH} - e^{-tH'}) \sim \sum_{k=0}^{\infty} t^{-\beta_k} b_k, \quad 0 = \beta_0 < \beta_1 < \beta_2 < \cdots \longrightarrow \infty.$$
(2.16)

b) *Suppose the hypothesis of 2.4 with m arbitrary and suppose for $t \to \infty$ an asymptotic expansion (2.16). Then the coefficients b_k in (2.16) are not local.*

Proof. of a) (2.16) follows from 2.3 and 2.4. b) Write for $|z| < \varepsilon_1$

$$\mathrm{tr}\left(S^*(z^m)\frac{dS(z^m)}{dz}\right) = \sum_{k=0}^{\infty} d_k z^k. \quad (2.17)$$

Then we see that the c_k in (2.13) and the d_k in (2.17) are related by $c_k = d_{k-1}/k$, $k \geq 1$. Hence the asymptotic expansion (2.16) determines $S(\lambda)$ in a neighbourhood of $\lambda = 0$ up to an additive constant, the b_k in (2.16) are nonlocal. □

We want to define a relative ζ-function also in the case $]0, \varepsilon[\subset \sigma_{ac}(H')$, i.e. very far from a spectral gap above zero.

Lemma 2.6 *Suppose the hypothesis of proposition 2.4. Then for $t \longrightarrow \infty$*

$$\mathrm{tr}(e^{-tH} - e^{-tH'}) = -\xi(0^+) + O(t^{-\frac{1}{m}}). \quad (2.18)$$

Proof.

$$\text{tr}(e^{-tH} - e^{-tH'}) = -t \int_0^\infty \xi(\lambda) e^{-t\lambda} d\lambda$$

$$= -t \left[\frac{e^{-t\lambda}}{-t} \xi(\lambda) \Big|_0^\infty - \int_0^\infty \frac{e^{-t\lambda}}{-t} \xi'(\lambda) d\lambda \right]$$

$$= -\xi(0^+) - \int_0^\infty e^{-t\lambda} \xi'(\lambda) d\lambda$$

$$= -\xi(0^+) - \int_0^\varepsilon \xi'(\lambda) d\lambda - \int_\varepsilon^\infty e^{-t\lambda} \xi'(\lambda) d\lambda$$

$$= \xi(0^+) + \frac{1}{2\pi i} \int_0^\varepsilon \left(e^{-t\lambda} \text{tr} S^*(\lambda) \frac{dS(\lambda)}{d\lambda} \right) d\lambda + O(e^{-t\frac{\varepsilon}{2}})$$

$$= \xi(0^+) + \frac{1}{2\pi i} \int_0^{\varepsilon_1} \left(e^{-t\lambda^m} \text{tr} S^*(\lambda) \frac{dS(\lambda^m)}{d\lambda} \right) d\lambda + O(e^{-t\frac{\varepsilon}{2}})$$

$$= -\xi(0^+) - O(t^{-\frac{1}{m}}). \tag{2.19}$$

$\square$

Now we would like to define again

$$\zeta(s, H, H') := \zeta_1(s, H, H') + \zeta_2(s, H, H')$$

$$:= \frac{1}{\Gamma(s)} \int_0^\varepsilon t^{s-1} \text{tr}(e^{-tH} - e^{-tH'}) dt$$

$$+ \frac{1}{\Gamma(s)} \int_\varepsilon^\infty t^{s-1} \text{tr}(e^{-tH} - e^{-tH'}) dt \tag{2.20}$$

and thereafter perform meromorphic extension.

Theorem 2.7 *Suppose the hypotheses of 2.4 and for $t \to \infty$ an* *asymptotic expansion*

$$\text{tr}(e^{-tH} - e^{-tH'}) \sim \sum_{j=0}^{\infty} a_j t^{\alpha_j}, \quad -\infty < \alpha_0 < \alpha_1 < \cdots \to \infty.$$
(2.21)

Then (2.20) defines a relative ζ-function which is meromorphic in $\mathbb{C}$ and holomorphic in $s = 0$.

Proof. We start with

$$\zeta_1(s, H, H') = \int_0^{\infty} t^{s-1} \text{tr}(e^{-tH} - e^{-tH'}) dt. \qquad (2.22)$$

Inserting (2.21) into (2.22) yields as in (2.5) a meromorphic extension with simple poles at $s = -\alpha_j$ which is holomorphic at $s = 0$ (since we multiply with $\frac{1}{\Gamma(s)}$). According to (2.19),

$$\begin{aligned}
\zeta_2(s, H, H') &= \int_{\varepsilon}^{\infty} t^{s-1} \text{tr}(e^{-tH} - e^{-tH'}) dt \\
&= -\frac{1}{\Gamma(s)} \int_{\varepsilon}^{\infty} t^{s-1} \xi(0^+) dt + \frac{1}{\Gamma(s)} \int_{\varepsilon}^{\infty} t^{s-1} O(t^{-\frac{1}{m}}) dt.
\end{aligned}$$
(2.23)

Here the first integral is absolutely convergent in the half-plane $\text{Re}(s) < 0$ and equals there to

$$-\frac{\varepsilon}{s \cdot \Gamma(s)} = -\frac{\varepsilon}{s(\frac{1}{s} + \gamma + s \cdot h(s))} = -\frac{\varepsilon}{1 + s \cdot \gamma \cdot s \cdot h(s)},$$

which admits a meromorphic extension to $\mathbb{C}$. The second integral yields a holomorphic function in the half-plane $\text{Re}(s) < \frac{1}{m}$. The meromorphic extension to $\mathbb{C}$ is given by integration of (2.16) with simple poles at $s = \beta_k$, $k > 0$. □

Remark 2.8 We see from the proof of theorem 2.7 that we only need (2.18) to have a well-established ζ_2, i.e. the stronger condition (2.21) is not necessary for that. Moreover, using

$$\operatorname{tr}(e^{-tH} - e^{-tH'}) = -t \int_0^\infty e^{-t\lambda} \xi(\lambda) d\lambda,$$

we see that

$$\operatorname{tr}(e^{-tH} - e^{-tH'}) = b_0 + O(t^{-\varrho}) \text{ for } t \longrightarrow \infty \qquad (2.24)$$

is equivalent to

$$\xi(\lambda) = -b_0 + O(\lambda^\varrho) \text{ for } \lambda \to 0^+. \qquad (2.25)$$

□

Corollary 2.9 *Suppose $e^{-tH} - e-tH'$ of trace class, (2.21) and (2.18) or (2.24). Then (2.20) defines a relative ζ-function which is meromorphic in the half-plane $\operatorname{Re}(s) < \varrho$ and holomorphic in $s = 0$, which is explicitely given by*

$$\zeta_2(s, H, H') = -\frac{b_0}{\Gamma(s+1)} + \frac{1}{\Gamma(s)} \int_1^\infty t^{s-1} (\operatorname{tr}(e^{-tH} - e^{-tH'}) - b_0) dt.$$

We apply this to our Clifford bundle situation.

Theorem 2.10 *Let $E \in \operatorname{CL}\mathcal{B}^{N,n}(I, B_k)$, $k \geq r+1 > n+3$, $E^1 \in \operatorname{gen comp}_{L,diff,F,rel}^{1,r+1}(E) \cap \operatorname{CL}\mathcal{B}^{N,n}(I, B_k)$. Suppose additionally $]0, \varepsilon[\subset \sigma_{ac}(D'^2)$ and $\det S(\lambda)$ extends to a holomorphic function on $\hat{D}_{\varepsilon,1}$. Then (2.20) defines a relative ζ-function*

$$\zeta(s, D^2, (U^*D'U)^2)$$

which is meromorphic in $\mathbb{C}$ and holomorphic in $s = 0$.

Proof. According to IV, theorem 3.2 for $t > 0$ $e-tD^2 P - e^{-t(U^*DU)^2} P'$ is of trace class. Hence the wave operators are defined, complete and S and ξ are well-defined. Lemma 1.3 yields an asymptotic expansion of type 2.21 and we get $\zeta_1(s, D^2, (U^*D*U)^2)$ as in (2.22) and thereafter. The assumptions provide (2.18) and we obtain $\zeta_2(s, D^2, (U^*D'U)^2)$ as above. □

3 Relative determinants and QFT

It is well–known that the evolution of a quantum system is described by the S–matrix of QFT. For the elements of this matrix there exist well–known formulas, given by the Feynman path integrals. The exact mathematical understanding and meaning has occupied up to now many mathematicians. There are several approaches. One essential part of these path integrals is the so–called partition function which can be written (perhaps after a so–called Wick–rotation) as

$$Z := \int e^{-S(A)} dA. \tag{3.1}$$

Here $S(A)$ is an action functional and A runs through an infinite dimensional space, e.g. a space of connections (in gauge theory), a space of Riemannian metrics and embeddings (in string theory). In many cases $S(A)$ is of the kind $\langle HA, A \rangle_{L_2}$, where H is an elliptic self–adjoint non–negative differential operator. The model, how to calculate (3.1) is now the Gauß integral

$$\int_{\mathbb{R}^N} e^{-(Ax,x)} dx = \left(\frac{\pi^N}{\det A} \right)^{\frac{1}{2}}. \tag{3.2}$$

This yields a hint, how to attack, better to define the integral in the infinite dimensional case. One simply replaces the determinant in (3.2) by the zeta function determinant, if the latter is defined.

Suppose the underlying space M^n to be compact. Then $\sigma(H)$ is purely discrete and one defines

$$\det(H) := \det_\zeta(H) := e^{-\frac{d}{ds}\zeta(s,H)|_{s=0}} = e^{-\zeta'(0)}, \tag{3.3}$$

$$\zeta(s, H) = \sum_{\substack{\lambda \in \sigma(H) \\ \lambda > 0}} \lambda^{-s} \tag{3.4}$$

with meromorphic extension.

In the case of string theory,

$$Z_p = \int_{\mathcal{M}(\Sigma_p)} \left(\frac{\det(\Delta_{g/2})}{(1,1)_g}\right)^{-\frac{n}{2}} D_g,$$

$Z = \sum_{p=0}^{\infty} Z_p$, Σ_p closed surface of genus p.

If the underlying manifold M^n is open then $\sigma(H)$ is not purely discrete and (3.3), (3.4) do not make sense.

We are now able to rescue this situation by considering relative determinants,

$$\det(H, H') := e^{-\zeta'(0,H,H')}. \tag{3.5}$$

If $E', E'' \in \operatorname{gen comp}_{L,diff,F}^{1,r+1}(E)$ then we denote as above $\tilde{D}'^2 = (U^*D'U)^2$, $\tilde{D}''^2 = (V^*D''V)^2$ for the transformed operators acting in $L_2((M,E),g,h)$.

Theorem 3.1 *Suppose* $E', E'' \in \operatorname{gen comp}_{L,diff,F}^{1,r+1}(E)$ *and* $\inf \sigma$ $(D^2|_{(\ker D^2)^\perp}) > 0$.

a) *Then* $\zeta(s, D^2, \tilde{D}'^2)$, $\zeta(s, D^2, \tilde{D}''^2)$, $\zeta(s, \tilde{D}'^2, \tilde{D}''^2)$ *are after meromorphic extension in* $\operatorname{Re}(s) > -1$ *well defined and holomorphic in* $s = 0$. *In particular* $\det(D^2, D'^2) = e^{-\zeta'(0,D^2,\tilde{D}'^2)}$, $\det(D^2, D''^2) = e^{-\zeta'(0,D^2,\tilde{D}''^2)}$, $\det(\tilde{D}'^2, D''^2) = e^{-\zeta'(0,\tilde{D}'^2,\tilde{D}''^2)}$ *are well defined.*

b) *There holds*

$$\det(\tilde{D}'^2, D^2) = \det(D^2, \tilde{D}'^2)^{-1} \tag{3.6}$$

etc. and

$$\det(D^2, \tilde{D}''^2) = \det(D^2, \tilde{D}'^2) \cdot \det(\tilde{D}'^2, \tilde{D}''^2). \tag{3.7}$$

Proof. a) follows from 2.1 and the fact that $E, E'' \in \operatorname{gen comp}(E)$ implies $E'' \in \operatorname{gen comp}(E')(= \operatorname{gen comp}(E))$.
b) immediately follows from the definitions and $\operatorname{tr}(e^{-tD^2} - e^{-t\tilde{D}''^2}) = \operatorname{tr}(e^{-tD^2} - e^{-t\tilde{D}'^2}) + \operatorname{tr}(e^{-t\tilde{D}'^2} - e^{-t\tilde{D}''^2})$. □

4 Relative analytic torsion

If we now restrict to the case $E = (\Lambda^* T^* M \otimes \mathbb{C}, g_\Lambda)$ then $g' \in \text{gen comp}^{1,r+1}(g)$ does not imply $E' = (\Lambda^* T^* M \otimes \mathbb{C}, g'_\Lambda) \in \text{gen comp}^{1,r+1}_{L,diff,F}(E)$ since the fibre metric changes, $g_\Lambda \longrightarrow g'_\Lambda$. Hence the above considerations for constructing the relative ζ–function are not immediately applicable, since they assume the invariance of the fibre metric. Fortunately we can define relative ζ–functions also in this case. We recall from [38], p. 65 – 74 the following well known fact which we used in (1.1), (1.2) already. Let P be a self adjoint elliptic partial differential operator of order 2 such that the leading symbol of P is positive definite, acting on sections of a vector bundle $(V, h) \longrightarrow (M^n, g)$. Let $W_P(t, p, m)$ be the heat kernel of e^{-tP}, $t > 0$. Then for $t \longrightarrow 0^+$

$$\operatorname{tr} W_P(t, m, m) \sim t^{-\frac{n}{2}} b_{-\frac{n}{2}}(m) + t^{-\frac{n}{2}+1} b_{-\frac{n}{2}+1}(m) + \cdots$$

and the $b_\nu(m)$ can be locally calculated as certain derivatives of the symbol of P according to fixed rules. As established by Gilkey, for $P = \Delta$ or $P = D^2$ the b's can be expressed by curvature expressions (including derivatives). This is (1.1), (1.2), (1.3). We apply this to $e^{-t\Delta}$ and $e^{-t(U^* i^* \Delta' iU)}$ but we want to compare the asymptotic expansions of $W_\Delta(t, m, m)$ and $W_{\Delta'}(t, m, m)$. The expansion of

$$W_{i^* \Delta' i}(t, m, m) \text{ in } L_2(g', g_{\Lambda^*}) \tag{4.1}$$

and

$$W_{U^* i^* \Delta' iU}(t, m, m) \text{ in } L_2(g, g_{\Lambda^*}) \tag{4.2}$$

coincide since

$$W_{U^* i^* \Delta' iU}(t, m, m) = \alpha^{-\frac{1}{2}}(m) W_{i^* \Delta' i}(t, m, m) \alpha^{\frac{1}{2}}(m). \tag{4.3}$$

The point is to compare the expansions of

$$W_{i^* \Delta' i}(t, m, m) \text{ in } L_2(g', g_{\Lambda^*}) \tag{4.4}$$

and

$$W_{\Delta'}(t, m, m) \text{ in } L_2(g', g'_{\Lambda^*}), \tag{4.5}$$

i.e. we have to compare the symbol of $i^*\Delta' i = i^*\Delta'$ and Δ'. For $q = 0$ they coincide. Let $q = 1$, $m \in M$, $\omega_1, \ldots, \omega_n$ a basis in $T_m^* M$, $\Phi \in \Omega^1(M)$, $\Delta'\Phi|_m = \xi^1\omega_1 + \cdots + \xi^n\omega_n$. Then, according to IV (2.55) $i^*(\Delta'\Phi_i)|_m = g^{kl}g'_{ik}\xi^i\omega_l$, i.e.

$$((i^* - 1)\Delta'_1\Phi)|_m = (g^{kl}g'_{ik}\xi^i - \xi^l)\omega_l. \tag{4.6}$$

Hence for the (local) coefficients of $i^*\Delta' i$ as differential operator holds

$$\text{coeff of } (i^*\Delta'_1 i) = (g^{kl}g'_{ik}) \text{ coeff of } (\Delta'_1). \tag{4.7}$$

Quite similar for $0 < q < n$, e.g.

$$\text{coeff of } (i^*\Delta'_2 i) = (g^{k_1 l_1}g^{k_2 l_2} g'_{i_1 k_1} g'_{i_2 k_2}) \text{ coeff of } (\Delta'_2).$$

Proposition 4.1 *Let $r > n + 2$, $g' \in \text{comp}^{1,r+1}(g)$, $l \leq \frac{n+3}{2}$, $b_{-\frac{n}{2}+l}(\Delta(g, g_{\Lambda^*}), g, g_{\Lambda^*}, m)$ and $b_{-\frac{n}{2}+l}(U^* i^* \Delta'(g', g'_{\Lambda^*}) i U, g, g_{\Lambda^*})$ the coefficients of the asymptotic expansion of $\text{tr}_{g_{\Lambda^*}} W_\Delta(t, m, m)$ and $\text{tr}_{g_{\Lambda^*}} W_{U^* i^* \Delta' i U}(t, m, m)$ in $L_2(M, g)$, repectively. Then*

$$b_{-\frac{n}{2}+l}(\Delta, g, g_{\Lambda^*}, m) - b_{-\frac{n}{2}+l}(U^* i^* \Delta' i U, g, g_{\Lambda^*}, m) \in L_1(M, g). \tag{4.8}$$

Proof. Write

$$b_{-\frac{n}{2}+l}(\Delta, g, g_{\Lambda^*}, m) - b_{-\frac{n}{2}+l}(U^* i^* \Delta' i U, g, g_{\Lambda^*}, m)$$
$$= b_{-\frac{n}{2}+l}(\Delta, g, g_{\Lambda^*}, m) - b_{-\frac{n}{2}+l}(\Delta', g', g'_{\Lambda^*}, m) + \tag{4.9}$$
$$+ b_{-\frac{n}{2}+l}(\Delta', g', g'_{\Lambda^*}, m) - b_{-\frac{n}{2}+l}(i^* \Delta' i, g', g_{\Lambda^*}, m) + \tag{4.10}$$
$$+ b_{-\frac{n}{2}+l}(i^* \Delta' i, g', g_{\Lambda^*}, m) - b_{-\frac{n}{2}+l}(U^* i^* \Delta' i U, g, g_{\Lambda^*}, m) \tag{4.11}$$

where $b_{-\frac{n}{2}+l}(\Delta', g', g'_{\Lambda^*}, m)$, $b_{-\frac{n}{2}+l}(i^*\Delta' i, g', g_{\Lambda^*})$ are explained in (4.5), (4.4), respectively. (4.11) vanishes according to (4.3). (4.9) $\in L_1(M, g)$ according to the expressions (1.3) and $g' \in \text{comp}^{1,r}(g)$. We conclude this as in the proof of 1.1. Finally (4.10) $\in L_1$ according to $i^*\Delta' i = (i^* - 1)\Delta' + \Delta'$, coeff $(i^*\Delta' i) = $ coeff $(i^* - 1) + $ coeff (Δ'), $(i^* - 1) \in \Omega^{0,r}(\text{End}(\Lambda^*))$, the rules for calculating the heat kernel expansion and according to the module structure theorem. $\square$

Theorem 4.2 Let (M^n, g) be open, satisfying (I), (B_k), $k \geq r+1 > n+3$, $g' \in \text{comp}^{1,r+1}(g)$, $\Delta = \Delta(g, g_{\Lambda^*})$, $\Delta' = \Delta(g', g'_{\Lambda^*})$ the graded Laplace operators, U, i as in (2.54), and assume $\inf \sigma_e(\Delta|_{(\ker \Delta)^\perp}) > 0$.

a) Then for $t > 0$
$$e^{-t\Delta} - e^{-tU^*i^*\Delta' iU}$$
is of trace class.

b) Denote $h = \text{tr}(\varphi_{\delta,\varepsilon}(\Delta) - \varphi_{\delta,\varepsilon}(U^*i^*\Delta' iU))$ for $0 < 3\varepsilon < \frac{\delta}{2}$, $\delta + 3\delta < \frac{\mu}{2}$, $\mu = \inf\{\text{nonzero spectrum of } \Delta, i^*\Delta' i\}$. Then

$$\zeta_q(s, \Delta, \Delta') := \frac{1}{\Gamma(s)} \int_0^\infty t^{s-1}[\text{tr}(e^{-t\Delta_q} - e^{-t(U^*i^*\Delta'_q iU)}) - h]\, dt$$

has a well defined meromorphic extension to $\text{Re}(s) > -1$ which is holomorphic in $s = 0$.

c) The relative analytic torsion $\tau^a(M^n, g, g')$,

$$\log \tau^a(M^n, g, g') := \frac{1}{2} \sum_{q=0}^n (-1)^q q \cdot \zeta'_q(0, \Delta, \Delta') \qquad (4.12)$$

is well-defined.

Proof. a) is just IV theorem 3.8. b) immediately follows from theorem 2.2 and the proof of theorem 2.2. c) is a consequence of b). □

We defined in II 4 gen $\text{comp}^{1,r+1}_{L,diff,rel}(g)$ which induces at the level of Λ^*T^*M gen $\text{comp}^{1,r+1}_{L,diff,rel}(\Lambda^*T^*, g_\Lambda)$. According IV 3.9, for $t > 0$

$$e^{-t\Delta}P - e^{t(U^*i^*\Delta' iU)}P'$$

is of trace class. Then we can apply the asymptotic expansion (1.16) and obtain as above relative ζ–functions

$$\zeta_q(s, \Delta, \Delta') = \zeta_q(x, (M, g, \Delta_g), (M', g', \Delta'_g)),$$

which are holomorphic at $s = 0$. Hence we got

Theorem 4.3 Let (M^n, g) be open, satisfying (I), (B_k), $k \geq r > n + 3$, $g' \in \text{gen comp}_{L,diff,rel}^{1,r+1}(M^n, g)$ and suppose $\inf \sigma_e(\Delta|_{(\ker \Delta)^\perp}) > 0$. Then there is a well–defined relative analytic torsion $\tau^a((M,g),(M',g'))$,

$$\log \tau^a((M,g),(M',g')) := \frac{1}{2} \sum_{q=0}^{n} (-1)^q q \cdot \zeta'_q(0, \Delta, \Delta'). \quad (4.13)$$

□

The last step ist to give up assumption $\inf \sigma_e(\Delta|_{(\ker \Delta)^\perp}) > 0$.

Theorem 4.4 Let (M^n, g) be open, satisfying (I), (B_k), $k \geq r > n + 3$, $g' \in \text{gen comp}_{L,diff,rel}^{1,r+1}(M^n, g)$ and suppose additionally $]0, \varepsilon[\subset \sigma_{ac}(\Delta')$ and that $\det S(\lambda)$ extends to a holomorphic function on $\hat{D}_{\varepsilon,m}$. Then there is a well defined relative analytic torsion $\tau^a((M,g),(M',g'))$,

$$\log \tau^a((M,g),(M',g')) := \frac{1}{2} \sum_{q=0}^{n} (-1)^q q \zeta'_q(0, \Delta, \Delta'), \quad (4.14)$$

where $\zeta_q(s, \Delta, \Delta')$ is defined by (2.20) and theorem 2.10. □

Remark 4.5 The assertion of theorem 4.4 remains valid if we replace the assumption $\det S(\lambda)$ extends to a holomorphic function on $\hat{D}_{\varepsilon,m}$ by (2.18). □

5 Relative η–invariants

Finally we turn to the relative η–invariant. On a closed manifold (M^n, g) and for a generalized Dirac operator the η–function is defined as

$$\eta_D(s) := \sum_{\substack{\lambda \in \sigma(D) \\ \lambda \neq 0}} \frac{\text{sign } \lambda}{|\lambda|^s} = \frac{1}{\Gamma\left(\frac{s+1}{2}\right)} \int_0^\infty t^{\frac{s-1}{2}} \text{tr}(De^{-tD^2}) \, dt.$$

(5.1)

$\eta_D(s)$ is defined for $\text{Re}(s) > n$, it has a meromorphic extension to $\mathbb{C}$ with isolated simple poles and the residues at all poles are locally computable. $\Gamma\left(\frac{s+1}{2}\right) \cdot \eta_D(s)$ has its poles at $\frac{n+1-\nu}{2}$ for $\nu \in \mathbb{N}$. One cannot conclude directly η is regular at $s = 0$ since $\Gamma(u)$ is regular at $u = \frac{1}{2}$, i.e. $\Gamma\left(\frac{s+1}{2}\right)$ is regular at $s = 0$. But one can show in fact using methods of algebraic topology that $\eta(s)$ is regular at $s = 0$. A purely analytical proof for this is presently not known (cf. [38], p. 114/115).

(5.1) does not make sense on open manifolds. But we are able to define a relative η-function and under an additional assumption the relative η-invariant. De^{-tD^2} is an integral operator with heat kernel $D_p W_D(t, m, p)$ which has at the diagonal a well defined asymptotic expansion (cf. [16], p. 75, lemma 1.9.1 for the compact case)

$$\text{tr} D_p W(t,m,p)|_{p=m} \underset{t \to 0^+}{\sim} \sum_{l \geq 0} b_{-\frac{n+l}{2}}(D^2, D, m) t^{\frac{-n+l}{2}}. \quad (5.2)$$

In [3] has been proved that the heat kernel expansion on closed manifolds also holds on open manifolds with the same coefficients (it is a local matter) independent of the trace class property. The (simple) proof there is carried out for e^{-tD^2}, $\text{tr} W(t,m,m)$, but can be word by word repeated for De^{-tD^2}, $DW(t,m,m)$. The rules for calculating the $b_{-\frac{n+l}{2}}(D^2, D, m)$ are quite similar to them for $b_{-\frac{n+l}{2}}(D, m)$ (cf. [38], Lemma 1.9.1). We sum up these considerations in

Proposition 5.1 *Let $E' \in \text{gen comp}_{L,diff,F}^{1,r+1}(E)$, $r + 1 > n + 3$. Then for $t > 0$*

$$e^{-tD^2}D - e^{-t(U^*D'U)^2}(U^*D'U) \quad (5.3)$$

is of trace class, for $t \to 0^+$ there exists an asymptotic expansion

$$\text{tr}(e^{-tD^2}D - e^{-t(U^*D'U)^2}(U^*D'U))$$
$$= \sum_{l=0}^{n+3} \int_M a_{-\frac{n+l}{2}}(m) \, dvol_m(g) t^{\frac{-n+l}{2}} + O(t^{\frac{4}{2}}). \quad (5.4)$$

Proof. The first assertion is just IV theorem 2.9. We recall from [38] the existence of an asymptotic expansion for the diagonal of the heat kernel

$$\tilde{D}' = (e^{-tD^2}D - e^{-t\tilde{D}'^2}\tilde{D}')(m,m)$$

$$= \sum_{l=0}^{n+3}\left[b_{-\frac{n+l}{2}}(D^2,D,m) - b_{-\frac{n+l}{2}}(\tilde{D}'^2,\tilde{D}',m)\right]t^{-\frac{n+l}{2}} + O(t^{\frac{4}{2}}). \tag{5.5}$$

It can be proved, absolutely parallel to lemma 1.1, that $[\ ] \in L_1$, $O(t^{\frac{4}{2}}) \in L_1$. Integration of (5.5) yields (5.4). □

We recall from [49] the following

Proposition 5.2 *Assume that D and $\tilde{D}' = U^*i^*D'iU$ satisfy (5.3), (5.4) and that the spectra of D and D' have a common gap $[a,b]$, $\left(\sigma(D) \cup \sigma(\tilde{D}')\right) \cap [a,b] = \emptyset$. Then there exists a spectral shift function $\xi(\lambda) = \xi\left(\lambda, D, \tilde{D}'\right)$ having the following properties.*

1) $\xi \in L_{1,loc}(\mathbb{R})$ *and* $\xi(\lambda) = 0$ *for* $\lambda \in [a,b]$. (5.6)
2) *For all* $\varphi \in C_c^\infty(\mathbb{R})$, $\varphi(D) - \varphi(\tilde{D}')$ *is a trace class operator and*

$$\operatorname{tr}\left(\varphi(D) - \varphi(\tilde{D}')\right) = \int_\mathbb{R} \varphi'(\lambda)\xi(\lambda)\,d\lambda. \tag{5.7}$$

3) $\operatorname{tr}\left(e^{-tD^2}D - e^{-t\tilde{D}'^2}\tilde{D}'\right) = \int_\mathbb{R} \frac{d}{d\lambda}(\lambda e^{-t\lambda^2})\xi(\lambda)\,d\lambda.$ (5.8)

□

Proposition 5.3 *Assume $E' \in \operatorname{gen\,comp}_{L,diff}^{1,r+1}(E)$ and $\inf \sigma_e (D^2|_{(\ker D^2)^\perp}) > 0$. Then there exists $c > 0$ s. t.*

$$\operatorname{tr}(e^{-tD^2}D - e^{-t(U^*i^*D'iU)^2}(U^*i^*D'iU)) = O(e^{-ct}). \tag{5.9}$$

Proof. We conclude as in the proof of 2.1 that there exists $\mu > 0$ s. t. $\sigma(D) \cup \sigma(i^*D'i) \cap ([-\mu, -\frac{1}{\nu}] \cup [\frac{1}{\nu}, \mu]) = \emptyset$ for all

$\nu \geq \nu_0$. Hence, according to (5.5), $\int_{-\mu}^{\mu} \frac{d}{dt}(\lambda e^{-t\lambda^2})\xi(\lambda)\,d\lambda = 0$ and

$$|\mathrm{tr}(e^{-tD^2}D - e^{-t(U^*i^*D'iU)^2})|$$

$$= \left| \int_{-\infty}^{-\mu} \frac{d}{d\lambda}(\lambda e^{-t\lambda^2})\xi(\lambda)\,d\lambda + \int_{\mu}^{\infty} \frac{d}{d\lambda}(\lambda e^{-t\lambda^2})\xi(\lambda)\,d\lambda \right|$$

$$\leq e^{-t\frac{\mu^2}{2}} \Big[\int_{-\infty}^{-\mu} e^{-t\frac{\lambda^2}{2}} |1 - 2t\lambda| |\xi(\lambda)|\,d\lambda$$

$$+ \int_{\mu}^{\infty} e^{-t\frac{\lambda^2}{2}} |1 - 2t\lambda| |\xi(\lambda)|\,d\lambda \Big]$$

$$= C \cdot e^{-t\frac{\mu^2}{2}}.$$

□

Theorem 5.4 *Assume* $E' \in \mathrm{gen\,comp}_{L,diff,F}^{1,r+1}(E)$, $k \geq r+1 > n+3$ *and* $\inf \sigma_e(D^2|_{(\ker D^2)^\perp}) > 0$. *Then there is a well defined relative η–function*

$$\eta(s, D, D') := \frac{1}{\Gamma\left(\frac{s+1}{2}\right)} \int_0^\infty t^{\frac{s-1}{2}} \mathrm{tr}(De^{-tD^2} - U^*D'Ue^{-t(U^*D'U)^2})\,dt \tag{5.10}$$

which is defined for $\mathrm{Re}(s) > \frac{n}{2}$ *and admits a meromorphic extension to* $\mathrm{Re}(s) > -5$. *It is holomorphic at $s = 0$ if the coefficient* $\int a_{-\frac{1}{2}}(m)\,dvol_m(g)$ *of* $t^{-\frac{1}{2}}$ *equals to zero. Then there is a well defined relative η–invariant of the pair* (E, E').

Proof. We write again $U^*D'U = \tilde{D}'$. Then according to proposition 5.1,

$$\eta(s, D, \tilde{D}') = \frac{1}{\Gamma\left(\frac{s+1}{2}\right)} \int_0^\infty \left[t^{\frac{s-1}{2}} \mathrm{tr}(e^{-tD^2} D - e^{-t\tilde{D}'^2} \tilde{D}') \right] dt$$

$$= \frac{1}{\Gamma\left(\frac{s+1}{2}\right)} \int_0^1 [\ldots] dt + \int_1^\infty [\ldots] dt$$

$$= \frac{1}{\Gamma\left(\frac{s+1}{2}\right)} \int_0^1 \left(t^{\frac{s-1}{2}} \left\{ \sum_{l=0}^{n+3} \int_M a_{-\frac{n+l}{2}} \, \mathrm{dvol}_m(g) + O(t^{\frac{4}{2}}) \right\} \right) dt$$

$$+ \frac{1}{\Gamma\left(\frac{s+1}{2}\right)} \int_1^\infty t^{\frac{s-1}{2}} \mathrm{tr}(e^{-tD^2} D - e^{-t\tilde{D}'^2} \tilde{D}') dt$$

$$= \frac{1}{\Gamma\left(\frac{s+1}{2}\right)} \sum_{l=0}^{n+3} \frac{1}{\frac{s}{2} - \frac{n}{2} + \frac{l}{2} + \frac{1}{2}} \int_M a_{-\frac{n+l}{2}} \, \mathrm{dvol}_m(g)$$

$$+ \frac{1}{\Gamma\left(\frac{s+1}{2}\right)} \int_0^1 t^{\frac{s-1}{2}} O(t^{\frac{4}{2}}) dt \tag{5.11}$$

$$+ \frac{1}{\Gamma\left(\frac{s+1}{2}\right)} \int_1^\infty t^{\frac{s-1}{2}} \mathrm{tr}(e^{-tD^2} D - e^{-t\tilde{D}'^2} \tilde{D}') dt \tag{5.12}$$

We infer from (5.9) that (5.13) is holomorphic in $\mathbb{C}$. (5.12) is holomorphic in $\mathrm{Re}(s) > -5$. (10.45) admits a meromorphic extension to $\mathbb{C}$. $\eta(s, D, D')$ is holomorphic at $s = 0$ if the coefficient $\int_M a_{-\frac{1}{2}}(m) \, \mathrm{dvol}_m(g)$ equals to zero. □

Theorem 5.4 immediately generalizes to the case of additional compact perturbations.

Theorem 5.5 *Assume* $E' \in \mathrm{gen\, comp}_{L,diff,F,rel}^{1,r+1}(E)$, $k \geq r+1 > n+3$

$$\sigma_e(D^2|_{(\ker D^2)^\perp}) > 0.$$

Then there is a well defined relative η-function

$$\eta(s, D, \tilde{D}') := \frac{1}{\Gamma\left(\frac{s+1}{2}\right)} \int_0^\infty t^{\frac{s-1}{2}} \operatorname{tr}(De^{-tD^2}P - \tilde{D}'e^{-t\tilde{D}'^2}P')dt,$$

which is defined for $\operatorname{Re}(s) > \frac{n}{2}$ and admits a meromorphic extension to $\operatorname{Re}(s) > -5$. Here $\tilde{D}' = U^*D'U$ as above. $\eta(s, D, \tilde{D}')$ is holomorphic at $s = 0$ if the integrated coefficient

$$a_{-\frac{1}{2}} = \int_K b_{-\frac{1}{2}}(D^2, D, m)\, \mathrm{dvol}_m(g)$$

$$- \int_{K'} b_{-frac12}(\tilde{D}'^2, \tilde{D}', m')\, \mathrm{dvol}_{m'}(g')$$

$$+ \int_{M\backslash K = M'\backslash K'} \left(b_{-\frac{1}{2}}(D^2, D, m) - b_{-\frac{1}{2}}(\tilde{D}'^2, \tilde{D}', m)\right) \mathrm{dvol}_m(g)$$

equals to zero.

We repeat for the proof the single arguments from the proof of 5.4 which remain valid in the case of 5.5. □

6 Examples and applications

In this section, we present examples of pairs of generalized Dirac operators which satisfy the assumptions of sections 1–5 and present applications of some theorems of these sections.

Let (M^n, g) be open with finitely many collared ends ε_i, the collar $[0, \infty[\times N_i^{n-1}$ of ε_i endowed with a warped product metric, i.e. $g|_{\varepsilon_i} \cong dr^2 + f_i(r)^2 d\sigma_{N_i}^2$, N_i closed, $h_i = d\sigma_{N_i}^2$, $i = 1, \ldots, m$.
We consider one end ε with collar $[0, \infty[\times N$ with the warped product metric $ds^2|_\varepsilon = dr^2 + f(r)^2 d\sigma^2$ and we first calculate the curvature.

Let $U_0, U_1, \ldots, U_{n-1}$ be an orthogonal basis in $T_{(r,u)}([0, \infty[\times N)$ with respect to ds^2, $U = \frac{\partial}{\partial r}$, $U_1, \ldots, U_{n-1}$ orthonormal in $T_U N$

with respect to $d\sigma^2 = h$. Then, in coordinates $(r, u^1, \ldots, u^{n-1})$, we get for the Christoffel symbols $\Gamma^\gamma_{\alpha,\beta}(g)$, $\alpha, \beta, \gamma = 0, \ldots, n-1$, the following expressions

$$\Gamma^0_{00} = 0, \qquad \Gamma^0_{0j} = 0 \text{ for } j > 0,$$

$$\Gamma^k_{00} = 0 \text{ for } k > 0, \qquad \Gamma^k_{0j} = \tfrac{1}{2}\tfrac{f}{f'}\delta^k_j \text{ for } j, k > 0, \qquad (6.1)$$

$$\Gamma^0_{ij} = -f'f h_{ij} \text{ for } i, j > 0, \quad \Gamma^k_{ij} = \Gamma^k_{ij}(h) \text{ for } i, j, k > 0.$$

For the curvature tensor and the sectional curvature holds

$$R(U_0, U_i)U_0 = \frac{f''}{f}U_i \qquad (6.2)$$

$$R(U_0, U_i)U_j = -f'' f h_{ij} U_0 \qquad (6.3)$$

$$R(U_i, U_j)U_0 = 0 \qquad (6.4)$$

$$R(U_i, U_j)U_k = -f'^2(h_{jk}U_i - h_{ik}U_j) + R_N(U_i, U_j)U_k, \quad (6.5)$$

which implies immediately

$$K(U_0, U_j) = -\frac{f''}{f}, \qquad (6.6)$$

$$K(U_i, U_j) = K_N(U_i, U_j)/f^2 - \frac{f'^2}{f^2}. \qquad (6.7)$$

Here $i, j, k = 1, \ldots, n-1$. The easy calcualtions are performed in [28]. It is now easy from (6.2) – (6.7) to calculate the general curvature $K(V, W)$.

Examples 6.1 1) *Take $f(r) = e^{-r}$, N flat, then $K \equiv -1$, ε satisfies (B_0) but $r_{\text{inj}}(\varepsilon) = 0$.*
2) *Choose $f(r) = e^{-r}$, $K_N \neq 0$, then ε does not satisfy (B_0) and again $r_{\text{inj}}(\varepsilon) = 0$.*
3) *If $f(r) = e^r$, N flat, then ε satisfies (B_0) and $r_{\text{inj}}(\varepsilon) > 0$.*
4) *Finally take $f(r) = e^{r^2}$, then ε does not satisfy (B_0) but (I). Hence all good and bad combinations of properties are possible.*

□

Proposition 6.2 *Suppose $f(r)$ such that*

$$\inf_r f(r) > 0 \text{ or } f \text{ monotone increasing} \tag{6.8}$$

and

$$|f^{(\nu)}| \leq c_\nu f, \nu = 1, 2, \ldots. \tag{6.9}$$

Then $g|_\varepsilon$ satisfies (I) and (B_∞).

Proof. $\inf_r f(r) > 0$ and $r_{\text{inj}}(N,h) > 0$ immediately imply $r_{\text{inj}}(\varepsilon) > 0$. (6.6) and (6.7) immediately imply (B_0). (B_1) is equivalent with (B_0) and

$$|\nabla_{e_\alpha}(R(e_\beta, e_\gamma)e_\delta)|_x \leq C \tag{6.10}$$

$$|\nabla_{e_\alpha} e_\beta|_x \leq C, \tag{6.11}$$

$e_0, \ldots, e_{n-1}$ tangential vector fields, orthonormal in $T_x M$. We apply this to $x = (v, y) \in [0, \infty[\times N$, $U_0 = \frac{\partial}{\partial r}$, $U_i = \frac{\partial}{\partial u^i}$, $e_0 = U_0$, $e_1 = \frac{U_1}{f}, \ldots, e_{n-1} = \frac{U_{n-1}}{f}$. Then according to (6.1),

$$\nabla_{e_0} e_0 = \nabla_{U_0} U_0 = \Gamma^0_{00} U_0 + \Gamma^i_{00} U_i = 0, \tag{6.12}$$

$$\nabla_{e_i} e_0 = \nabla_{\frac{U_i}{f}} U_0 = \frac{1}{f}\nabla_{U_i} U_0 = \frac{1}{f}[\Gamma^0_{i0} U_0 + \Gamma^j_{i0} U_j]$$

$$= \frac{1}{f}\Gamma^j_{i0} U_j = \frac{1}{f}\frac{1}{2}\frac{f'}{f}\delta^j_i U_j = \frac{1}{2}\frac{f'}{f} e_i, \tag{6.13}$$

$$\nabla_{e_0} e_i = -\frac{1}{2}\frac{f'}{f} e_i, \tag{6.14}$$

$$\nabla_{e_i} e_j = -\frac{f'}{f} h_{ij} U_0 + \frac{1}{f}\Gamma^l_{ij}(h) e_l = -\frac{f'}{f} h_{ij} e_0 + \frac{1}{f}\Gamma^l_{ij}(h) e_l. \tag{6.15}$$

We see, each term on the r.h.s. of (6.12) – (6.15) is – up to a constant or bounded function – a sum of terms

$$\frac{f'}{f} e_i, \frac{f'}{f} e_0, \frac{1}{f} e_l \tag{6.16}$$

with pointwise norms (w.r.t. $g|_\varepsilon$)

$$\left|\frac{f'}{f}\right|, \frac{1}{f}, \tag{6.17}$$

i.e. (6.11) is satisfied. Next we establish (6.10)

$$\nabla_{e_0}(R(e_0, e_i)e_0) = \nabla_{U_0}(R(U_0, \frac{U_i}{f}), U_0) = \nabla_{U_0}\frac{1}{f}R(U_0, U_i)U_0$$

$$= \nabla_{U_0}\frac{1}{f}\frac{f''}{f}U_i \left(\frac{f''}{f^2}\right)' U_i + \frac{f''}{f^2}\nabla_{U_0}U_i$$

$$= \frac{ff''' - \frac{3}{2}f'f''}{f^2}e_i, \qquad (6.18)$$

$$\nabla_{e_j}(R(e_0, e_i), e_0) = \frac{1}{f^2}\nabla_{U_j}(R(U_0, U_i)U_0) = \frac{1}{f^2}\nabla_{U_j}\frac{f''}{f}U_i$$

$$= \frac{f''}{f^3}\nabla_{U_j}U_i = \frac{f''}{f^3}(-f'fh_{ij}U_0 + \Gamma_{ij}^k(h)U_k)$$

$$= -\frac{f''f'}{f^2}h_{ij}e_0 + \frac{f''}{f^2}\Gamma_{ij}^k(h)e_k, \qquad (6.19)$$

$$\nabla_{e_0}(R(e_0, e_i)e_j) = \nabla_{U_0}\frac{1}{f^2}R(U_0, U_i)U_j$$

$$= -\frac{2f'}{f^3}R(U_0, U_i)U_j) + \frac{1}{f^2}\nabla_{U_0}(-f''fh_{ij}U_0)$$

$$= -\frac{2f'}{f^3}(-f''fh_{ij}U_0) + \frac{1}{f^2}(-f'''f - f''f')h_{ij}U_0$$

$$= \frac{2f'f''}{f^2}h_{ij}e_0 - \left(\frac{f'''}{f} + \frac{f''f'}{f^2}\right)h_{ij}e_0$$

$$= \left(\frac{f'f''}{f^2} - \frac{f'''}{f}\right)h_{ij}e_0, \qquad (6.20)$$

$$\nabla_{e_k}(R(e_0, e_i)e_j) = \frac{1}{f^3}\nabla_{U_k}(R(U_0, U_i)U_j) = \frac{1}{f^3}\nabla_{U_k}(-f''fh_{ij}U_0)$$

$$= -\frac{f''}{f^2}h_{ij}\Gamma_{k0}^l U_l = -\frac{f''}{f^2}h_{ij}\frac{1}{2}\frac{f'}{f}\partial_k^l U_l$$

$$= -\frac{1}{2}\frac{f''f'}{f^3}h_{ij}U_k, \qquad (6.21)$$

$$\nabla_{e_0}(R(e_i,e_j),e_0) = \frac{1}{f^2}\nabla_{U_0}(R(U_i,U_j)U_0) = 0$$

$$= \frac{1}{f^3}\nabla_{U_k}(R(U_i,U_j)U_0)$$

$$= \nabla_{e_k}(R(e_i,e_j)e_0), \qquad (6.22)$$

$$\nabla_{e_l}(R(e_i,e_j)e_k) = \frac{1}{f^4}\nabla_{U_l}(R(U_i,U_j),U_k)$$

$$= \frac{1}{f^4}\nabla_{U_l}[-f'^2(h_{jk}U_i - h_{ik}U_j$$

$$+ R_N(U_i,U_j)U_k]$$

$$= -\frac{f'^2}{f^4}(h_{jk}\nabla_{U_l}U_i - h_{ik}\nabla_{U_l}U_j)$$

$$+ \frac{1}{f^4}\nabla_{U_l}(R_N(U_i,U_j)U_k)$$

$$= -\frac{f'^2}{f^4}(h_{jk}\Gamma^m_{li}(h)U_m - h_{ik}\Gamma^m_{lj}(h)U_m)$$

$$+ \frac{1}{f^4}\nabla_{U_l}(R_N(U_i,U_j)U_k). \qquad (6.23)$$

If we take the pointwise norm of the r.h.s. of (6.18) – (6.23) and apply the triangle inequality, then we obtain on the r.h. sides a finite number of terms, each of which is – up to a constant or a bounded function – of the type

$$\frac{1}{f^\alpha}\frac{|f^{(\nu_1)}|\cdot|f^{(\nu_2)}|}{f^2}, \quad \alpha \geq 0. \qquad (6.24)$$

But according to (6.9), each term of the kind (6.24) is bounded on ε, i.e. we established (B_1).

To establish (B_2), we have at the end to estimate expressions of the kind

$$\left(\frac{f'}{f}\right)\cdot\left(\frac{f'}{f}\right)', \quad \left(\frac{f'}{f}\right)\cdot\frac{1}{f}, \quad \left(\frac{f'}{f}\right)'\cdot\frac{1}{f}, \quad \frac{f'}{f}\cdot\left(\frac{1}{f}\right)',$$

$$\frac{1}{f^\alpha}\cdot\frac{f^{(\nu_1)}f^{(\nu_2)}}{f^2}\cdot\left(\frac{1}{f^\beta}\cdot\frac{f^{(\mu_1)}f^{(\mu_2)}}{f^2}\right)'. \qquad (6.25)$$

Again, according to (6.9), each term of the kind (6.25) is bounded.

A very easy induction now proves (B_k) for all k, i.e. (B_∞). □

Collared ends are isolated ends. Hence, if all ends of an open manifold M^n are collared, then M^n can have only a finite number of ends. If an open manifold has an infinite number of ends, then at least one end is not isolated.

Theorem 6.3 *Let (M^n, g) be open. If each end ε of M^n is collared then M has only a finite number of ends, $\varepsilon_1, \ldots, \varepsilon_m$. Suppose $g|_{\varepsilon_i} \cong dr^2 + f_i^2(r) d\sigma_{N_i}^2$ such that each f_i satisfies (6.8) and (6.9). Then (M^n, g) satifies (I) and (B_0).*

This follows immediately from proposition 6.2. □

Interesting examples for the fs are $f(r) = e^{g(r)}$, $g(r) > 0$ and $g^{(\nu)}(r)$ bounded for all ν.

We consider here the special case $g(r) = b \cdot r$, $b > 0$ i.e.

$$g|_{\varepsilon_i} = dr^2 + e^{2b_i r} d\sigma_{N_i}^2. \tag{6.26}$$

In the sequel, we need the knowledge of the essential spectrum σ_e of such manifolds.

Theorem 6.4 *Suppose (M^n, g) has only collared ends ε_i, $i = 1, \ldots, m$, each of them endowed with a metric of type (6.26). Then there holds*

$$\sigma_e(\Delta_q(M^n, g)) \setminus \{0\}$$
$$= \Big[\min_i \Big(\min\Big\{\Big(\frac{n-2q-1}{2}\Big)^2 b_i^2, \Big(\frac{n-2q+1}{2}\Big)^2 b_i^2\Big\}\Big),$$
$$\infty\Big[\setminus \{0\} \text{ for } q \neq \frac{n}{2} \tag{6.27}$$

and

$$\sigma_e(\Delta_q(M^n, g)) = \{0\} \cup \Big[\frac{\min(b_i)^2}{4}, \infty\Big[\text{ for } q = \frac{n}{2} \tag{6.28}$$

We refer to [3] for the proof which essentially relies on [28]. □

Corollary 6.5 *Suppose the hypotheses of 6.4, n even and*

$$\min_i b_i^2 > 0. \tag{6.29}$$

Then the graded Laplace operator $\Delta = (\Delta_0, \ldots, \Delta_n)$ *has a spectral gap above zero.* □

A special case of theorem 6.4 is the case of a rotationally symmetric metric at infinity, i.e. $(M^n \setminus K_M, g|_{M \setminus K}) \cong (\mathbb{R}^n \setminus K_{\mathbb{R}^n}, dr^2 + e^{2br} d\sigma_{S^{n-1}}^2)$. Then for $b > 0$

$$\sigma_e(\Delta_q) = \left[\left\{\left(\frac{n-2q-1}{2}\right)^2 b^2, \left(\frac{n-2q+1}{2}\right)^2 b^2\right\}, \infty\right[,$$

$$\text{for } q \neq \frac{n}{2}. \tag{6.30}$$

If we replace e^{2br} by $(\sinh r)^2$ and set $K = \emptyset$, then we get the real hyperbolic space H_{-1}^n and

$$\sigma_e(\Delta_q(H_{-1}^n)) = \begin{cases} \left[\left\{\left(\frac{n-2q-1}{2}\right)^2, \left(\frac{n-2q+1}{2}\right)^2 b^2\right\}, \infty\right[& \text{for } q \neq \frac{n}{2} \\ \{0\} \cup [\frac{1}{4}, \infty[& \text{for } q = \frac{n}{2} \end{cases}$$
$$\tag{6.31}$$

Corollary 6.6 *In the case* $(M^n, g) \cong (\mathbb{R}^{2k}, dr^2 + e^{2br} d\sigma_{S^{n-1}}^2)$ *or* $(M^n, g) = H_{-1}^{2k}$, *the graded Laplace operator* $\Delta = (\Delta_0, \ldots, \Delta_n)$ *has a spectral gap above zero.* □

Corollary 6.7 *In the following cases, the graded Laplace operator* $\Delta = (\Delta_0, \ldots, \Delta_n)$ *has a spectral gap above zero.*

a) (M^n, g) *is a finite connected sum of manifolds with collared ends, warped product metrics (6.26) satisfying (6.29) and n is even,*

b) *any compact perturbation of manifolds of a),*

c) *any finite connected sum of manifolds of type* $(M^n \setminus K_M, g|_{M \setminus K}) \cong (\mathbb{R}^n \setminus K_{\mathbb{R}^n}, dr^2 e^{2br} d\sigma_{S^{n-1}}^2)$, $b > 0$ *and n even,*

d) *any compact perturbation of c),*

e) *any finite connected sum of the hyperbolic space H^{2k}_{-1},*

f) *any compact perturbation of e),*

g) *any (M^{2k}, g'), $g' \in \text{comp}^{1,r+1}_{rel}(g)$, g of type a) – f).* □

Remark 6.8 Compact perturbations and connected sums of collared manifolds with (6.26) are again of this type. We introduced 6.6. a), b) to indicate how to enlarge step by step a given set of such warped product metrics at infinity by forming connected sums and compact perturbations. □

We apply the facts above to the case

$$E = (\Lambda^* T^* M \otimes \mathbb{C}, g_{\Lambda^*}), \nabla^{g_{\Lambda^*}}), D = d + d^*, D^2 = \Delta = (\Delta_0, \ldots, \Delta_n).$$

Theorem 6.9 *Let (M^{2k}, g) be one of the manifolds 6.6 a) – f), $g' \in \text{comp}^{1,r+1}_{rel}(g)$, g' smooth, $r + 1 > 2k + 3$. Then the relative ζ-function $\zeta_q(s, \Delta, \Delta')$ as in section 4 and the relative analytic torsion, $\tau^a((M, g), M', g'))$,*

$$\log \tau^a((M, g), (M', g')) = \sum_{q=0}^{2k} (-1)^q q \cdot \zeta'_q(0, \Delta, \Delta')$$

are well defined.

Proof. According to proposition 6.2, (M^n, g) and (M', g') satisfy (I) and (B_∞), and the general Laplace operator has a spectral gap above zero. The assertion then follows from theorem 4.4. □

Corollary 6.10 *Let (M, g) be as is 6.6 a) – g). Then the attachment $(M'^{2k}, g') \longrightarrow \tau^a((M, g), (M', g'))$ yields a contribution to the classification of the elements of $\text{gen comp}^{1,r+1}_{L,diff,rel}(\Lambda^* T^*, g_{\Lambda^*})$.*

Remark 6.11 If $n = 2k+1$ and (M^n, g) belongs to one of the classes 6.6 a) – g) then the relative ζ–functions $\zeta(s, \Delta_q, \Delta'_q)$ are for $(M'^n, g') \in \text{gen comp}^{1,r+1}_{L,diff,rel}(M^n, g) \cap C^{r+1}$ and $q \neq k, k+1$ well defined. □

Another very special case is given by $b = 0$ in (6.26), i.e. cylindrical ends ε,

$$ds^2 = dr^2 + d\sigma_N^2. \tag{6.32}$$

Suppose, we have m cylindrical ends ε_i, $i = 1, \ldots, m$,

$$g|_{\varepsilon_i} \cong dr^2 + d\sigma_{N_i}^n,$$

and let $\{\lambda_k^q(i)\}_k$ be the (purely discrete) spectrum of Δ_q (N_i^{n-1}, h_i), $i = 1, \ldots, m$.

Proposition 6.12 *Then*

$$\sigma_e(\Delta_q(M, g)) = \bigcup_i \bigcup_k ([\lambda_k^q(i), \infty[\cup [\lambda_k^{q-1}(i), \infty[). \tag{6.33}$$

We refer to [3] and [28] for the proof. □

Corollary 6.13 a) *If* $H^q(N_i) = H^{q-1}(N_i) = (0)$, $i = 1, \ldots, m$, *then* $\sigma(\Delta_q(M))$ *has a spectral gap above zero.*
b) *If for at least one i $H^q(N_i) \neq 0$, then* $\sigma(\Delta_{q+1}(M)) = \sigma(\Delta_q(M)) = [0, \infty[$.
c) *In the case of cylindrical ends, the graded Laplace operator never has a spectral gap above zero.*

Proof. a) and b) immediately follow from (6.34). c) follows from $H^0(N_i) \neq 0$ for all i, hence $\sigma(\Delta_0(M)) = \sigma(\Delta_1(M)) = [0, \infty[= \sigma(\Delta_0, \ldots, \Delta_n)$. □

Corollary 6.14 *Suppose* (M^n, g) *with cylindrical ends* $\varepsilon_1, \ldots, \varepsilon_m$ *and let* $(M'^n, g') \in \text{gen comp}_{L,diff,rel}(M^n, g)$. *If* $H^q(N_i) = H^{q-1}(N_i) = (0)$, $i = 1, \ldots, m$, *then* $\zeta(s, \Delta_q, \Delta'_q)$ *and* $\det(\Delta_q, \Delta'_q)$ *are well defined.* □

A special case is given by the pair

(M^n, g) and

$$\left(\bigcup_{i=1}^{m} N_i \times [0, \infty[= N \times [0, \infty[, \bigcup_{i=1}^{m} dr^2 + f_i^2(r) d\sigma_{N_i}^2 \right) = (M'^n, g').$$

Here M'^n is a manifold with boundary $\partial M'^n = N = \bigcup N_i$, and the latter falls out from our considerations. But if we consider the case $q = 0$ and $\Delta_0(M'^n, g')$ with Dirichlet boundary conditions at $\partial M' = N = \bigcup N_i$, then we get an essentially self–adjoint operator Δ_0', $\sigma(\Delta_0') = [0, \infty[= [0, \infty[\cup \bigcup_{\lambda_j > 0} [\lambda_j, \infty[$, and

$e^{-t\Delta_0} - e^{-t\Delta_0' p'}$ is of trace class. The latter fact is an immediate consequence of the proof of theorem 3.9. Hence the wave operator $W_\pm(\Delta_0, \Delta_0')$ exist, are complete and the absolutely continuous parts of Δ_0 and Δ_0' are unitarily equivalent. We intend to present an explicit representation of the scattering matrix $S(\lambda)$, of $\text{tr}(e^{-t\Delta_0} - e^{-t\Delta_0'})$ and of the relative ζ–function. Here we essentially follow [49]. Then

$$\sigma(\Delta_0) = \sigma_p(\Delta_0) \cup \sigma_{ac}(\Delta_0),$$

and $\sigma_p(\Delta_0)$ consists of eigenvalues

$$0 < \mu_1 \leq \mu_2 \leq \mu_3 \leq \cdots \longrightarrow \infty$$

of finite multiplicity without finite accumulation point and

$$N_p(\mu) := \#\{j | \mu_j \leq \mu\} \leq C \cdot (1+\mu)^{\frac{n}{2}}. \quad (6.34)$$

(6.35) immediately implies for $t > 0$

$$\sum_j e^{-t\mu_j} < \infty \quad (6.35)$$

and

$$\text{tr}(e^{-t\Delta_0}|_{L_{2,\sigma_p}}) = \sum_j e^{-t\mu_j}, \quad (6.36)$$

where $\Delta_0|_{L_{2,\sigma_p}}$ is the restriction of Δ_0 to the subspace spanned by the L_2–eigenfunctions. If we apply (6.34) to the case $q=0$ then we obtain

$$\sigma(\Delta_0) = \sigma_{ac}(\Delta_0) = \bigcup_i \bigcup_k [\lambda_k^0(i), \infty[= \bigcup_j [\lambda_j, \infty[, \qquad (6.37)$$

where the λ_js are the eigenvalues of $\Delta_0(N)$ and simultaneously the thresholds of the (absolutely) continuous spectrum.

We describe the continuous spectrum in terms of generalized eigenfunction. Consider $\Delta_0(N)(= (\Delta_0(N_1), \ldots, \Delta_0(N_m)))$, an eigenvalue $\lambda_j \in \sigma(\Delta_0(N))$ and let $E(\lambda_j)$ be the corresponding eigenspace.

For $\mu > \lambda_j$ and different from all thresholds and for $\varphi \in E(\lambda_j)$ there exists a unique eigenfunction $E(\varphi,\mu) \in C^\infty(M)$ of Δ_0 for the eigenvalue λ, the restriction to $\mathbb{R}_+ \times N$ of which has the representation

$$E(\varphi,\mu)(r,y) = e^{ir\sqrt{\mu-\lambda_j}}\varphi(y) + \sum_{\lambda_l<\mu} e^{-ir\sqrt{\mu-\lambda_l}}(T_{jl}(\mu)\varphi)(y) + \psi,$$
(6.38)

$\psi \in L_2(\mathbb{R}_+ \times N)$ and $T_{jl}(\mu): E(\lambda_j) \longrightarrow E(\lambda_l)$ a linear operator. Let Σ be the minimal Riemann surface to which all functions $\mu \longrightarrow \sqrt{\mu - \lambda_j}$, $j \in \mathbb{N}$ extend as holomorphic functions.

Lemma 6.15 *Each generalized eigenfunction $E(\varrho,\mu)$ extends to a meromorphic function on Σ.*

See [49] for a proof. □

Define now

$$S_{jl}(\mu) = \left(\frac{\mu-\lambda_j}{\mu-\lambda_l}\right)^{\frac{1}{4}} T_{jl}(\mu)$$

and set

$$S(\mu) = \bigoplus_{\lambda_j,\lambda_l<\mu} S_{jl}(\mu).$$

$S(\mu)$ is the scattering matrix of $S(\mu, \Delta_0, \Delta_0')$. The $E(\varrho,\mu)$ are meromorphic functions on Σ, which implies the same for $T_{jl}(\mu)$ and $S_{jl}(\mu)$. This yields (cf. [49])

Proposition 6.16 *The scattering matrix $S(\mu)$ is a unitary operator of finite rank, smoothly depending on $\mu \in]\lambda_{j-1}, \lambda_j[$. The rank changes when μ crosses a threshold. The coefficients of $S(\mu)$ extend to meromorphic functions on Σ.* □

Proposition 6.17 *For $t > 0$ there holds*

$$\begin{aligned}\operatorname{tr}(e^{-t\Delta_0} - e^{-t\Delta_0'} P') &= \sum_j e^{-t\mu_j} + \frac{1}{2} \sum_{k=1}^{\infty} \dim E(\lambda_k) e^{-t\lambda_k} \\ &+ \frac{1}{4} \operatorname{tr} S(0) \\ &+ \frac{1}{2\pi i} \int_0^{\infty} e^{-t\lambda} \frac{d}{d\lambda} \log \det S(\lambda) d\lambda. \quad (6.39)\end{aligned}$$

We refer to [49] for the proof. □

Next we want to establish an explicite expression for the relative ζ–function $\zeta(s, \Delta_0 + z, \Delta_0' + z)$, which is well defined for $\Re(z) > 0$.

Proposition 6.18 *Under the above assumptions, the relative ζ-function $\zeta(s, \Delta_0 + z, \Delta_0' + z)$ for $\Re(z) > 0$ is given by*

$$\begin{aligned}\zeta(s, \Delta_0 + z, \Delta_0' + z) &= \sum_j (\mu_j + z)^{-s} + \sum_k (\lambda_k + z)^{-s} \\ &+ \frac{1}{2\pi i} \int_0^{\infty} (\lambda + z)^{-s} \frac{d}{d\lambda} \log \det S(\lambda) d\lambda.\end{aligned}$$

(6.40)

Proof. At first we remark that for $0 < \lambda < \lambda_1$, $S(\lambda)$ is a scalar, $S(\lambda^2)$ extends to a meromorphic function on $\{z \in \mathbb{C} | \ |z| < \lambda_1\}$, which is holomorphic at $\lambda = 0$ and, in particular, $S(\lambda)$ is holomorphic in $U_\varepsilon(0)$, ε small. Using this and equation (6.39), we see that for $t \longrightarrow \infty$ we have a representation $\operatorname{tr}(e^{-t\Delta_0} - e^{-t\Delta_0'} P') = b_0 + O(t^{-\varrho})$, $\varrho > 0$, which is equivalent to $\xi(\lambda) = -b_0 - O(\lambda^\varrho)$ for $\lambda \longrightarrow 0^+$. Hence (2.24) and both assumptions of proposition 2.4 are satisfied. We obtain from corollary 2.5 a) and (6.39) an asymptotic expansion

$$\operatorname{tr}(e^{-t\Delta_0} - e^{-t\Delta_0'} P') \sim \dim \ker \Delta + \frac{1}{4}\operatorname{tr} S(0) + \sum_{j=1}^{\infty} c_j t^{-\frac{j}{2}}, \quad t \longrightarrow \infty.$$

Moreover, for $t \longrightarrow 0^+$, we have the standard expansion

$$\operatorname{tr}(e^{-t\Delta_0} - e^{-t\Delta_0'} P') \sim \sum_{j=0}^{\infty} a_j t^{-\frac{n}{2}+j}, \quad t \longrightarrow 0^+,$$

and we get the existence of the relative ζ–function $\zeta(s, \Delta_0, \Delta_0')$ as a meromorphic function. According to [49]

$$\int_0^\Lambda \frac{d}{dz} \log \det S(\lambda) d\lambda = O(\Lambda^n) \text{ as } \Lambda \longrightarrow \infty. \qquad (6.41)$$

Using this and inserting (6.39) into (,), we obtain finally (6.40). □

It is clear that the product geometry of cylindrical ends is an extremly special case of possible geometries on $\varepsilon = N \times [0, \infty[$, and one should admit much more general bounded geometries on M, e.g. bounded geometries of the type $g_\varepsilon = g|_{N \times [0,\infty[} = dr^2 + (e^{g(r)})^2 d\sigma_N^2$, $g(r) > 0$, $g^{(\nu)}(r)$ bounded for all ν. Again we get a pair $(\Delta_0 = \Delta_0(g), \Delta_0' = \Delta_0'(g_\varepsilon))$, where $(\varepsilon, g_\varepsilon) \in$ gen $\operatorname{comp}_{L,diff,rel}^{1,r}(M,g)$, $e^{-t\Delta_0} - e^{-t\Delta_0'} P'$ of trace class. If $\inf \sigma_e(\Delta_0) > 0$ or for $t \longrightarrow \infty$

$$\operatorname{tr}(e^{-t\Delta_0} - e^{-t\Delta_0'} P') = b_0 + O(t^{-\varrho}), \quad \varrho > 0 \qquad (6.42)$$

then the relative ζ–function $\zeta(s, \Delta_0, \Delta_0')$ and relative determinant are defined. Explicit $g(r)$ leads to explicit calculations. Similarly for $q > 0$. If e.g. $n = 2m$, g on M is such that $g|_\varepsilon = dr^2 + e^r d\sigma_N^2$ and $g = m$ then $\inf \sigma_e(\Delta_m|_{(\ker \Delta_m)^\perp}) > 0$. Here we take for Δ_m' the Friedrichs' extension of Δ_m on $M \times [0, \infty[$

with zero boundary condition on $M \times \{0\}$. Hence $\zeta(s, \Delta_m, \Delta'_m)$ and $\det(\Delta_m, \Delta'_m)$ are well defined. In the case of a manifold $M = M' \cup N \times [0, \infty[, g)$ with cylindrical ends, the main and interesting part of the geometry is contained in the compact part M', $\partial M' = N$. At the boundary we assume product geometry. $X = (M' \underset{N}{\cup} M', g_X = g_{M'} \cup g_{M'}$ is then a closed manifold. It is now a natural and interesting question, how are the Δ–determinant for X and the relative Δ–determinant for M related? The answer would also give a meaning, an interpretation of the relative determinant. A certain answer is contained in [53], and we give an outline of the corresponding result.

Consider the following situation. Given (M^n, g) closed, oriented, connected, $Y \subset M$ a hypersurface, separating M into two components $\overset{\circ}{M}_1, \overset{\circ}{M}_2$. Set $M_i = \overset{\circ}{M}_i$. $i = 1, 2$. M_i are compact with boundary Y, $M = M_1 \underset{Y}{\cup} M_2$, $Y = \partial M_1 = \partial M_2$. Moreover, let $E \longrightarrow M$ be a Hermitean vector bundle and $\Delta = \Delta_M : C^\infty(M, E) \longrightarrow C^\infty(M, E)$ a Laplace type operator, i.e. Δ is symmetric, non–negative with principal symbol $\sigma_\Delta(x, \xi) = -|\xi|^2 \cdot \text{id}$. Suppose moreover, that Y has a product collar neighbourhood $N \cong Y \times [-1, 1]$ such that all geometries over N are product geometries, i.e. $g|_N = dr^2 + g_Y$, $E|_N \cong \pi^*(E_0 \longrightarrow Y)$ and $\Delta_M|_N = -\frac{\partial^2}{\partial r^2} + \Delta_y$, $\Delta_y : C^\infty(Y, E_0) \longrightarrow C^\infty(Y, E_0)$ again of Laplace type. Denote by $\Delta_{M_i, D}$ the Friedrich's extension of $\Delta_M|_{M_i}$ with Dirichlet boundary conditions on $\partial M_i = Y$.

We define a family $\{(M_r, g_r)\}_{r>0}$ by setting $M_r = M_1 \underset{Y}{\cup} N_r \underset{Y}{\cup} M_2$, where $N_r = Y \times [-r, r]$ and $g_r|_{M_i}$ as before, $g_r|_{N_r} = dr^2 + g_y$. Similarly, we extend the bundle $E \longrightarrow M$ to $E_r \longrightarrow M_r$. Then Δ_{M_r}, $\zeta(s, \Delta_{M_r})$ and $\det \Delta_{M_r} = e^{-\zeta'(0, \Delta_{M_r})}$ are well defined. Set $M_{i,\infty} = M_i \underset{Y}{\cup} Y \times [0, \infty[$, extend $g|_{M_i}$ and $E|_{M_i}$ canonically as product structures and consider $\Delta_{i,\infty}$, $\Delta_{i,\infty}|_{M_i} = \Delta_M|_{M_i}$, $\Delta_{i,\infty}|_{Y \times [0,\infty[} = -\frac{\partial^2}{\partial r^2} + \Delta_Y$, $i = 1, 2$. $\Delta_{i,\infty}$ has its closure as unique self–adjoint extension which we denote by the same letter. Let Δ_0 be the Friedrichs' extension of $-\frac{\partial^2}{\partial r^2} + \Delta_Y$ on $C^\infty(Y \times [0, \infty[, E_{i,\infty}|_{Y \times [0,\infty[})$ with Dirichlet boundary conditions on $Y \times \{0\}$. Then $(\pi^* E_0 \longrightarrow (Y \times [0, \infty[, dr^2 + g_y)) \in$

gen $\text{comp}_{L,diff,rel}(E_{i,\infty} \longrightarrow (M_{i,\infty}, g_{i,\infty}))$) and, according to the above considerations the relative ζ–function $\zeta(s, \Delta_{i,\infty}, \Delta_0)$ and the relative determinant $\det(\Delta_{i,\infty}, \Delta_0) = e^{-\zeta'(0,\Delta_{i,\infty},\Delta_0)}$ are well defined.

Denote

$$\xi_Y(s) := \frac{\Gamma(s-\frac{1}{2})}{\sqrt{\pi}\Gamma(s)} \zeta(s-\frac{1}{2}, \Delta_y).$$

Theorem 6.19 Assume $\ker \Delta_y = \{0\}$ and $\ker \Delta_{i,\infty} = \{0\}$, $i = 1, 2$. Then Δ_{M_r} is invertible for $r \geq r_0$ and

$$\lim_{r \to \infty} e^{r\xi_y'(0)} \det \Delta_{M_r} = (\det \Delta_y)^{-\frac{1}{2}} \prod_{i=1}^{2} \det(\Delta_{i,\infty}, \Delta_0). \quad (6.43)$$

We refer to [53] for the proof. □

Remark 6.20 (6.43) establishes a connection of the desired kind. We cannot expect a "simple formula" which relates $\det(\Delta_M)$, $\det(\Delta_{M_1})$, $\det(\Delta_{M_2})$ since this would contradict the global character of $\zeta'(0)$ and $\det(\cdot)$. □

For $E' \in \text{gen comp}_{L,diff,rel}^{1,r+1}(E)$, $\inf \sigma_e(D^2|_{(\ker D^2)^\perp}) > 0$, we have the asymptotic expansion (1.16),

$$\text{tr}(e^{-tD^2} P - e^{-t\tilde{D}'^2} P') = t^{-\frac{n}{2}} c_{-\frac{n}{2}} + \cdots + t^{-\frac{n}{2}+[\frac{n+3}{2}]} c_{-\frac{n}{2}+[\frac{n+3}{2}]}$$
$$+ O(t^{-\frac{n}{2}+[\frac{n+3}{2}]})$$

and can express the logarithm of the determinant more explicitely. First we remark that in $U(s=0)$

$$\Gamma(s) = \frac{1}{s} + \gamma \cdot s \cdot h(s),$$

where $h(s)$ is a holomorphic function near $s = 0$ and γ denotes the Euler constant. Hence

$$\frac{d}{ds}\Big|_{s=0} \frac{1}{\Gamma(s)} = 1,$$

$$\frac{d}{ds}\Big|_{s=0} \left(\frac{1}{\Gamma(s)} \frac{1}{s}\right) = \frac{d}{ds}\Big|_{s=0} \frac{1}{\Gamma(s+1)} = -\Gamma'(1),$$

$$\frac{1}{\Gamma(s)}\Big|_{s=0} = 0.$$

(1.16), (2.4) and (2.11) then imply

$$-\log \det(D^2, D'^2) = \frac{d}{ds}\Big|_{s=0}\zeta_1(s, D^2, D'^2) - h \cdot \Gamma'(1)$$

$$+ \int_1^\infty t^{-1}[\operatorname{tr}(e^{-tD^2}P - e^{-t\tilde{D}'^2}P') - h]dt$$

$$= \sum_{l=0}^{[\frac{n+3}{2}]} c_{-\frac{n}{2}+l} - c_0\Gamma'(1) + \int_0^1 t^{-1}[\operatorname{tr}(e^{-tD^2}P - e^{-t\tilde{D}'^2}P')$$

$$- \sum_{l=0}^{[\frac{n+3}{2}]} c_{-\frac{n}{2}+l}t^{-\frac{n}{2}+l}]dt - h\Gamma'(1)$$

$$+ \int_1^\infty t^{-1}[\operatorname{tr}(e^{-tD^2}P - e^{-t\tilde{D}'^2}P' - h]dt. \tag{6.44}$$

Let as in section 1 $P_d(D)$ be the projector on the discrete subspace and $P_c = 1 - P_d$ the projector on the continuous subspace and denote by D_d and D_c the corresponding restrictions of. These subspaces are orthogonal and remain invariant under D_d or D_c, respectively. Suppose that $e^{-tD_d^2}$ is of trace class. Then $e^{-tD_c^2} - e^{-t\tilde{D}'^2}$ is of trace class too. If for $e^{-tD_d^2}$ there exist an asymptotic expansion of the type (1.1) then there exists a well defined zeta function $\zeta(s, D_d)$ and $\det D_d^2$ can be defined by

$$\det D_d^2 := e^{-\zeta'(0, D_d^2)}. \tag{6.45}$$

Then (3.7), (6.45) and

$$\operatorname{tr}(e^{-tD^2}P - e^{-t\tilde{D}'^2}P') = \operatorname{tr}(e^{-tD_d^2}P + e^{-tD_c^2}P - e^{-t\tilde{D}'^2}P')$$
$$= \operatorname{tr} e^{-tD_d^2}P + \operatorname{tr}(e^{-tD_c^2}P - e^{-t\tilde{D}'^2}P')$$

yield

$$\det(D^2, D'^2) = \det(D_d^2) \cdot \det(D_c^2, D'^2). \tag{6.46}$$

If $a > 0$, then $\inf \sigma_e(D^2 + a) > 0$ then $h = 0$ in (2.9), (2.24) for the pair $D^2 + a$, $\tilde{D}' + a$ is satisfied and the corresponding relative

ζ–function is given by

$$\zeta(s, D^2 + a, \tilde{D}'^2 + a) = \frac{1}{\Gamma(s)} \int_0^\infty t^{s-1} e^{-ta} \mathrm{tr}(e^{-tD^2} - e^{-t\tilde{D}'^2}) dt,$$

(6.47)

$\mathrm{Re}(s) > -1$. The r.h.s. of (6.47) is also well defined if we replace a by any $z \in \mathbb{C}$ with $\mathrm{Re}(z) > 0$. Then the corresponding function $\zeta(s, z, D^2, \tilde{D}'^2)$ admits as function of s a meromorphic extension to $\mathbb{C}$ which is holomorphic at $s = 0$ and we finally define

$$\det(D^2 + z, \tilde{D}'^2 + z) := e^{-\frac{\partial}{\partial s}|_{s=0} \zeta(s, z, D^2, \tilde{D}'^2)}.$$

An important property concerning the z–dependence is expressed by

Proposition 6.21 *Suppose* $E' \in \mathrm{gen\,comp}_{L,diff,rel}^{1,r+1}(E)$ *and (2.24). Then* $\det(D^2 + z, \tilde{D}'^2 + z$ *is a holomorphic function of* $z \in \mathbb{C} \setminus]0, \infty[$.

Proof. Here we essentially follow [49]. According to V, lemma 1.10 a),

$$\mathrm{tr}(e^{-tD^2} P e^{-t\tilde{D}'^2} P') = -t \int_0^{2c} e^{-t\lambda} \xi(\lambda) d\lambda - t \int_{2c}^\infty e^{-t\lambda} \xi(\lambda) d\lambda.$$

(6.48)

The second integral on the r.h.s. is $O(e^{-tc})$ for $t \longrightarrow \infty$ since $e^{-t\lambda} \xi(\lambda) \in L_1 \mathbb{R}$ for $t > 0$. Moreover,

$$\int_0^{2c} e^{-t\lambda} \xi(\lambda) d\lambda = \sum_{k=0}^N c_k t^k + O(t^{N+1}) \text{ for } t \longrightarrow 0,$$

(6.49)

$N \in \mathbb{N}$ arbitrary. (1.16) and (6.49) imply (by taking the difference) that $\int_{2c}^\infty e^{-t\lambda \xi(\lambda)} d\lambda$ has a similar asymptotic expansion as (1.16). We infer from this that the integral

$$F(s, z) = \frac{1}{\Gamma(s)} \int_0^\infty t^s e^{-tz} \int_{2c}^\infty e^{-t\lambda} \xi(\lambda) d\lambda \, dt$$

in the half planes $\text{Re}(s) > -\frac{n}{2}$ and $\text{Re}(z) > -c$ absolutely converges and, as function of s, it admits a meromorphic extension to $\mathbb{C}$ which is holomorphic at $s = 0$.

The first integral on the r.h.s. of (6.48) can be discussed as follows. We obtain for $\text{Re}(z) > 0$

$$-\frac{1}{\Gamma(s)} \int_0^\infty t^s e^{-tz} \int_0^{2c} e^{-t\lambda} \xi(\lambda) d\lambda dt$$

$$= \frac{1}{\Gamma(s)} \int_0^{2c} \xi(\lambda) \int_0^\infty t^s e^{-t(z+\lambda)} dt d\lambda$$

$$= -s \int_0^{2c} (z+\lambda)^{-(s+1)} \xi(\lambda) d\lambda. \tag{6.50}$$

Hence, for $\text{Re}(z) > 0$,

$$\det(D^2 + z, \tilde{D}'^2 + z) = e^{\int_0^{2c}(z+\lambda)^{-1}\xi(\lambda)d\lambda + \frac{\partial F}{\partial s}(0,z)}. \tag{6.51}$$

The r.h.s. of (6.51) has an obvious extension to an analytic function of $z \in \mathbb{C}\setminus]-\infty, 0]$, $\frac{\partial F}{\partial s}(0, z)$ is holomorphic in $\text{Re}(z) > -c$. $c > 0$ was arbitrary, hence we get an analytic extension to $\mathbb{C}\setminus]-\infty, 0]$. $\square$

It is an interesting question, how the relative determinant and the relative torsion change under 1–parameter change of the metric.

We could consider e.g. the most natural evolution of the metrix which is given by the Ricci flow,

$$\frac{\partial}{\partial \tau} g(\tau) = -2\text{Ric}\,(g(\tau)), \quad g(0) = g_0. \tag{6.52}$$

If (M^n, g_0) is complete and has bounded curvature then, according to [69], [25], there exists for $0 \leq \tau \leq T$ in the class of metrics with bounded curvature a unique solution of (6.51).

Denote $\Delta(\tau) \equiv \Delta_q(g(\tau))$. $\tilde{\Delta}(\tau)$ is defined as before by the diagram

$$
\begin{array}{ccc}
L_2(g(0)) \supset \mathcal{D}_{\Delta(0)} & \xrightarrow[\tilde{\Delta}(\tau)]{\Delta(0)} & L_2(g()) \\
\cdot \frac{\mathrm{dvol}(0)}{\mathrm{dvol}(\tau)} \downarrow & & \uparrow \cdot \frac{\mathrm{dvol}(\tau)}{\mathrm{dvol}(0)} \\
L_2(g(\tau)) \supset \mathcal{D}_{\Delta(\tau)} & \xrightarrow{\Delta(\tau)} & L_2(g(\tau))
\end{array}
$$

It is not yet clear, whether $g(\tau) \in \mathrm{comp}^{1,r+1}(g(0))$. We proved in [25] that $g(\tau) \in {}^b\mathrm{comp}^2(g(0))$, but $g(\tau) \in \mathrm{comp}^{1,r+1}(g(0))$ is still open. Therefore we make the following

Assumption. $E' \in \mathrm{comp}^{1,r+1}_{L,diff}(E) \subset \mathrm{gen}\,\mathrm{comp}^{1,r+1}_{L,diff}(E)$. Here $\mathrm{comp}(\cdot)$ denotes the arc component. Hence there exists an arc $\{E(\tau)\}_{-\varepsilon \le \tau \le \varepsilon}$ connecting E and E', we assume in the sequel to the arc to be at least C^1.

Then we get a C^1-arc $\{D(\tau)\}_{\tau'}$ and we will study the behaviour of the relative determinant $\det(D^2(0), \tilde{D}^2(\tau))$ under variation of τ. Additionally, we suppose again (2.24). As before and in the sequel, $\tilde{D}^2(\tau)$ denotes the transformed to the Hilbert space for $\tau = 0$ $D^2(\tau)$.

Denote $\dot{\tilde{D}}^2(\tau) = \frac{d}{d\tau}\tilde{D}^2(\tau)$. By Duhamels principle,

$$\frac{d}{d\tau}e^{-t\tilde{D}^2(\tau)} = \int_0^t e^{-s\tilde{D}^2(\tau)} \dot{\tilde{D}}^2(\tau) e^{-(t-s)\tilde{D}^2(\tau)} ds. \qquad (6.53)$$

If $\dot{\tilde{D}}^2(\tau)e^{-t\tilde{D}^2(\tau)}$ is for $t > 0$ of trace class with trace norm uniformly bounded on compact t–intervals $[a_0, a_1]$, $a_0 > 0$, then according to (6.53), $\frac{d}{d\tau}e^{-t\tilde{D}^2(\tau)}$ is also of trace class for $t > 0$ and

$$\frac{d}{d\tau}\mathrm{tr}(e^{-t\tilde{D}^2(\tau)} - e^{-tD^2(0)}) = -t\,\mathrm{tr}(\dot{\tilde{D}}^2(\tau)e^{-t\tilde{D}^2(\tau)}). \qquad (6.54)$$

To establish in the sequel substantial results, we must make two additional assumptions.

Assumption 1. $D^2(\tau)$ is invertible for $\tau \in [-\varepsilon, \varepsilon]$. (6.55)

Assumption 2. There exists for $t \longrightarrow 0^+$ an asymptotic expansion

$$\operatorname{tr}(\dot{\tilde{D}}^2(\tau)\tilde{D}^{-2}(\tau)e^{-t\tilde{D}^2(\tau)}) \sim \sum_{j=0}^{\infty}\sum_{k=0}^{k(j)} c_{jk}(\tau)t^{\gamma_j}\log^k t + C_1(\tau) + C_2(\tau)\log t$$
(6.56)

with $-\infty < \gamma_1 < \gamma_2 < \cdots$, $\gamma_j \neq 0$, $j \in \mathbb{N}$ and $\gamma_j \longrightarrow \infty$. (6.55) implies the existence of $c > 0$ such that

$$\operatorname{tr}(\dot{\tilde{D}}^2(\tau)\tilde{D}^{-2}(\tau)e^{-t\tilde{D}^2(\tau)}) = O(e^{-ct}) \text{ for } t \longrightarrow \infty. \quad (6.57)$$

We infer from (6.54)–(6.57) that for $\operatorname{Re}(s) \gg 0$

$$\frac{d}{d\tau}\zeta(s, D^2(0), \tilde{D}^2(\tau))$$

$$= \frac{1}{\Gamma(s)} \int_0^\infty t^s \frac{\partial}{\partial t} \operatorname{tr}(\dot{\tilde{D}}^2(\tau)\tilde{D}^{-2}(\tau)e^{-t\tilde{D}^2(\tau)})dt$$

$$= -\frac{s}{\Gamma(s)} \int_0^\infty \operatorname{tr}(\dot{\tilde{D}}^2(\tau)\tilde{D}^{-2}(\tau)e^{-t\tilde{D}^2(\tau)})dt. \quad (6.58)$$

Proposition 6.22 *Suppose that the arc* $\{D(\tau)\}_{-\varepsilon \leq \tau \leq \varepsilon}$ *satisfies the assumptions (6.55) and (6.56). Then* $\det(D^2(0), \tilde{D}^2(\tau))$ *is a differentiable function of* τ *and*

$$\frac{d}{d\tau}\log\det(D^2(0), \tilde{D}^2(\tau)) = C_1(\tau) + \Gamma'(1)C_2(\tau), \quad (6.59)$$

$C_1(\tau)$, $C_2(\tau)$ *as in (6.56).*

Proof. Consider

$$\int_0^\infty t^{s-1}\operatorname{tr}(\dot{\tilde{D}}^2(\tau)\tilde{D}^{-2}(\tau)e^{-t\tilde{D}^2(\tau)})dt.$$

This integral has an analytic extension to a meromorphic function of $s \in \mathbb{C}$ with poles determined by the expansion (6.56). The pole at $s = 0$ has order ≤ 2 with Laurent expansion

$$\frac{C_2(u)}{s^2} + \frac{C_1(u)}{s} + \cdots.$$

Inserting this into (6.58) and differentiating at $s = 0$ yields (6.59). □

Corollary 6.23 *Suppose the hypothesis of 6.22. Then*

$$\det(D^2(0), \tilde{D}^2(\varepsilon)) = e^{\int_0^\infty C_1(\tau) + \Gamma'(1)C_2(\tau))d\tau}. \qquad (6.60)$$

□

Until now we defined the relative ζ-determinant only for pairs D^2, D'^2. But there is a reasonable way also to introduce $\det(D, D')$. Let (M^n, g) be closed. Then Singer already defined in [70] by very natural considerations

$$\det D := \det |D| \cdot e^{\frac{i\pi}{2}(\eta_D(0) - \zeta_{|D|}(0))}. \qquad (6.61)$$

Consider now the open case, $E' \in \text{comp}_{L,diff,F,rel}^{1,r+1}(E)$, and assume for a moment $\inf \sigma(D^2), \inf \sigma(D'^2) > 0$. Then D and D' are invertible and $\inf \sigma_e(D'^2|_{(\ker D'^2)^\perp}) > 0$. We can apply 2.2 and 5.4 with $h = 0$ and obtain

$$\zeta(s, D^2, D'^2) = \text{tr}((D^2)^{-s} - (\tilde{D}'^2)^{-s}) = \text{tr}(|D|^{-2s} - |\tilde{D}'|^{-2s}),$$
$$\text{Re}(s) > -5. \qquad (6.62)$$

(6.62) yields

$$\zeta(s, D^2, D'^2) = \zeta(2s, |D|, |\tilde{D}'|) \qquad (6.63)$$

and (6.63) has a meromorphic extension to $\mathbb{C}$ which is holomorphic at $s = 0$. Hence

$$\det(|D|, |\tilde{D}'|) := e^{-\zeta(0, |D|, |\tilde{D}'|)}$$

is well defined and we set in analogy to (6.61)

$$\det(D, D') := \det(|D|, |\tilde{D}'|) \cdot e^{\frac{\pi i}{2}(\eta(0, D, D') - \zeta(0, |D|, |\tilde{D}'|))}. \quad (6.64)$$

But to define (6.64) we don't need D, D' invertible. The former assumption $\inf \sigma_e(D^2|_{(\ker D^2)^\perp}) > 0$ is sufficient to define (6.64) since this assumption is sufficient for the l.h.s. of (6.63) and hence for the r.h.s. of (6.64). □

Finally we draw some conclusions for the Schrödinger operator $\Delta_q + V$, V as in IV, proposition 3.11.

Theorem 6.24 *Suppose $g \in \mathcal{M}(I, B_k)$, $k \geq r > n + 2$, $V \in \Omega^{1,r}(M^n, g)$ a real–valued function. Then the absolutely continuous parts of Δ_q and $\Delta_q + (V \cdot)$ are unitarily equivalent.*

Proof. This immediately follows from IV 3.11 and the completeness of the wave operators. □

Corollary 6.25 *Suppose $V \in \Omega^{1,r}(\mathbb{R}^n, g_{standard})$, real–valued, $r > n + 2$, $\Delta = -\sum_{i=1}^{n} \frac{\partial^2}{\partial x_i^2}$. Then the absolutely continuous parts of Δ and $\Delta + (V \cdot)$ are unitarily equivalent.* □

VII Scattering theory for manifolds with injectivity radius zero

In chapters II – VI we always assumed $r_{\text{inj}}(M, g) > 0$. The background for this assumption was the fact that to establish our uniform structures, we used the module structure theorem for Sobolev spaces and this theorem has been proved until now under the assumption $r_{\text{inj}} > 0$. An extension to the case $r_{\text{inj}} = 0$ for weighted Sobolev spaces is in preparation.

In [54] W. Müller and G. Salomonsen introduced a scattering theory for manifolds with bounded curvature, admitting $r_{\text{inj}}(M, g) = 0$. They restrict themselves to the case of the scalar Laplacian $\Delta_0 = (\nabla^g)^* \nabla^g$. We partially follow here an extended to arbitrary Riemannian vector bundles version of their approach but using our language of components of uniform spaces and our procedures of chapter IV to establish trace class properties.

1 Uniform structures defined by decay functions

We apply the procedure of [27], [32], our chapter II and reformulate and extend the approach of [54].

Let $V \in C^0([1, \infty[)$ be non–increasing. It is called of moderate decay if

$$1) \quad \sup_{x \in [1, \infty[} x \cdot V(x) < \infty \qquad (1.1)$$

and

$$2) \quad \text{there exists } C = C(V)$$
$$\text{such that } V(x + y) \geq C(V) \cdot V(x) \cdot V(y). \qquad (1.2)$$

It is called of subexponential decay if for any $c > 0$

$$e^{cx}V(x) \xrightarrow[x\to\infty]{} \infty. \qquad (1.3)$$

Examples 1.1 1) $V(x) = e^{-tx}$ is for $t > 0$ of moderate decay.
2) If V, V_1, V_2 are of moderate decay then this holds for V^δ, $\delta > 0$, $V_1 \cdot V_2$ too. The same holds for subexponential decay.
3) $V(x) = x^{-1}$ and e^{-x^α}, $0 < \alpha < 1$, are of sub–exponential decay. □

Denote by $\mathcal{M} = \mathcal{M}(M)$ the set of all complete (smooth) metrics on M and let $g, h \in \mathcal{M}$. Then we define

$$^m|g - h|_g(x) := |g - h|_g(x) + \sum_{i=0}^{m-1} |(\nabla^g)^i(\nabla^g - \nabla^h)|_g(x) \quad (1.4)$$

and

$$^{b,m}|g - h|_g := \sup_{x \in M} {}^m|g - h|_g(x).$$

Remark 1.2 The conditions $^{b,0}|g - h|_g \equiv {}^b|g - h|_g < \infty$ and $^b|g - h|_h < \infty$ are equivalent to g and h quasi–isometric. □

Lemma 1.3 Suppose $^b|g-h|_g, {}^b|g-h|_g < \infty$. Then there exists for every $m \geq 0$ a polynomial

$$P_m(|g - h|_g(x), |\nabla^g - \nabla^h|_g(x), \ldots, |(\nabla^g)^{m-1}(\nabla^g - \nabla^h)|_g(x))$$

with non–negative coefficients and without constant term such that

$$^m|g - h|_h(x) \leq P_m(|g - h|_g(x), \ldots, |(\nabla^g)^{m-1}(\nabla^g - \nabla^h)|_{g(x)}). \qquad (1.5)$$

Proof. The proof is essentially contained in that of II, proposition 2.16. There we work with $^{b,i}|\ |$, but all steps remain valid without "b". □

300

Lemma 1.4 *Suppose g_1, g_2, g_3 quasi–isometric. Then there exists for every $m \geq 0$ a polynomial*

$$Q_m(^i|g_1 - g_2|_{g_1}(x), {}^j|g_2 - g_3|_{g_2}(x), i, j = 0, \ldots, m)$$

with non–negative coefficients and without constant term such that

$$^m|g_1 - g_3|_{g_1}(x) \leq Q_m(^i|g_1 - g_2|_g(x), {}^j|g_2 - g_3|_{g_2}(x)). \quad (1.6)$$

□

Proof. This proof is also contained in that of II, 2.16, replacing there $^{b,i}|\ |$ by $^i|\ |(x)$. □

Lemma 1.5 *Let V be a function of moderate decay, $g, h \in \mathcal{M}$, $x_0 \in M$, $\varepsilon > 0$, and suppose*

$$|g - h|_g(x) \leq \varepsilon V(1 + d_g(x, x_0)). \quad (1.7)$$

Then there holds
a) *g and h are quasiisometric,*
b) *there exist constants $\varepsilon_1, \varepsilon_2 > 0$ such that*

$$\varepsilon_1 d_g(x, y) \leq d_h(x, y) \leq \varepsilon_2 d_g(x, y), x, y \in M \quad (1.8)$$

and

$$\varepsilon_1 V(1 + d_g(x, x_0)) \leq V(1 + d_h(x, x_0)) \leq \varepsilon_2 V(1 + d_g(x, x_0)), x \in M. \quad (1.9)$$

We refer to [54] for the simple proof. □

Remark 1.6 *If $V(x)$ is of moderate decay then there exist constants c, C_1, C_2 such that*

$$C_1 \cdot e^{-cx} \leq V(x) \leq C_2 \cdot x^{-1}, x \geq 1. \quad (1.10)$$

We refer to [54] for the proof. □

Define now for $m \geq 0$, V a function of moderate decay, $g, h \in \mathcal{M}$

$$^{m,V}|g - h|_g = \inf\{\varepsilon > 0 |\, ^m|g - h|_g(x) \leq \varepsilon \cdot V(1 + d_g(x, x_0))$$
$$\text{for all } x \in M\}$$

if $\{\cdots\} \neq 0$ and $^{m,V}|g - h|_g = \infty$ otherwise.

Let $m \geq 0$, $\delta > 0$, $C(n, \delta) = 1 + \delta + \delta\sqrt{2n(n-1)}$ function of moderate decay and set

$$V_\delta = \{(g, h) \in \mathcal{M}^2 | C(n, \delta)^{-1} g \leq h \leq C(n, \delta) g$$
$$\text{and } ^{m,V}|g - h|_g < \delta\}. \tag{1.11}$$

Proposition 1.7 $\mathfrak{B} = \{V_\delta\}_{\delta > 0}$ *is a basis for a metrizable uniform structure* $^{m,V}\mathfrak{U}(\mathcal{M})$.

Proof. Certainly holds $\bigcap_\delta V_\delta = \text{diagonal} = \{(g, g) | g \in \mathcal{M}\}$.
For the symmetry, we have to show $^{m,V}|g - h|_g < \delta$ implies $^{m,V}|g - h|_h < \delta'(\delta)$ such that $\delta'(\delta) \longrightarrow 0$. But (1.5), (1.10), (1.11) and $C(n, \delta)^{-1} g \leq h \leq C(n, \delta) g$ immediately imply

$$^m|g - h|_h(x) \leq \varepsilon V(1 + d_h(x, x_0)) \text{ for all } x \in M,$$

$\varepsilon = \varepsilon(C(n, \delta), \varepsilon_1(\delta), \varepsilon_2(\delta), \delta)$, $\lim_{\delta \to 0} C(n, \delta) = 1$, $\lim_{\delta \to 0} \varepsilon_1(\delta), \varepsilon_2(\delta) = 1$ and

$$\lim_{\delta \to 0}(\inf\{\text{all } \varepsilon(C_1, C_2, \varepsilon_1, \varepsilon_2, \delta)\}) = 0.$$

This yields the symmetry. Similarly we get the transitivity from (1.6). □

Denote by $^V_m\mathcal{M}$ the pair $(\mathcal{M}, ^{m,V}\mathfrak{U}(\mathcal{M}))$ and by $^{m,V}\mathcal{M}$ the completion $\overline{^V_m\mathcal{M}}^\mathfrak{U}$. The elements of $^{m,V}\mathcal{M}$ are C^m–metrics (cf. [67]). There is an equivalent metrizable uniform structure, based on $|g - h|, |\nabla^g(g - h)|, \ldots, |(\nabla^g)^m(g - h)|$ instead of $|g - h|, |\nabla^g - \nabla^h|, |\nabla^g(\nabla^g - \nabla^h)|, \ldots, |(\nabla^g)^{m-1}(\nabla^g - \nabla^h)|$, i.e. a uniform structure giving the same (metrizable) topology.

Suppose $|g-h|_g(x) \le C_g \cdot V(1+d_g(x,x_0))$, hence $|g-h|_h(x) \le C_h V(1+d_h(x,x_0))$. According to lemma 1.5, for any $(p,q) \in (\mathbf{Z}_+)^2$, there exist constants $C_1(p,q), C_2(p,q) > 0$, such that for any (p,q)–tensor t

$$C_1(p,q)|t|_h(x) \le |t|_g(x) \le C_2(p,q)|t|_h(x). \tag{1.12}$$

The equivalence of the uniform structures would follow from an inequality

$$\begin{aligned}C_{1,i}|(\nabla^g)^i(g-h)|_g(x) &\le |(\nabla^g)^{i-1}(\nabla^g-\nabla^h)|_g(x) \\ &\le C_{2,i}|(\nabla^g)^i(g-h)|_g(x),\end{aligned} \tag{1.13}$$

since then

$$|(\nabla^g)^i(g-h)|_g(x) \le \varepsilon_1 V(1+d_g(x,x_0)) \tag{1.14}$$

if and only if

$$|(\nabla^g)^{i-1}(\nabla^g-\nabla^h)|_g(x) \le \varepsilon_2 V(1+d_g(x,x_0)). \tag{1.15}$$

We prove (1.13) and set $B = h-g$, $D = \nabla^h - \nabla^g$. Then

$$\begin{aligned}|(\nabla^g)^i B|_g(x) &= |(\nabla^g)^i h|_g(x), \\ |(\nabla^h)^i B|_h(x) &= |(\nabla^h)^i g|_h(x),\end{aligned} \tag{1.16}$$

$$h(D(X,Y),Z) = \frac{1}{2}\{\nabla^g_X B(Y,Z) + \nabla^g_Y B(X,Z) \\ -\nabla^g_Z B(X,Y)\} \tag{1.17}$$

$$\nabla^h_X B(Y,Z) = g(D(X,Y),Z) + g(Y,D(X,Z)). \tag{1.18}$$

If we insert into (1.18) for X, Y, Z local (with respect to h) orthonormal vector fields e_i, e_j, e_k, square, sum up and take the square root, then we get

$$\begin{aligned}|\nabla^h(h-g)|_h(x) &\le \sqrt{2}\left(\sum_{i,j,k} g((\nabla^h_{e_i}-\nabla^g_{e_i})e_j, e_k)^2\right)^{\frac{1}{2}} \\ &= \sqrt{2}|g((\nabla^h-\nabla^g)_{(\cdot)}(\cdot),(\cdot)|_h(x). \end{aligned} \tag{1.19}$$

According to (1.12)

$$|g((\nabla^h - \nabla^g)_{(\cdot)}(\cdot),(\cdot))|_h(x)$$
$$\leq \frac{1}{C_1(3,0)}|g((\nabla^h - \nabla^g)_{(\cdot)}(\cdot),(\cdot))|_g(x)$$
$$= \frac{1}{C_1(3,0)}|\nabla^h - \nabla^g|_g(x)$$
$$\leq \frac{C_1(2,1)}{C_1(3,0)}|\nabla^h - \nabla^g|_h(x), \tag{1.20}$$

together with (1.19)

$$|\nabla^h(h-g)|_h(x) \leq d'_{1,0}|\nabla^h - \nabla^g|_h(x),$$

or, exchanging the role of g and h,

$$|\nabla^g(g-h)|_g(x) \leq d_{1,0}|\nabla^g - \nabla^h|_g(x),$$

i.e. we have the first inequality (1.13) for $i = 1$. We get in the same manner from (1.17)

$$|\nabla^h - \nabla^g|_h(x) \leq d_1|\nabla^g(h-g)|_h(x) \leq d_2|\nabla^g(h-g)|_g(x),$$

together with

$$|\nabla^g - \nabla^h|_g(x) \leq C_2(2,1)|\nabla^g - \nabla^h|_h(x)$$

finally
$$|\nabla^g - \nabla^h|_g(x) \leq c_{2,i}|\nabla^g(h-g)|_g(x)$$

for $i = 1$, (1.13) is done.

For $i > 1$, the proof follows by a simple but extensive induction, differentiating (1.17), (1.18) and applying (1.16). □

We now set

$$^{m,met}|g-h|_g(x) := \sum_{i=0}^{m} |(\nabla^g)^i(g-h)|_g(x),$$
$$^{m,met,V}|g-h|_g := \inf\{\varepsilon > 0 |^{m,met}|g-h|_g(x)$$
$$\leq \varepsilon V(1+d_g(x,x_0)) \text{ for all } x \in M\}$$

if $\{\cdots\} \neq \emptyset$ and $^{m,met,V}|g-h|_g = \infty$ otherwise and define

$$\begin{aligned}V_\delta &= \{(g,h) \in \mathcal{M}^2 | C(1,\delta)^{-1}g \leq h \leq C(n,\delta)g \\ &\quad \text{and } {}^{m,met,V}|g-h| < \delta\}.\end{aligned}$$

Proposition 1.8 $\mathfrak{B} = \{V_\delta\}_{\delta>0}$ *is a basis for a metrizable uniform structure* $^{m,met,V}\mathfrak{U}(\mathcal{M})$. □

Denote by $^{m,met,V}\mathcal{M}$ the corresponding completion. Then, as metrizable topological spaces

$$^{m,V}\mathcal{M} = {}^{m,met,V}\mathcal{M}. \tag{1.21}$$

Proposition 1.9 $^{m,V}\mathcal{M}$ *is locally contractible and hence locally arcwise connected.*

Proof. According to (1.21), we would be done if we could prove the assertion for $^{m,met,V}\mathcal{M}$. Locally a neighbourhood basis is given by

$$\begin{aligned}U_\varepsilon(g) &= \{h \in {}^{m,met,V}\mathcal{M} | C(n,\varepsilon)^{-1}g \leq h \leq C(n,\varepsilon) \cdot g \\ &\quad \text{and } {}^{m,met,V}|g-h|_g < \varepsilon\}. \tag{1.22}\end{aligned}$$

Let $h \in U_\varepsilon(g)$, $g_t = th + (1-t)g$, $|g-g_t|_g(x) = t|g-h|_g(x)$, $^{m,met,V}|g-g_t| = t \cdot {}^{m,met,V}|h-g|_g < t \cdot \varepsilon \leq \varepsilon$, i.e. all of the arc is contained in $U_\varepsilon(g)$ and this arc deforms to g. □

Remark 1.10 Quite analogous to (1.22) on defines

$$\begin{aligned}{}^{m,V}U_\varepsilon(g) &\equiv U_\varepsilon(g) = \{g' | C(n,\varepsilon)^{-1}g \leq g' \leq C(n,\varepsilon)g \\ &\quad \text{and } {}^{m,V}|g-g'|_g < \varepsilon\}. \tag{1.23}\end{aligned}$$

□

Theorem 1.11 **a)** *In* $^{m,V}\mathcal{M}$ *coincide components with arc components.*

b) *There exists a topological sum representation*

$$^{m,V}\mathcal{M} = \sum_{i \in I} \mathrm{comp}^{m,V}(g_i). \tag{1.24}$$

c) *For $g \in {}^{m,V}\mathcal{M}$,*

$$\mathrm{comp}^{m,V}(g) = \{g' \in {}^{m,V}\mathcal{M} | {}^{m,V}|g - g'|_g < \infty\} \equiv {}^{m,V}\mathcal{U}(g).$$

Proof. a) and b) immediately follow from proposition 1.9. We remark further that

$$g' \in {}^{m,V}\mathcal{U}(g) \text{ if and only if } g \in {}^{m,V}\mathcal{U}(g'). \tag{1.25}$$

This follows from the symmetry in proposition 1.7. In particular, g and g' are quasi–isometric. Moreover,

$$^{m,V}\mathcal{U}(g) \cap {}^{m,V}\mathcal{U}(g') \neq \emptyset \text{ implies } {}^{m,V}\mathcal{U}(g) = {}^{m,V}\mathcal{U}(g'). \tag{1.26}$$

Let $g'' \in {}^{m,V}\mathcal{U}(g) \cap {}^{m,V}\mathcal{U}(g')$ and $g''' \in {}^{m,V}\mathcal{U}(g)$. Write $g' - g''' = (g' - g'') + (g'' - g) + (g - g''')$ and apply (1.6) to get $^{m,V}|g' - g'''|_{g'} < \infty$. This yields $^{m,V}\mathcal{U}(g) \subseteq {}^{m,V}\mathcal{U}(g')$. From symmetry considerations $^{m,V}\mathcal{U}(g') \subseteq {}^{m,V}\mathcal{U}(g)$.

Suppose now $g' \in \mathrm{comp}^{m,V}(g)$ and let $\{g_t\}_{0 \leq t \leq 1}$ be an arc between g and g', $g_0 = g$, $g_1 = g'$. Then for even $\varepsilon > 0$, the arc can be covered by a finite number of neighbourhoods $^{m,V}\mathcal{U}_\varepsilon(g_0), {}^{m,V}\mathcal{U}_\varepsilon(g_{t_1}), \ldots, {}^{m,V}\mathcal{U}_\varepsilon(g_{t_r}) = {}^{m,V}\mathcal{U}_\varepsilon(g')$, $^{m,V}\mathcal{U}_\varepsilon(g_{t_{i-1}}) \cap {}^{m,V}\mathcal{U}_\varepsilon(g_{t_i}) \neq \emptyset$. If $g_{i,i-1} \in {}^{m,V}\mathcal{U}_\varepsilon(g_{t_{i-1}}) \cap {}^{m,V}\mathcal{U}_\varepsilon(g_{t_i})$, then

$$^{m,V}\mathcal{U}(g_{t_{i-1}}) \supset {}^{m,V}\mathcal{U}_\varepsilon(g_{t_{i-1}}) \ni g_{i,i-1} \in {}^{m,V}\mathcal{U}_\varepsilon(g_{t_i}) \subset {}^{m,V}\mathcal{U}(g_{t_i}),$$

i.e., according to (1.26)

$$^{m,V}\mathcal{U}(g_{t_{i-1}}) = {}^{m,V}\mathcal{U}(g_{t_i}).$$

Repeating this conclusion, we obtain

$$^{m,V}\mathcal{U}(g) = {}^{m,V}\mathcal{U}(g'),$$

$g' \in {}^{m,V}\mathcal{U}(g)$, $\mathrm{comp}^{m,V}(g) \subseteq {}^{m,V}\mathcal{U}(g)$. We have to show that $^{m,V}\mathcal{U}(g) \subseteq \mathrm{comp}^{m,V}(g)$. Suppose now $g' \in {}^{m,V}\mathcal{U}(g)$. Set $g_t =$

$tg' + (1-t)g$. Then we conclude immediately from the proof of proposition 1.9 that $\{g_t\}_{0 \leq t \leq 1}$ is an arc in $^{m,V}\mathcal{U}(g)$ which connects g and g', i.e. $^{m,V}\mathcal{U}(g) \subseteq \mathrm{comp}^{m,V}(g)$. □

Proposition 1.12 *The following properties are invariants of* $\mathrm{comp}^{m,V}(g)$, $m \geq 2$.

a) $|\nabla^i R| \leq C_i$, $0 \leq i \leq m-2$,

b) *sectional curvature bounded from below or bounded from above, respectively,*

c) *Ricci curvature bounded from below or from above, respectively.* □

2 The injectivity radius and weighted Sobolev spaces

We assumed in chapter II – VI the positivity of the injectivity radius, $r_{\mathrm{inj}}(M,g) > 0$. The background for this assumption was the fact that we often used the Sobolev embedding theorem II 1.6, II 1.7 and the module structure theorem II 1.17. In the proof of these theorems, the assumption $r_{\mathrm{inj}} > 0$ plays an essential role. Unfortunately, open complete manifolds with finite volume fall out from these considerations. Nevertheless, a big part of the necessary Sobolev analysis can be rescued if we work with weighted Sobolev spaces. We present in a forthcoming paper a comparatively comprehensive investigation on weighted Sobolev spaces for open complete manifolds with $r_{\mathrm{inj}} = 0$. Brought all discussions to a point, one has to work with the weight $(r_{\mathrm{inj}}(x))^{-\frac{n}{p}}$ or a weight function with at least this growth. For the trace class investigations in this chapter, we need much less of the corresponding Sobolev analysis and we essentially follow [18] and [54].

Lemma 2.1 *Let* (M^n, g) *be open, complete, g and h quasi isometric, $|K_g|, |K_h| \leq K$. Then there exist constants $c, c' > 0$ such*

that
$$r_{\text{inj}}(h, x) \geq \min\{c \cdot r_{\text{inj}}(g, x), c'\}, \quad x \in M.$$

We refer to [54] for the proof. □

A special case is given by $h \in \text{comp}^{0,V}(g)$, V of moderate decay. Set for $x \in M$
$$\tilde{r}_{\text{inj}}(x) := \min\left\{\frac{\pi}{12\sqrt{K}}, r_{\text{inj}}(x)\right\}.$$

Then, under the assumptions of lemma 2.1,
$$\tilde{r}_{\text{inj}}(h, x) \geq c_2 \tilde{r}_{\text{inj}}(g, x).$$

Using the standard volume comparison theorem
$$\frac{2\pi^{\frac{n}{2}}}{\Gamma(\frac{n}{2})} \int_0^r \left(\frac{\sinh t\sqrt{K}}{\sqrt{(K)}}\right)^{n-1} dt \leq \text{vol}(B_r(x_0))$$
$$\leq \frac{2\pi^{\frac{n}{2}}}{\Gamma(\frac{n}{2})} \int_0^r \left(\frac{\sinh t\sqrt{K}}{\sqrt{K}}\right)^{n-1} dt \quad (2.1)$$

and theorem 4.7 from [18],
$$\tilde{r}_{\text{inj}}(x) \geq \frac{r}{2} \frac{1}{1 + (V_{r_0+s}^K/\text{vol}(B_r(y)))(V_{d(x,y)+r}^K/V_s^K)}, \quad (2.2)$$

$r_0 + 2s < \pi/\sqrt{K}$, $r_0 \leq \pi/4\sqrt{K}$, V_ϱ^K = volume of a ϱ-ball in hyperbolic space H_{-K}^n, then one immediately gets
$$\tilde{r}_{\text{inj}}(x) \geq C\tilde{r}_{\text{inj}}(y)^n e^{-(n-1)\sqrt{K}d(x,y)}, \quad (2.3)$$

$C = C(K)$. (2.3) implies the existence of a constant $C = C(K, y)$ such that
$$\tilde{r}_{\text{inj}}(x) \geq Ce^{-(n-1)\sqrt{K}d(x,y)}, \quad x \in M. \quad (2.4)$$

Lemma 2.2 *There exists a constant $C = C(K)$, such that*
$$\tilde{r}_{\text{inj}}(y) \geq C\tilde{r}_{\text{inj}}(x)e^{-\frac{(n-1)\pi}{12}\frac{d(x,y)}{\tilde{r}_{\text{inj}}(x)}}. \quad (2.5)$$

We refer to [54] for the proof. □

We recall from II, lemma 1.10 the existence of appropriate uniformly locally finite covers of (M^n, g) if $\mathrm{Ric}\,(g) \geq k$.

Set for $s > \varepsilon \geq 0$, $\kappa_\varepsilon(M, g, s) \in \mathbb{N} \cup \{\infty\}$ = smallest number such that there exists a sequence $\{x_i\}_{i=1}^\infty$ such that $\{B_{s-\varepsilon}(x_i)\}_{i=1}^\infty$ is an open cover of M satisfying

$$\sup_{x \in M} \#\{i \in \mathbb{N} | x \in B_{3s+\varepsilon}(x_i)\} \leq \kappa_\varepsilon(M, g, s) \qquad (2.6)$$

and set $\kappa(M, g, s) := \kappa_0(M, g, s)$, $\kappa(M, g, 0) := 1$.

We need for norm estimates of $\cos\sqrt{\Delta}$ in the forthcoming section 3 the following

Lemma 2.3 $\kappa_\varepsilon(M, g, s)$ *is finite for all* $s > \varepsilon$ *and there exist constants* $C, c > 0$ *depending on* K *such that for* $s > \frac{2\pi}{\sqrt{K}} + \varepsilon$

$$\kappa_\varepsilon(M, g, s) \leq C \cdot e^{cs}. \qquad (2.7)$$

We refer to [54] for the proof. □

Next we collect some facts concerning weighted Sobolev spaces. Let $(E, h_E, \nabla^E) \longrightarrow (M^n, g)$ be a Riemannian vector bundle, ξ a positive, measurable function on M which is finite a.e.. Define the associated weighted Sobolev spaces as follows

$$\begin{aligned}\Omega_\xi^{p,r}(E, \nabla) &= \{\varphi \in L_{p,loc}(E) \mid |\varphi|_{p,r,\nabla,\xi} \\ &= \left(\int \sum_{i=0}^r (\xi^{\frac{1}{p}}|\nabla^i \varphi|_x)^p \, \mathrm{dvol}_x(g)\right)^{\frac{1}{p}} < \infty\},\end{aligned}$$

where ∇^i is iteratively applied in the distributional sense. In the case $p = 2$ we write

$$\Omega_\xi^{2,r}(E, \nabla) = W_\xi^r(E, \nabla) \equiv (W_\xi^r(E, \nabla), ||_{W_\xi^r}).$$

Set $\Delta = \nabla^*\nabla$,

$$\begin{aligned}\Omega_\xi^{2,2r}(E, \Delta) &= \{\varphi \in L_{p,loc}(E) \mid |\varphi|_{p,2r,\Delta,\xi} \\ &= \left(\int \sum_{i=0}^r (\xi^{\frac{1}{p}}|\Delta^i \varphi|_x)^p \, \mathrm{dvol}_x(g)\right)^{\frac{1}{p}} < \infty\}\end{aligned}$$

and for $p = 2$

$$\Omega^{2,2r}_\xi(E,\Delta) = H^{2r}_\xi(E,\Delta) \equiv (H^{2r}_\xi(E,\Delta), ||_{H^{2r}_\xi}).$$

We refer to [27], [54] for further general information on weighted Sobolev spaces. For $\xi = 1$ we get back standard Sobolev spaces.

Lemma 2.4 *The natural inclusion* $W^{2r}_\xi(E) \hookrightarrow H^{2r}_\xi(E)$ *is bounded.*

Proof. Let $\varphi \in W^{2r}_\xi$, $|\varphi|_{2,2r,\nabla,\xi} < \infty$. $\Delta\varphi = -\mathrm{tr}_g(\nabla^2 \varphi)$ and $\nabla \mathrm{tr}_g = 0$ imply for the pointwise norms

$$|\Delta^i \varphi|_x \leq C |\nabla^{2i} \varphi|_x, \quad 0 \leq i \leq r, x \in M,$$

hence

$$|\varphi|_{2,2i,\Delta,\xi} < \infty$$

and

$$|\varphi|_{2,2r,\Delta,\xi} \leq C \cdot |\varphi|_{2,2r,\nabla,\xi}.$$

$\square$

In the sequel, we will establish some Sobolev estimates by locally finite covers. We proved in [29] the following

Proposition 2.5 a) *If* (M^n, g) *satisfies* (B_k) *and* $\mathfrak{U} = \{(U_\alpha, \Phi_\alpha)\}_\alpha$ *is a locally finite cover by normal charts, then there exist constants* $C_\beta, C'_\beta, C'_\gamma$, *multiindexed by* β, γ *such that*

$$|D^\beta g_{ij}| \leq C_\beta, \quad |D^\beta g^{ij}| \leq C'_\beta, \quad \text{for } |\beta| \leq k \qquad (2.8)$$

and

$$|D^\gamma \Gamma^m_{ij}| \leq C'_\gamma, \quad |\gamma| \leq k-1, \qquad (2.9)$$

all constants are independent of α.

b) *If* $(E, h_E, \nabla^E) \longrightarrow (M^n, g)$ *is a Riemannian vector bundle satisfying* $(B_k(M,g))$, $(B_k(E, \nabla^E))$, *then additionally to (2.8), there holds for the connection coefficients* $\Gamma^\lambda_{i\mu}$ *defined by*

$\nabla_{\frac{\partial}{\partial u^i}} \varphi_\lambda = \Gamma^\mu_{i\lambda} \varphi_\mu$, $\{\varphi_\mu\}_\mu$ a local orthonormal frame defined by radial parallel translation,

$$|D^\beta \Gamma^\mu_{i\lambda}| \leq D_\beta, \quad |\beta| \leq k-1. \tag{2.10}$$

□

Consider $\varrho > 0$, $B_\varrho = B_\varrho(0) \subset \mathbb{R}^n$, $m, k \in \mathbb{N}$, $K, \lambda > 0$ and denote by $\mathrm{Ell}^m_k(\varrho, K, \lambda)$ the set of elliptic operators

$$P = \sum_{|\alpha| \leq m} a_\alpha(x) D^\alpha, \quad a_\alpha(x) = (a_{\alpha,i,j})_{1 \leq i,j \leq k}$$

satisfying
1) $a_{\alpha,i,j} \in C^m(B_\varrho)$,
2) $\sum_{|\alpha|<m} |a_\alpha|_{C^0(B_\varrho)} \leq K$, $\sum_{|\alpha|<m} |a_\alpha|_{C^1(B_\varrho)} \leq K$,
3) $\lambda^{-1}|\xi|^m \leq \sum_{|\alpha|=m} \left(\sum_{1 \leq i,j \leq k} a_{\alpha,i,j}(x) \right) \xi^\alpha \leq \lambda |\xi|^m$,

where $|a_\alpha|_{C^l(B_\varrho)} = \sup_{i,j} \sup_{|\beta| \leq l} \sup_{x \in B_\varrho} |D^\beta a_{\alpha,i,j}(x)|$.

We recall a standard elliptic inequality for Euclidean balls B_ϱ and refer to [27], II (1.52), p. 75.

Lemma 2.6 *For given k, K, λ, there exist $\varrho = \varrho(K, \lambda) > 0$ and $C(\lambda) > 0$ such that for all $\varrho \leq \varrho_0$, $P \in \mathrm{Ell}^m_k(\varrho, K, \lambda)$*

$$|u|_{W^m(B_\varrho, \mathbb{C}^k)} \leq C(|Pu|_{L_2(B_\varrho, \mathbb{C}^k)} + |u|_{L_2(B_\varrho, \mathbb{C}^k)})$$

for all $u \in C_c^\infty(B_\varrho)$.

□

Using 2.5 and 2.6, we immediately get a generalization to balls in Riemannian manifolds and bundles satisfying (B_k) (as in [27], II (1.52)).

Lemma 2.7 *Suppose $(E, h_E, \nabla^E) \longrightarrow (M^n, g)$ being a Riemannian vector bundle satisfying $B_{2k}(M, g)$, $B_{2k}(E, \nabla^E)$, $|\nabla^i R^g|$,*

$|\nabla^j R^E| \leq K$, $i = 0, \ldots, 2k$. Then there exist constants $\varrho_0(K) > 0$, $C(K) > 0$ such that for all $x_0 \in M$ and $\varrho \leq \min\{\varrho_0, \tilde{r}_{\text{inj}}(x_0)\}$ there holds

$$|u|_{W^{2k}(B_\varrho(x_0),E)} \leq C |u|_{H^{2k}(B_\varrho(x_0),E)} \qquad (2.11)$$

for all $u \in C_c^\infty(B_\varrho(x_0), E)$.

We refer to [27] or to [54]. □

If $V : [1, \infty[\longrightarrow \mathbb{R}_+$ is of moderate decay, then we define the associated $V : M^n \longrightarrow \mathbb{R}_+$ of moderate decay by $V(x) = V(1 + d_g(x, x_0))$.

Lemma 2.8 *Let* $(E, h_E, \nabla^E) \longrightarrow (M^n, g)$ *be a Riemannian vector bundle satisfying* $(B_k(M^n, g))$, $B_{2k}(E, \nabla^E)$, *k even, and let* $\beta : M \longrightarrow \mathbb{R}_+$ *be of moderate decay. Then there exist bounded inclusions*

$$H^k_{V\tilde{r}_{\text{inj}}^{-2kn}}(E, \Delta) \hookrightarrow W^k_V(E, \nabla) \qquad (2.12)$$

and

$$H^k_V(E, \Delta) \hookrightarrow W^k_{V\tilde{r}_{\text{inj}}^{2kn}}(E, \nabla). \qquad (2.13)$$

Proof. We apply (2.6) to get a cover $\left\{ B_{\frac{r_{\text{inj}}(x_i)}{2^k}}(x_i) \right\}_{i=1}^\infty$ by balls satisfying for all $x \in M$

$$\#\{x_i \mid x \in B_{\tilde{r}_{\text{inj}}(x_i)}(x_i)\} \leq C. \qquad (2.14)$$

Set for $u \in C^\infty(\mathbb{R})$ with $u = 1$ on $[0,1]$ and $= 0$ on $[2, \infty[$ and $1 \leq j \leq k$

$$u_{j,x}(y) = \begin{cases} u(2^{j \cdot \frac{d(x,y)}{\tilde{r}_{\text{inj}}(x)}}), & y \in B_{\tilde{r}_{\text{inj}}(x)}(x), \\ 0 & \text{otherwise.} \end{cases}$$

Hence $u_{j,x} \in C_c^\infty(M)$ and for $\varphi \in H^k(E, \Delta)$,

$$u_{j,x}\varphi \in H^k(B_{\tilde{r}_{\text{inj}}(x)}, E).$$

We infer from (2.11) $u_{j,x}\varphi \in W^k(B_{\tilde{r}_{\text{inj}}(x)}(x), E$ and

$$\nabla^j(u_{k,x}\varphi) = \sum_{l=0}^{j} \binom{j}{l} ((\nabla^g)^l u_{k,x})((\nabla^h)^{j-l}\varphi).$$

Moreover, we infer from (2.11)

$$\begin{aligned}
|u_{k,x}\varphi|_{W^k(E)} &\leq C_1 |\varphi|_{W^k(B_{\frac{\tilde{r}_{\text{inj}}}{2^{k-1}}(x)}(x), E)} \\
&\quad + C_2 \sum_{l=1}^{k} \binom{k}{l} (\tilde{r}_{\text{inj}}(x))^{-l} \cdot |u_{k-1,x}\varphi|_{W^{k-l}(E)} \\
&\leq C_1 |\varphi|_{H^k(B_{\frac{\tilde{r}_{\text{inj}}}{2^{k-1}}(x)}(x))} \\
&\quad + C_3 \sum_{l=1}^{k} \binom{k}{l} (\tilde{r}_{\text{inj}}(x))^{-l} |u_{k-1,x}\varphi|_{H^{k-l}(E)}.
\end{aligned} \tag{2.15}$$

We conclude from (2.14) by an easy induction

$$|u_{k,x_i}\varphi|_{W^k(E)} \leq C \tilde{r}_{\text{inj}}^{-k}(x_i) |\varphi|_{H^k(B_{\tilde{r}_{\text{inj}}(x_i)}(x_i), E)}. \tag{2.16}$$

□

Lemma 2.9 *Let $V : [1, \infty[\longrightarrow \mathbb{R}_+$ be a function of moderate decay. Then for all $x, y, q \in M$ there holds*

$$C_V \cdot V(1 + d(x,y)) \leq \frac{V(1+d(x,q))}{V(1+d(y,q))} \leq \frac{1}{C_V \cdot V(1+d(x,y))},$$

and for every $q' \in M$ there exists $C = C(q, q') > 0'$ such that

$$C^{-1} \cdot V(1 + d(x, q')) \leq V(1 + d(x,q)) \leq C \cdot V(1 + d(x, q')).$$

We refer to [54] for the simple proof. □

Let now $\varphi \in H^k(E, \Delta)$. Then lemma 2.9, lemma 2.7, (2.14) and (2.16) imply

$$\begin{aligned}
|\varphi|_{W_V^k(E,\nabla)} &\leq C \sum_{i=1}^\infty V^{\frac{1}{2}}(x_i) |u_{k,x_i}\varphi|_{W^k(E,\nabla)} \cdot \tilde{r}_{\mathrm{inj}}(x_i)^k \\
&\leq C \sum_{i=1}^\infty V^{\frac{1}{2}}(x_i) |u_{k,x_i}\varphi|_{H^k(E,\Delta)} \\
&\leq C \sum_{i=1}^\infty V^{\frac{1}{2}}(x_i) \tilde{r}_{\mathrm{inj}}(x_i)^{-k} \cdot |\varphi|_{H^k(B_{\tilde{r}_{\mathrm{inj}}(x_i)},E)}.
\end{aligned}$$
(2.17)

According to (2.4), there exists $D_1 > 0$ such that

$$\tilde{r}_{\mathrm{inj}}(x_i)^{-k} \tilde{r}_{\mathrm{inj}}(x)^{kn} \leq D_1,$$

hence

$$\sum_{i=1}^\infty V^{\frac{1}{2}}(x_i) \cdot \tilde{r}_{\mathrm{inj}}(x_i)^{-k} |\varphi|_{H^k(B_{\tilde{r}_{\mathrm{inj}}(x_i)}(x_i),E)} \leq D_2 |\varphi|_{H^k_{\tilde{r}_{\mathrm{inj}}^{-2kn} V(E)}},$$

which yields together with (2.17) the inclusion (2.12). The proof of (2.13) is quite analoguous. □

Lemma 2.10 *Let ξ be continuous. Then $W_\xi^r(E, \nabla) \cap C^\infty(E)$ or $H_\xi^r(E, \Delta) \cap C^\infty(E)$ are dense in $W_\xi^r(E, \nabla)$ or $H_\xi^r(E, \Delta)$, respectively.*

We refer to [54], lemma 3.1 for the simple proof.

We conclude this section with the invariance of weighted Sobolev spaces $W_\xi^r(E, \nabla^E, g)$, $H_\xi^r(E, \Delta^E, g)$ under change of the metric g inside $\mathrm{comp}^{r,V}(g)$.

Proposition 2.11 *Let V be of moderate decay, $(E, h_E, \nabla^E) \longrightarrow (M^n, g)$ be a Riemannian vector bundle satisfying $(B_0(M,g))$ and suppose $h \in \mathrm{comp}^{r,V}(g)$. Then*

$$W_\xi^\varrho(E, \nabla^E, g) \cong W_\xi^\varrho(E, \nabla^E, h), \quad 0 \leq \varrho \leq r, \quad (2.18)$$

as equivalent Sobolev spaces.

Proof. For $r = 0$, the assertion is clear since, according to (1.8), g and h are quasi–isometric. The case $r = 1$ is also clear, since into the derivative ∇^E the derivatives ∇^g or ∇^h do not enter. In the case $r = 2$, (M, h) satisfies $(B_0(M, h))$ too, and h is a C^r–metric. We set

$$\nabla^2 = (\nabla^E \otimes \nabla^g) \circ \nabla^E, \tag{2.19}$$

$$\vdots \tag{2.20}$$

$$\nabla^\varrho = (\nabla^E \otimes \bigotimes_1^{\varrho-1} \nabla^g) \circ \cdots \circ (\nabla^E \otimes \nabla^g) \circ \nabla^E, \tag{2.21}$$

$$\nabla'^2 = (\nabla^E \otimes \nabla^h) \circ \nabla^E, \tag{2.22}$$

$$\vdots \tag{2.23}$$

$$\nabla'^\varrho = (\nabla^E \otimes \bigotimes_1^{\varrho-1} \nabla^h) \circ \cdots \circ (\nabla^E \otimes \nabla^h) \circ \nabla^E. \tag{2.24}$$

According to II (1.34), for $\varphi \in W^\varrho_\xi(E, \nabla^E, g) \cap C^\infty(E)$

$$\nabla'^\varrho \varphi = \sum_{i=1}^\varrho \nabla^{i-1}(\nabla' - \nabla) \nabla'^{\varrho-i} \varphi + \nabla^\varrho \varphi. \tag{2.25}$$

By assumption $\xi^{\frac{1}{2}}|\nabla^\varrho \varphi| \in L_2(M, g)$. There remains to consider the terms

$$\nabla^{i-1}(\nabla' - \nabla) \nabla'^{\varrho-i} \varphi. \tag{2.26}$$

Iterating the procedure, i.e. applying it to $\nabla'^{\varrho-i}$ and so on, we have to estimate expressions of the kind

$$\nabla^{i_1}(\nabla' - \nabla)^{i_2} \cdots \nabla^{i_{\varrho-2}}(\nabla' - \nabla)^{i_{\varrho-1}} \nabla^{i_\varrho} \varphi \tag{2.27}$$

with $i_1 + \cdots + i_\varrho = \varrho$, $i_\varrho < \varrho$. Consider a typical expression

$$\nabla^k (\nabla' - \nabla)^j \nabla^i \varphi.$$

In a local bundle chart

$$\nabla^i \varphi|_u = \sum_{\text{finite}} \psi \otimes t_1 \otimes \cdots \otimes t_i.$$

We calculate $(\nabla' - \nabla)\nabla^i\varphi$ at a typical term $\psi \otimes t_1 \otimes \cdots \otimes t_i$:

$$(\nabla' - \nabla)(\psi \otimes t_1 \otimes \cdots \otimes t_i)$$
$$= (\nabla^E \otimes \bigotimes_1^i \nabla^h - \nabla^E \otimes \bigotimes_1^i \nabla^g)(\psi \otimes t_1 \otimes \cdots \otimes t_i)$$
$$= \psi \otimes (\nabla^h - \nabla^g)t_1 \otimes t_2 \otimes \cdots \otimes t_i + \cdots$$
$$+ \psi \otimes t_1 \otimes \cdots \otimes t_{i-1} \otimes (\nabla^h - \nabla^g)t_i. \qquad (2.28)$$

Linear extension of (2.26) and Schwartz' inequality for the pointwise norm yield

$$|(\nabla' - \nabla)\nabla^i\varphi|_{h_E,g,x} \leq i \cdot |\nabla^i\varphi|_{h_E,g,x} \cdot {}^1|g - h|_g(x),$$

together with (1.8)

$$|(\nabla' - \nabla)\nabla^i\varphi|_{h_E,h,x} \leq C \cdot i \cdot |\nabla^i\varphi|_{h_E,g,x} \cdot {}^1|g - h|_g(x).$$

Iterating the calculation and estimate, we obtain

$$|(\nabla' - \nabla)^2 \nabla^i\varphi|_x \leq C(i,2) \cdot |\nabla^i\varphi|({}^1|g - h|_g^2(x) + {}^2|g - h|_g(x)),$$

$$\vdots$$

$$|(\nabla' - \nabla)^j \nabla^i\varphi|_x \leq C(i,j) \cdot |\nabla^i\varphi|({}^1|g - h|_g^j(x) + \cdots + {}^j|g - h|_g(x)),$$

$$|\nabla(\nabla' - \nabla)^j \nabla^i\varphi|_x \leq P_{i,j,1}({}^1|g - h|_g(x), \ldots, {}^{j+1}|g - h|_g(x)) \cdot |\nabla^{j+1}\varphi|_x,$$

$$\vdots$$

$$|\nabla^k(\nabla' - \nabla)^j \nabla^i\varphi|_x \leq P_{i,j,k}({}^1|g - h|_g(x), \ldots, {}^{j+k}|g - h|_g(x)) \cdot |\nabla^{j+k}\varphi|_x,$$

where $P_{i,j,k}$ is a polynomial in the indicated variables without constant terms. Continuing this way, we finally get

$$|\nabla^{i_1}(\nabla' - \nabla)^{i_2}\nabla^{i_3}\cdots\nabla^{i_\varrho-2}(\nabla' - \nabla)^{i_{\varrho-1}}\nabla^{i_\varrho}\varphi|_{h_E,h,x}$$
$$\leq P_{i_1,i_2,\ldots,i_\varrho}(\ldots,{}^j|g - h|_g(x),\ldots)|\nabla^{i_1+i_3+\cdots+i_\varrho}\varphi|_{h_E,g,x}$$
$$\leq P_{i_1,i_2,\ldots,i_\varrho}(V(x),\ldots,V(x)) \cdot |\nabla^{i_1+i_3+\cdots i_\varrho}\varphi|_{h_E,g,x}. \qquad (2.29)$$

By assumption,

$$\xi^{\frac{1}{2}}|\nabla^{i_1+i_3+\cdots+i_\varrho}\varphi|_{h_E,g,x} \in L_2(M,g) \cong L_2(M,h).$$

$P_{i_1,i_2,\ldots i_\varrho}(V(x),\ldots,V(x))$ is bounded, even
$P_{i_1,i_2,\ldots i_\varrho}(V(x),\ldots,V(x)) \xrightarrow[x\to\infty]{} 0$ uniformly,

hence

$$\xi^{\frac{1}{2}}\cdot|\nabla^{i_1}(\nabla'-\nabla)^{i_2}\nabla^{i_3}\cdots\nabla^{i_{\varrho-2}}(\nabla'-\nabla)^{i_{\varrho-1}}\nabla^{i_\varrho}\varphi|_{h_E,h,x} \in L_2(M,h),$$

$\varphi \in W_\xi^\varrho(E,\nabla,h)$, $W_\xi^\varrho(E,\nabla,g) \subseteq W_\xi^\varrho(E,\nabla,h)$, $0 \le \varrho \le r$.
Exchanging the role of g and h, we obtain the other inclusion. $\square$

Corollary 2.12 *Suppose the hypotheses of 2.11 and $B_r(M,g)$, $B_r(M,h)$, $B_r(E,\nabla^E)$. Then*

$$H^\varrho(E,\Delta,q) \cong W^\varrho(E,\nabla,g) \cong W^\varrho(E,\nabla,h) \cong H^\varrho(E,\nabla,h)$$

as equivalent Hilbert spaces.

Proof. This immediately follows from 2.11 with $\xi = 1$ and [27], [66]. $\square$

3 Mapping properties $e^{-t\Delta}$

In the decisive next section, we need mapping properties of the operator $e^{-t\Delta}$ which we establish now. Here we densely follow section 4 of [54]. On a complete manifold the operator $\Delta = \nabla^*\nabla : C_c^\infty(E) \longrightarrow L_2(M,E)$ is essentially self–adjoint and for an admissible function $f(\cdot)$ the operator $f(\Delta)$ is defined by the spectral theorem. The same holds for $\sqrt{\Delta}$, i.e.

$$f(\sqrt{\Delta}) = \int_0^\infty f(\lambda)dE_\lambda,$$

where $\{E_\lambda\}_\lambda$ is the spectral family for $\sqrt{\Delta}$. For $f \in L_1(\mathbb{R})$ we denote by $\hat{f}$ the cosine transformation of f

$$\hat{f}(\lambda) = \int_{-\infty}^{+\infty} f(x)\cos(\lambda x)dx$$

and $f(\sqrt{\Delta})$ has a representation as

$$f(\sqrt{\Delta}) = \frac{1}{2\pi} \int_{-\infty}^{+\infty} \hat{f}(\lambda)\cos(\lambda\sqrt{\Delta})d\lambda. \quad (3.1)$$

Theorem 3.1 *Let* $(E, h_E, \nabla^E) \longrightarrow (M^n, g)$ *be a Riemannian vector bundle satisfying* $(B_0(M^n, g))$ *and let* $V : M \longrightarrow \mathbb{R}_+$ *be a function of moderate decay. Then* $\cos(s\sqrt{\Delta})$ *extends to a bounded operator in* $L_{2,V}(M, E) = \{\varphi \mid V^{\frac{1}{2}}|\varphi| \in L_2(M, g)\}$ *for all* $s \in \mathbb{R}$ *and there exist* $C, c > 0$ *such that*

$$|\cos(s\sqrt{\Delta})|_{L_{2,V}, L_{2,V}} \leq C \cdot e^{c|s|}, \quad s \in \mathbb{R},$$

and $\cos(s\sqrt{\Delta}) : L_{2,V}(M, E) \longrightarrow L_{2,V}(M, E)$ *is strongly continuous in* s.

Proof. Recall $\kappa(M, g, s)$ from (2.6), let $s > 0$, $\{x_k\}_{k=1}^{\infty}$ be a sequence in M which minimizes $\kappa(M, g, s)$ and let $P_k(\cdot) = \chi_{B_s(x_k) \setminus \bigcup_{i=1}^{k-1} B_s(x_i)}$. P_k is an orthonormal projection in $L_2(M, E)$ and $L_{2,V}(M, E)$ satisfying $P_k P_{k'} = 0$ for $k \neq k'$ and $\sum_{k=1}^{\infty} P_k = 1$ strongly. $\cos(t\sqrt{\Delta})$ has unit propagation speed which implies for $\varphi \in L_2(M, E)$

$$\operatorname{supp}\,\cos(s\sqrt{\Delta})P_k\varphi \subset B_{2s}(x_k)$$

and

$$\operatorname{supp}\,\cos(s\sqrt{\Delta})((1 - \chi_{B_{3s}})\varphi) \subset M \setminus B_{2s}(x_k).$$

This yields

$$|\cos(s\sqrt{\Delta})\varphi|^2_{L_2,V}$$
$$= \sum_{k=1}^{\infty} \langle \cos(s\sqrt{\Delta})P_k\varphi, \cos(s\sqrt{\Delta})\varphi \rangle_{L_2,V}$$
$$= \sum_{k=1}^{\infty} \langle \cos(s\sqrt{\Delta})P_k\varphi, \cos(s\sqrt{\Delta})\chi_{B_{3s}(x_k)}\varphi \rangle_{L_2,V}. \quad (3.2)$$

As operator $\cos(s\sqrt{\Delta}) : L_2(M, E) \longrightarrow L_2(M, E)$, $\cos(s\sqrt{\Delta})$ has the operator norm ≤ 1, hence

$$|\langle \cos(s\sqrt{\Delta}P_k\varphi, \cos(s\sqrt{\Delta})(\chi_{B_{3s}}(x_k)\varphi) \rangle_{L_2,V}|$$
$$\leq \sup_{y \in B_{3s}(x_k)} V(y) \cdot |P_k\varphi|_{L_2(M,E)} \cdot |\chi_{B_{3s}}(x_k)\varphi|_{L_2(M,E)}. \quad (3.3)$$

The firs two factors in (3.2) can be expressed as

$$\sup_{y \in B_{3s}(x_k)} V(y) |P_k\varphi|^2_{L_2(M,E)}$$
$$= \int_M |P_k\varphi(x)|^2 \sup_{y \in B_{3s}(x_k)} \left(\frac{V(y)}{V(x)} \right) V(x) \, d\mathrm{vol}_x(g)$$
$$\leq C_V^{-1} \frac{1}{V(1+4s)} |P_k\varphi|^2_{L_2,V(M,E)},$$

where we used supp $P_k\varphi \subseteq B_s(x_k)$ and lemma 2.9. Performing the analogous estimate for $|\chi_{B_{3s}(x_k)}\varphi|_{L_2(M,E)}$, we obtain altogether

$$\langle \cos(s\sqrt{\Delta})P_k\varphi, \cos(s\sqrt{\Delta})(\chi_{B_{2s}(x_k)}\varphi) \rangle_{L_2,V}$$
$$\leq C_V^{-1} \frac{1}{V(1+6s)} |P_k\varphi|_{L_2,V(M,E)} \cdot |\chi_{B_{3s}(x_k)}\varphi|_{L_2,V(M,E)}.$$
$$(3.4)$$

According to lemma 2.3, $\kappa(M, g, s) < \infty$, and we obtain

$$\sum_{k=1}^{\infty} |\chi_{B_{3s}(x_k)}\varphi|^2_{L_2,V(M,E)} \leq \kappa(M,g,s) |\varphi|^2_{L_2,V(M,E)}.$$

We conclude from (3.1) and (3.3)

$$|\cos(s\sqrt{\Delta})\varphi|^2_{L_{2,V}(M,E)}$$
$$\leq C_V^{-1} \frac{1}{V(1+6s)} |\varphi|^2_{L_{2,V}(M,E)} \sum_{k=1}^{\infty} |\chi_{B_{3s}(x_k)}\varphi|$$
$$\leq C_V^{-1} \frac{1}{V(1+6s)} \kappa(M,g,s)^{\frac{1}{2}} |\varphi|^2_{L_{2,V}(M,V)}. \qquad (3.5)$$

We have $L_2(M,E) \subset L_{2,V}(M,E)$ as a dense subspace since, according to (1.11), $V(x) \leq C(1+d(x,p))^{-1}$, $x \in M$. Hence $\cos(s\sqrt{\Delta})$ extends to a bounded operator $L_{2,V}(M,E) \longrightarrow L_{2,V}(M,E)$. (1.11) and (2.7) imply

$$|\cos(s\sqrt{\Delta})|_{L_{2,V},L_{2,V}} \leq Ce^{cs}, \quad s \in [0,\infty[,$$

which extends to $s \in \mathbb{R}$ since $\cos(s\sqrt{\Delta}) = \cos(-s\sqrt{\Delta})$,

$$|\cos(s\sqrt{\Delta})|_{L_{2,V},L_{2,V}} \leq Ce^{|s|}.$$

The local bound of the norm and the strong continuity on the dense subspace $L_2(M,V) \subset L_{2,V}(M,E)$ yield the strong continutity of $\cos(s\sqrt{\Delta}) : L_{2,V}(M,E) \longrightarrow L_{2,V}(M,E)$. $\square$

Define for given $c \geq 0$

$$\mathcal{F}^1(c) = \{f \in L_1(\mathbb{R}) | \int_{-\infty}^{+\infty} |\hat{f}(\lambda)| e^{c|\lambda|} d\lambda < \infty\}.$$

Proposition 3.2 *Let $(E, h_E, \nabla^E) \longrightarrow (M^n, g)$ be a Riemannian vector bundle satisfying $(B_0(M,g))$ and let V be a function of moderate decay. Then there exists a constant $c = c(M,g,V)$ such that for all even $f \in \mathcal{F}^1(c)$ the operator $f(\sqrt{\Delta})$ extends to a bounded operator $L_{2,V}(M,E) \longrightarrow L_{2,V}(M,E)$, and there exists a constant $C_1 = C_1(M,g,V)$ such that*

$$|f(\sqrt{\Delta})|_{L_{2,V},L_{2,V}} \leq C_1 |\hat{f}|_{L_{1,e^{c|\cdot|}}(\mathbb{R})} \qquad (3.6)$$

for all even $f \in \mathcal{F}^1(c)$. If $\kappa(M,g,s)$ is at most subexponentially increasing, then $c(M,g,V) > 0$ can be choosen arbitrarily.

Proof. According to theorem 3.1, there exist constants $C, c > 0$ depending on M, g, V such that

$$|\cos(s\sqrt{\Delta})|_{L_{2,V}(M,E), L_{2,V}(M,E)} \leq Ce^{c|s|}$$

for all $s \in \mathbb{R}$. Let $\varphi \in L_2(M, E)$. We infer from (3.1)

$$
\begin{aligned}
|f(\sqrt{\Delta})\varphi|_{L_{2,V}(M,E)} &= \left| \frac{1}{2\pi} \int_{-\infty}^{+\infty} \hat{f} \cos(\lambda\sqrt{\Delta}) d\lambda \varphi \right|_{L_{2,V}(M,E)} \\
&\leq \left| \frac{1}{2\pi} \int_{-\infty}^{+\infty} |\hat{f}| \cdot C \cdot e^{c|\lambda|} d\lambda \varphi \right|_{L_{2,V}(M,E)} \\
&= C_1 |\hat{f}|_{L_{1,e^{c|\cdot|}}} |\varphi|_{L_{2,V}}, \quad (3.7)
\end{aligned}
$$

from which we infer the bounded extension to $L_{2,V}(M, E)$. The last statement follows immediately from the meaning of $\kappa(M, g, s)$ and the proof of theorem 3.1. $\square$

Corollary 3.3 *Suppose $(E, h_E, \nabla^E) \longrightarrow (M^n, g)$ and V as above. Then for every $t > 0$, the operator $e^{-t\Delta}$ extends to a bounded operator on $L_{2,V}(M, E)$, and its norm is uniformly bounded in t on compact intervalls $[a_0, a_1]$, $a_0 > 0$.*

Proof. This immediately follows from proposition 3.2 and

$$\int_{-\infty}^{+\infty} e^{-tx^2} \cos(xy) dx = \sqrt{\frac{\pi}{t}} e^{-\frac{y^2}{4t}},$$

$$\int_{-\infty}^{+\infty} \frac{1}{\lambda + x^2} \cos(xy) dx = \sqrt{\frac{\pi}{\sqrt{\lambda}}} e^{-\sqrt{\lambda}|y|}.$$

$\square$

4 Proof of the trace class property

In this section, we will prove the trace class property of $e^{-t\Delta_g} - e^{-t\tilde{\Delta}_h}$, $h \in \text{comp}^{2,V}(g)$, $\Delta_g = -g^{ij}\nabla_i^E \nabla_j^E$. For this, we essentially reduce the task to the case of the scalar Laplacian, using the

Semi–group domination principle. *Let $(E, h_E, \nabla^E) \longrightarrow (M^n, g)$ be a Riemannian vector bundle, P an operator of the kind $P = \Delta^E + \mathcal{R} = (\nabla^E)^*\nabla^E + \mathcal{R}$, $\mathcal{R}$ a 0-order Weitzenboeck term. Set $b = \inf_{x \in M} b(x)$, $b(x) = \min_{\substack{\psi \in E_x \\ |\psi|=1}} h_E(\psi, \mathcal{R}\psi)$. Let $W_P(t, x, y) \equiv e^{-tP}(x, y)$ or $W_{\Delta_0(g)}(t, x, y) = e^{-t\Delta_0(g)}(x, y)$ the heat kernels of e^{-tP} or $e^{-t\Delta_0(g)}$ repectively. Then there holds for the pointwise norms*

$$|W_P(t, x, y)| \leq e^{-bt} W_{\Delta_0(g)}(t, x, y) \tag{4.1}$$

for all $t > 0$, $x, y \in M$.

We refer to [23] for details. □

In this section we must distinguish between $\Delta_g = (\nabla^E)^*\nabla^E = -g^{ij}\nabla_i^E \nabla_j^E$, $* = *(g)$, Δ_g acting on sections of E, and $\Delta_0(g) = \nabla^*\nabla = -g^{ij}\nabla_i\nabla_j$, where $\nabla_i = \nabla_i^g$ is the Levi–Civita connection of g and $\Delta_0(g)$ acts on a function on M.

Suppose (M^n, g) with $(B_0(M, g))$, $|$sectional curvature $K_g| < K$. Let $W_{\Delta_0(g)}(t, x, y) = e^{-t\Delta_0(g)}(x, y)$ be the heat kernel of $\Delta_0(g)$, $0 < a_0 < a_1$.

Lemma 4.1 *There exist $C_1, c_1 > 0$ such that*

$$e^{-t\Delta_0(g)}(x, y) \leq C_1 \tilde{r}_{\text{inj}}(x)^{-\frac{n}{2}} \tilde{r}_{\text{inj}}(y)^{-\frac{n}{2}} e^{-c_1 d^2(x,y)}, \quad t \in [a_0, a_1]. \tag{4.2}$$

We refer to propostion 1.3 of [18].
For $c < c_1$ there exists according to (4.2) and (2.4) a $C > 0$ such that

$$e^{-t\Delta_0(g)}(x, y) \leq C\tilde{r}_{\text{inj}}(x)^{-\frac{n(n+1)}{2}} e^{-cd^2(x,y)}, \quad t \in [a_0, a_1]. \tag{4.3}$$

Lemma 4.2 Let V be a function of moderate decay. Suppose that $a, b \in \mathbb{R}$ are such that

a) $a + b = 2$,

b) $V^b \in L_1(M, g)$,

c) $V^a \tilde{r}_{\text{inj}}(x)^{-\frac{n(n+1)}{2}} \in L_\infty(M, g)$.

Let M_V the operator $V\cdot$, $p \in \mathbb{N}$. Then the operator $M_V \Delta_0(g)^p e^{-t\Delta_0(g)}$ is Hilbert–Schmidt, and for any compact intervall $[a_0, a_1]$, $a_0 > 0$, the Hilbert–Schmidt norm is uniformly bounded.

Proof. Write

$$M_V \Delta_0(g)^p e^{-t\Delta} = (M_V e^{-\frac{t}{2}\Delta_0(g)})(\Delta_0(g)^p e^{-\frac{t}{2}\Delta_0(g)}). \quad (4.4)$$

The second factor on the r.h.s. of (4.4) has bounded operator norm on any intervall $[a_0, a_1]$, $a_0 > 0$. Hence we can restrict to the case $p = 0$. According to corollary 3.3, $e^{-t\Delta_0(g)}$ extends to a bounded operator in $L_{2,V^b}(M, g)$ with uniformly bounded norm in $0 < a \le t \le b$. We infer from $V^b \in L_1(M, g)$ that $1 \in L_{2,V^b}(M, g)$ and $e^{-t\Delta_0(g)} 1 \in L_{2,V^b}(M, g)$. Hence

$$\langle 1, e^{-t\Delta_0(g)} 1 \rangle_{L_{2,V^b}} = \int_M \int_M V(x)^b e^{-t\Delta_0(g)}(x, y) dy dy$$

is well defined, and we obtain for the Hilbert–Schmidt norm of $M_V e^{-t\Delta_0(g)}$

$$|M_V e^{-t\Delta_0(g)}|^2_{HS} = \int_M \int_M |V(x) e^{-t\Delta_0(g)}(x, y)|^2 dy dx$$

$$= \int_M \int_M V(x)^2 e^{-2t\Delta_0(g)}(x, y) dy dx$$

$$\le \sup_{u,v \in M} |V(u)^a e^{-t\Delta_0(g)}(u, v)| \cdot \int_M \int_M V(x)^b e^{-t\Delta_0(g)}(x, y) dy dx$$

$$\le C \cdot \sup_{u \in M} |V(u)^a |\tilde{r}_{\text{inj}}^{-\frac{n(n+1)}{2}}(u)| \int_M |V(x)^b (e^{-t\Delta_0(g)} 1)(x)| dx$$

$$\le C_1 \cdot |e^{-t\Delta_0(g)} 1|_{L_{2,V^b}(M,g)} < \infty. \qquad \square$$

Lemma 4.3 *Let (M^n, g) and V be as above and suppose that a, b are real numbers such that*

a) $b \geq 1$, $a + b = 2$,

b) $V^{\frac{b}{3}} \in L_1(M, g)$,

c) $V^{\frac{a}{3}} \tilde{r}_{\text{inj}}^{-\frac{n(n+1)}{2}} \in L_\infty(M, g)$.

Then the operator $M_{\tilde{r}_{\text{inj}}^{-2n}} M_V \Delta_0(g)^p e^{-t\Delta_0(g)}$ is for $p \in \mathbb{N}$ of trace class, and the trace norm is bounded on any compact t-intervall $[a_0, a_1]$, $a_0 > 0$.

Proof. Write

$$M_{\tilde{r}_{\text{inj}}^{-2n}} M_V \Delta_0(g)^p e^{-t\Delta_0(g)}$$
$$= [M_{\tilde{r}_{\text{inj}}^{-2n}} M_V e^{-\frac{t}{2}\Delta_0(g)} M_{V^{-\frac{1}{3}}}] \cdot [M_{V^{\frac{1}{3}}} \Delta_0(g)^p e^{-t\Delta_0(g)}]. \quad (4.5)$$

The properties of V and $b \geq 1$ immediately imply $V^{\frac{1}{3}} \leq C \cdot V^{\frac{b}{3}}$, hence according to assumption b) $V^{\frac{1}{3}} \in L_1(M, g)$. According to c) and the proof of lemma 4.2, the second factor on the r.h.s. of (4.5) is Hilbert–Schmidt, and the HS–norm is bounded for $t \in [a_0, a_1]$, $a_0 > 0$. We have to show this for the first factor. c) immediately implies $V^a \tilde{r}_{\text{inj}}^{-\frac{n(n+1)}{2} - 2n} \in L_\infty(M, g)$. Hence, together with (4.3),

$$\int_M \int_M |\tilde{r}_{\text{inj}}(x)^{-2n} V(x) e^{-t\Delta_0(g)}(x, y) V(y)^{-\frac{1}{3}}|^2 \, dx \, dy$$
$$\leq C \sup_{u \in M} |\tilde{r}_{\text{inj}}(u)^{-\frac{n(n+1)}{2} - 2n} V(u)^a|$$
$$\cdot \int_M \int_M V(x)^b e^{-t\Delta_0(g)}(x, y) V(y)^{-\frac{2}{3}} \, dx \, dy.$$

By assumption b), $V^{\frac{b}{3}} \in L_1$, i.e. $V^{-\frac{2}{3}} V^{-\frac{2}{3}} V^{\frac{b+4}{3}} \in L_1$, $V^{-\frac{2}{3}} \in L_{2, V^{\frac{b+4}{3}}}(M, g)$. Moreover, we get from $V^{\frac{b+4}{3}} \leq C \cdot V^{\frac{b}{3}}$ and

b) that $V^{\frac{b+4}{3}} \in L_1$. According to corollary 3.3, $e^{-t\Delta_0(g)}$ has an extension to a bounded operator in $L_{2,V^{\frac{b+4}{3}}}(M,g)$, and we get $\int_M e^{-t\Delta_0(g)}(x,y)V(y)^{-\frac{2}{3}}dy \in L_{2,V^{\frac{b+4}{3}}}(M,g)$ with uniformly bounded norm in $t \in [a_0, a_1]$, $a_0 > 0$. Write $V^b \cdot V^b V^{-\frac{b+4}{3}} = V^{\frac{b}{3}} \cdot V^{\frac{4}{3}(b-1)}$, from which follows $V^b \in L_{2,V^{-\frac{b+4}{3}}}(M,g)$. Between $L_{2,V}(M,g)$ and $L_{2,V^{-1}}(M,g)$ there is a standard pairing $\langle,\rangle$

$$\langle \varphi, \psi \rangle = \int \varphi(x)\psi(x)dx, \quad \varphi \in L_{2,V}, \psi \in L_{2,V^{-1}}. \tag{4.6}$$

(4.6) allows to rewrite

$$\int_M \int_M V(x)^b e^{-t\Delta_0(g)}(x,y)V(y)^{-\frac{3}{2}}dxdy = \langle V^b, e^{-t\Delta_0(g)}V^{-\frac{2}{3}}\rangle < \infty.$$

(4.7)

□

Now we prepare our main result. Let $(E, h_E, \nabla^E) \longrightarrow (M^n, g)$ be a Riemannian vector bundle, (M^n, g) complete, satisfying $(B_2(M,g))$, V a function of moderate decay, $h \in \text{comp}^{2,V}(g)$ satisfying $(B_2(M,h))$. We denote $\nabla = \nabla^E$, $\Delta_g = \nabla^*\nabla = -g^{ij}\nabla_i\nabla_j$, $\Delta_h = \nabla^{*(h)}\nabla = -h^{ij}\nabla_i\nabla_j$, $\Delta_0(g)$, $\Delta_0(h)$ are the scalar Laplace operators. We see, h_E remains unchanged and we consider only perturbations of the metric on M, g and h are quasi–isometric. We denote by $\text{dvol}_x g(x)$, $\text{dvol}_x h(x)$ the corresponding volume forms. It is really very easy and quite natural to define a metrizable uniform structure $^{m,V}\mathfrak{U}(vol)$ quite analogous to $^{m,V}\mathfrak{U}(\mathcal{M})$ and to establish the following

Lemma 4.4 $h \in \text{comp}^{2,V}(g)$ *implies* $\text{dvol}(h) \in \text{comp}^{2,V}(\text{dvol}(g))$. □

The metrics h_E, g, h define Hilbert spaces $L_2((M,g),(E,h_E)) \equiv L_2(g,h_E)$, $L_2((M,h),(E,h_E)) \equiv L_2(g,h_E)$ and $L_2(g,h_E) \cong L_2(h, h_E)$ as equivalent Hilbert spaces, but Δ_h must not be self–adjoint in $L_2(g, h_E)$, even not symmetric.

325

But we are exactly in the situation which we described in section IV, 2, between (2.1) and (2.3). We have to graft Δ_h to $L_2(g, h_E)$. We write $\operatorname{dvol}_x(g) \equiv dx(g) = \alpha(x) dx(h) \equiv \alpha(x) \operatorname{dvol}_x(h)$ and define $U : L_2(g, h_E) \longrightarrow L_2(h, h_E)$ by $U\varphi := \alpha^{\frac{1}{2}}\varphi$. Then U is a unitary equivalence between $L_2(g, h_E)$ and $L_2(h, h_E)$. $\tilde{\Delta}_h = U^*\Delta_h U$ acts in $L_2(g, h)$, is self-adjoint on $U^{-1}(\mathcal{D}_{\Delta_h})$ since U is a unitary equivalence.

Lemma 4.5
$$(|(\nabla^g)^2(\alpha - 1)|_g + |\nabla^g(\alpha - 1)|_g + |\alpha - 1|_g)(x) \leq c \cdot V(x).$$

Proof. Write $(\alpha - 1) \operatorname{dvol}_x(h) = \operatorname{dvol}_x(g) - \operatorname{dvol}_x(h)$ and apply lemma 4.4. □

Proposition 4.6 *Suppose* $(E, h_E, \nabla^E) \longrightarrow (M^n, g)$, V *and* $h \in \operatorname{comp}^{2,V}(g)$ *as above. Then for* $t > 0$, $e^{-t\Delta_g}(\alpha - 1)$ *is of trace class and the trace norm is uniformly bounded on compact t-intervalls* $[a_0, a_1]$, $a_{>0}$.

Proof. We write
$$e^{-t\Delta_g}(\alpha - 1) = (e^{-\frac{t}{2}\Delta_g} M_{V^{\frac{1}{3}}}) \cdot (M_{V^{-\frac{1}{3}}} e^{-t\Delta_g}(\alpha - 1)) \quad (4.8)$$

and estimate, using (4.1), the proofs of 4.2, 4.3., lemma 4.5, (4.6)

$$|e^{-\frac{t}{2}\Delta_g} M_{V^{\frac{1}{3}}}|^2_{HS} = \int_M \int_M |(e^{-t\Delta_g}(x, y), \cdot) V^{\frac{1}{3}}(x)|^2 dx dy$$

$$\leq C_1 \int_M \int_M |e^{-t\Delta_0(g)}(x, y)|^2 V^{\frac{2}{3}}(x) dx dy < \infty,$$

$$|M_{V^{-\frac{1}{3}}} e^{-\frac{t}{2}\Delta_g}(\alpha - 1)|^2_{HS}$$
$$= \int_M \int_M |V(y)^{-\frac{1}{3}} (e^{-\frac{t}{2}\Delta_g}(x, y), \cdot)(\alpha - 1)(x)|^2 dx dy$$

$$\leq C_2 \int_M \int_M V(x)^2 e^{-t\Delta_0(g)}(x, y) V(y)^{-\frac{2}{3}} dx dy < \infty.$$

Now we are able to state and to prove our

Main Theorem 4.7 Let $(E, h_E, \nabla^E) \longrightarrow (M^n, g)$ be a Riemannian vector bundle, (M^n, g) being complete and satisfying $B_2(M^n, g)$. Let V be of moderate decay and $h \in \mathrm{comp}^{2,V}(g) \cap C^4$, satisfying $(B_2(M, h))$ and set $\alpha(x) = \mathrm{dvol}_x(g)/\mathrm{dvol}_x(h)$, $U : L_2(g, h_E) \longrightarrow L_2(h, h_E)$ defined by $U\varphi = \alpha^{\frac{1}{2}}\varphi$. Then

$$e^{-t\Delta_g} - e^{-tU^*\Delta_h U}$$

is for $t > 0$ of trace class and the trace norm is uniformly bounded on compact t-intervalls $[a_0, a_1]$, $a_1 > 0$.

Proof. According to Duhamels principle IV (2.5),

$$e^{-t\Delta_g} - e^{-tU^*\Delta_h U} = e^{-t\Delta_g} - e^{-t(\alpha^{-\frac{1}{2}}\Delta_h \alpha^{\frac{1}{2}})}$$

$$= e^{-t\Delta_g}(\alpha - 1) - \int_0^t e^{-s\Delta_g}(\Delta_g - \alpha^{-\frac{1}{2}}\Delta_h \alpha^{\frac{1}{2}})e^{-(t-s)\alpha^{-\frac{1}{2}}\Delta_h \alpha^{\frac{1}{2}}} ds$$

$$= e^{-t\Delta_g}(\alpha - 1) - \int_0^t e^{-s\Delta_g}(\Delta_g - \alpha^{-\frac{1}{2}}\Delta_h \alpha^{\frac{1}{2}})\alpha^{-\frac{1}{2}} e^{-(t-s)\Delta_h} \alpha^{\frac{1}{2}} ds.$$

According to proposition 4.6, the term $e^{-t\Delta_g}(\alpha - 1)$ is done. We write

$$\Delta_g - \alpha^{-\frac{1}{2}}\Delta_h \alpha^{\frac{1}{2}} = \Delta_g(1 - \alpha^{\frac{1}{2}}) + (\Delta_g - \Delta_h)\alpha^{\frac{1}{2}} + (1 - \alpha^{-\frac{1}{2}})\Delta_h \alpha^{\frac{1}{2}},$$

$$e^{-s\Delta_g}[\Delta_g(1 - \alpha^{\frac{1}{2}}) + (\Delta_g - \Delta_h)\alpha^{\frac{1}{2}}$$
$$+ (1 - \alpha^{-\frac{1}{2}})\Delta_h \alpha^{\frac{1}{2}}]\alpha^{-\frac{1}{2}} e^{-(t-s)\Delta_h} \alpha^{\frac{1}{2}}$$
$$= \Delta_g e^{-s\Delta_g}(1 - \alpha^{\frac{1}{2}})\alpha^{-\frac{1}{2}} e^{-(t-s)\Delta_h} \alpha^{\frac{1}{2}} \quad (4.9)$$
$$+ e^{-s\Delta_g}(\Delta_g - \Delta_h)e^{-(t-s)\Delta_h} \alpha^{\frac{1}{2}} \quad (4.10)$$
$$+ e^{-s\Delta_g}(1 - \alpha^{-\frac{1}{2}})\Delta_h e^{-(t-s)\Delta_h} \alpha^{\frac{1}{2}} \quad (4.11)$$

and decompose $\int_0^t = \int_0^{\frac{t}{2}} + \int_{\frac{t}{2}}^t$. This yields altogether 6 integrals,

$$\int_0^{\frac{t}{2}}(4.9)ds, \int_0^{\frac{t}{2}}(4.10)ds, \int_0^{\frac{t}{2}}(4.11)ds, \int_{\frac{t}{2}}^t(4.9)ds, \int_{\frac{t}{2}}^t(4.10)ds, \int_{\frac{t}{2}}^t(4.11)ds.$$

The definition of α implies $0 < c_1 \le \alpha \le c_2$, and we obtain from lemma 4.5

$$|1 - \alpha^{-\frac{1}{2}}| = \left|1 - \frac{1}{\sqrt{\alpha}}\right| = \left|\frac{1 - \frac{1}{\alpha}}{1 + \frac{1}{\sqrt{\alpha}}}\right| = \left|\frac{\frac{1}{\alpha}(1 - \alpha)}{1 + \frac{1}{\sqrt{\alpha}}}\right| \le C_2 \cdot V.$$

Now we decompose $\int_{\frac{t}{2}}^t (4.11)ds$ as follows,

$$\int_{\frac{t}{2}}^t (4.11)ds$$

$$= \int_{\frac{t}{2}}^t [\{e^{-\frac{s}{2}\Delta_g} M_{V^{\frac{1}{3}}}\}\{M_{V^{-\frac{1}{3}}} e^{-\frac{s}{2}\Delta_g}(1 - \alpha^{-\frac{1}{2}})\}](\Delta_h e^{-(t-s)\Delta_h}\alpha^{\frac{1}{2}})ds.$$

But $\{\ldots\}$, $\{\ldots\}$ in $[\ldots]$ are Hilbert–Schmidt operators:

$$|(e^{-\frac{s}{2}\Delta_g}, \cdot)M_{V^{\frac{2}{3}}}| \le C_3 |e^{-\frac{s}{2}\Delta_0(g)}(x,y) M_{V^{\frac{2}{3}}}|, \quad (4.12)$$

$$|M_{V^{-\frac{2}{3}}}(e^{-\frac{s}{2}\Delta_g}, \cdot)(1 - \alpha^{-\frac{1}{2}})| \le C_4 |V^{-\frac{2}{3}}(y) e^{-\frac{s}{2}\Delta_0(g)}(x,y) V(x)|. \quad (4.13)$$

According to lemma 4.2 and (4.7), the right hand sides of (4.12), (4.13) are square integrable, and the value of the HS–norm is uniformly bounded on the interval $[\frac{t}{2}, t]$.

Finally

$$|\Delta_h e^{-(t-s)\Delta_h}\alpha^{\frac{1}{2}}|_{L_2 \to L_2} \le C_5(t-s)^{-\frac{1}{2}}.$$

We conclude as in IV (1.29)–(1.34) that $\int_{\frac{t}{2}}^t (4.11)ds$ is a trace class operator and its trace norm is bounded in any compact t–interval

$[a_0, a_1]$, $a_0 > 0$. Quite analogous, we handle the other 5 integrals and indicate this by the decompositionn of the integrands and their estimate:

$\int_0^{\frac{t}{2}}(4.11)ds:$

$$e^{-s\Delta_g}(1-\alpha^{-\frac{1}{2}})\Delta_h e^{-(t-s)\Delta_h}\alpha^{\frac{1}{2}}$$
$$= e^{-s\Delta_g}[\{(1-\alpha^{-\frac{1}{2}})e^{-\frac{t-s}{4}\Delta_h}M_{V^{-\frac{2}{3}}}\}\{M_{V^{\frac{2}{3}}}e^{-\frac{t-s}{4}\Delta_h}\}]$$
$$\cdot \Delta_h e^{-\frac{t-s}{2}\Delta_h}\alpha^{\frac{1}{2}},$$

$$|(1-\alpha^{-\frac{1}{2}})e^{-\frac{t-s}{4}\Delta_h}M_{V^{-\frac{2}{3}}}|$$
$$\leq C\cdot|V(x)e^{-\frac{t-s}{4}\Delta_0(g)}(x,y)V(y)^{-\frac{2}{3}}|$$
$$|M_{V^{\frac{2}{3}}}e^{-\frac{t-s}{4}\Delta_h}|\leq C\cdot|V(x)^{\frac{2}{3}}e^{-\frac{t-s}{4}\Delta_0(g)}(x,y)|,$$

$\int_{\frac{t}{2}}^{t}(4.10)ds:$

$$e^{-s\Delta_g}(\Delta_g-\Delta_h)e^{-(t-s)\Delta_h}\alpha^{\frac{1}{2}}$$
$$= [\{e^{-\frac{s}{2}\Delta_g}M_{V^{\frac{2}{3}}}\}\{M_{V^{-\frac{2}{3}}}e^{-\frac{s}{2}\Delta_g}(h^{ij}-g^{ij})\}]$$
$$\cdot \nabla_j\nabla_j e^{-(t-s)\Delta_h}\alpha^{\frac{1}{2}}$$

$e^{-\frac{s}{2}\Delta_g}M_{V^{\frac{2}{3}}}$ is done,

$$|M_{V^{-\frac{2}{3}}}e^{-\frac{s}{2}\Delta_g}(h^{ij}-g^{ij})|\leq C\cdot|V(x)^{-\frac{2}{3}}e^{-\frac{s}{2}\Delta_0(g)}(x,y)V(x)|,$$

$\int_0^{\frac{t}{2}}(4.10)ds:$

$$e^{-s\Delta_g}[\{(\Delta_g-\Delta_h)e^{-\frac{t-s}{4}\Delta_h}M_{V^{-\frac{2}{3}}}\}\{M_{V^{\frac{2}{3}}}e^{-\frac{t-s}{4}\Delta_h}\}]\cdot\Delta_h e^{-\frac{t-s}{2}\Delta_h}\alpha^{\frac{1}{2}},$$

everything is done, $\int_{\frac{t}{2}}^{t}(4.9)ds:$

$$\Delta_g e^{-\frac{s}{2}\Delta_g}[\{e^{-\frac{s}{4}\Delta_g}M_{V^{\frac{2}{3}}}\}\{M_{V^{-\frac{2}{3}}}e^{-\frac{s}{4}\Delta_g}(1-\alpha^{\frac{1}{2}})\}]\cdot\alpha^{-\frac{1}{2}}e^{-(t-s)\Delta_h}\alpha^{\frac{1}{2}},$$

everything is done, $\int_0^{\frac{t}{2}}(4.9)ds$:

$$\Delta_g e^{-s\Delta_g}\alpha^{-\frac{1}{2}}[\{(1-\alpha^{\frac{1}{2}})e^{-\frac{t-s}{4}\Delta_h}M_{V^{-\frac{2}{3}}}\}\{M_{V^{\frac{2}{3}}}e^{-\frac{t-s}{4}\Delta_h}\}]\cdot e^{-\frac{t-s}{2}}\alpha^{\frac{1}{2}},$$

everything is done.
This finishes the proof of 4.7. □

Remarks 4.8 1) Theorem 4.7 and the theorems in chapter IV are neither disjoint nor there is an inclusion.
2) In 4.7 we only permitted a perturbation of the metric of M^n. It would be also interesting to consider additionally appropriate perburbations of the fibre metric and the fibre connection. The advantage of this chapter is that we do not restrict to the case $r_{\text{inj}}(M,g) > 0$. Hence we admit e.g. locally symmetric spaces of finite volume. □

As an immediate consequence of theorem 4.7 we have

Theorem 4.9 *Suppose the hypotheses of theorem 4.7. Then there exist the wave operators*

$$W_{\pm}(\Delta_g, \Delta_h) = s - \lim_{t\to\pm\infty} e^{it\Delta_h}e^{-it\Delta_g}P_{ac}(\Delta_g)$$

and they are complete. Hence the absolutely continuous parts of Δ_g and Δ_h are unitarily equivalent. □

We have in mind still other admitted perturbations to establish a scattering theory. This will be the topic of a forthcoming treatise.

References

[1] Abresch, U., Über das Glätten Riemannscher Metriken, Habilitationsschrift, Universität Bonn, 1988.

[2] Anghel, N., An abstract index theorem on non–compact Riemannian manifolds, *Houston J. Math.* 19 (1993), 223–237.

[3] Antoci, F., On the spectrum of the Laplace–Beltrami operator for p–forms for a class of warped product metrics, *Adv. Math.* 188 (2004), 247–293.

[4] Atiyah, M.F., Elliptic operators, discrete groups and von Neumann algebras, *Asterisque* 32/33 (1976), 43–72.

[5] Atiyah, M.F., Patodi, V., Singer, I.M., Spectral asymmetry and Riemannian geometry I, *Math. Proc. Cambr. Philosoph. Soc.* 77 (1975), 43–69.

[6] Attie, O., Quasi–isometry classification of some manifolds of bounded geometry, *Math. Z.* 216 (1994), 501–527.

[7] Barbasch, P., Moscovici, H., L_2–index and the Selberg trace formula, *J. Functional Analysis* 53 (1983), 151–201.

[8] Bunke, U., Spektraltheorie von Diracoperatoren auf offenen Mannigfaltigkeiten, PhD thesis, Greifswald, 1991.

[9] Bunke, U., Relative index theory, *J. Funct. Anal.* 105 (1992), 63–76.

[10] Buser, P., A note on the isoperimetric constant, *Ann. Sci. Ecole Norm. Sup.* 15 (1982), 213–230.

[11] Carron, G., Index theorems for noncompact manifolds *J. Reine Angew. Math.* 541 (2001), 81–115.

[12] Carron, G., L_2–cohomology and Sobolev inequalities, *Math. Ann.* 314 (1999), 613–639.

[13] Carron, G., L_2–cohomology of manifolds with flat ends, *GAFA* 13 (2003), 366–395.

[14] Carron, G., Un theoreme de l'indice relativ, *Pacific J. Math.* 198 (2001), 81–107.

[15] Cheeger, J., Gromov, M., Bounds on the von Neumann dimension of L_2–cohomology and the Gauß–Bonnet theorem for open manifolds, *J. Diff. Geom.* 21 (1985), 1–34.

[16] Cheeger, J., Gromov, M., On the characteristic numbers of complete manifolds of bounded geometry and finite volume, Differential Geometry and Complex Analysis, Springer Berlin, 1985, 115—154.

[17] Cheeger, J., Gromov, M., Chopping Riemannian manifolds, Differential Geometry, Pitman Monogr. Surveys Pure Appl. Math. 52 (1991), 85–94.

[18] Cheeger, J., Gromov, M., Taylor, M., Finite propagation speed, kernel estimates for functions of the Laplacian, and geometry of complete Riemannian manifolds, *J. Diff. Geom.* 17 (1983), 15–53.

[19] Chen, B.L., Zhu, X.P., Uniqueness of the Ricci flow on complete noncompact manifolds, arXiv:math.DG/05054473v3.

[20] Chernoff, P., Essential self–adjointness of powers of generators of hyperbolic equations, *J. Funct. Anal.* 12 (1973), 401–414.

[21] Cohn–Vossen, S., Kürzeste Wege und Totalkrümmung auf Flächen, *Compositio Math.* 2 (1935), 69–133.

[22] Connes, A., Moscovici, H., The L_2–index theorem for homogeneous spaces of Lie groups, *Ann. Math.* 115 (1982), 291–330.

[23] Donnelly, H., Essential spectrum and heat kernel, *J. Funct. Anal.* 75 (1987), 362–381.

[24] Donnelly, H., On the point spectrum for finite volume symmetric spaces of negative curvature, *Comm. Part. Diff. Eq.* 6 (1981), 963–992.

[25] Eichhorn, J., Global properties of the Ricci flow on open manifolds, WSPC Conference Proceedings, DGA 2007, 99–118.

[26] Eichhorn, J., Characteristic numbers, bordism theory and the Novikov conjecture for open manifolds, Banach Center Publ. 76 (2007), 269–312.

[27] Eichhorn, J., Global analysis on open manifolds, Nova Science Publ., New York 2007, 664 pp.

[28] Eichhorn, J., Riemannsche Mannigfaltigkeiten mit einer zylinderähnlichen Endenmetrik, *Math. Nachr.* 114 (1983), 23–51.

[29] Eichhorn, J., The boundedness of connection coefficients and their derivatives, *Math. Nachr.* 152 (1991), 145–158

[30] Eichhorn, J., Gauge theory on open manifolds of bounded geometry, *Int. Journ. Mod. Physics* 7 (1992), 3927–3977.

[31] Eichhorn, J., The manifold structure of maps between open manifolds, *Ann. of Global Analysis and Geom.* 11 (1993), 253–300.

[32] Eichhorn, J., Spaces of Riemannian metrics on open manifolds, *Results in Mathematics* 27 (1995), 256–283.

[33] Eichhorn, J., Uniform structures of metric spaces and open manifolds, *Results in Math.* 40 (2001), 144–191.

[34] Eichhorn, J., Invariants for proper metric spaces and open Riemannian manifolds, *Math. Nachr.* 253 (2003), 8–34.

[35] Eichhorn, J., The Euler characteristic and signature for open manifolds, Suppl. ai Rendiconti del Circolo Matematico di Palermo, Serie II, no. 16 (1987), 29–41.

[36] Eichhorn, J., Fricke, J., The module structure theorem for Sobolev spaces on open manifolds, *Math. Nachr.* 194 (1998), 35–47.

[37] Eschenburg, J.-H., Comparison theorems in Riemannian geometry, Lecture Notes Series Dipartimento di Matematica Universita degli Studi di Trento, No. 3, Trento, 1994.

[38] Gilkey, P.J., Invariance Theory, the Heat Equation and the Atiyah–Singer Index Theorem, Studies in Advanced Math., Boca Raton, 1995.

[39] Gromov, M., Lafontaine, J., Pansu, P., Structures metriques pour les varietes Riemanniennes, Paris, 1981.

[40] Hebey, E., Sobolev spaces on Riemannian manifolds, *Lecture Notes in Mathematics* 1635, Berlin, 1996.

[41] Hebey, E., Nonlinear analysis on manifolds, *Courant–Lecture Notes in Mathematics* 5, Providence, 1999.

[42] Heitsch, J. L., Lazarov, C., A Lefschetz theorem on open manifolds, *Contemp. Math.* 105 (1990), 33–45.

[43] Huber, A., On subharmonic functions and differential geometry in the large, *Comm. Math. Helv.* 32 (1957), 13–72.

[44] Januszkiewicz, T., Characteristic invariants of noncompact Riemannian manifolds, *Topology* 23 (1984), 289–301.

[45] Kucerovsky, D., Averaging operators and open manifolds, *Contemp. Math.* 148 (1993), 151–153.

[46] Lück, W., L_2–invariants: theory and applications to geometry and K–theory, Ergebnisse der Mathematik und ihrer Grenzgebiete, 3rd Series. A Series of Modern Surveys in Mathematics 44, Springer Publ., Berlin, 2002.

[47] Milnor, J., Stasheff, J., Characteristic classes, *Ann. Math. Studies* 76, Princeton 1974.

[48] Moscovici, H., L_2–index of elliptic operators on locally symmetric spaces of finite volume, *Contemp. Math.* 10 (1982), 129–138.

[49] Müller, W., Relative zeta functions, relative determinants and scattering theory, *Comm. Math. Phys.* 192 (1998), 309–347.

[50] Müller, W., L_2–index theory, eta invariants and values of L_2–functions, *Contemp. Math.* 105 (1990), 145–189.

[51] Müller, W., Signature defects of cusps of Hilbert modular varieties and values of L–series at $s = 1$, *J. Diff. Geom.* 20 (1984), 55–119.

[52] Müller, W., Manifolds with cusps of rank one, spectral theory and L_2–index theorem, *Lecture Notes in Mathematics* 1339, Berlin 1988.

[53] Müller, J., Müller, W., Regularized determinants of Laplace–type operators, analytic surgery, and relative determinants, *Duke Math. J.* 1

[54] Müller, W., Salomonsen, G., Scattering theory for the Laplacian on manifolds with bounded curvature, *J. Funct. Anal.* 253 (2007), 158–206.

[55] Narasimhan, M.S., Ramanan, S., Existence of universal connections, *Amer. J. Math.* 83 (1961), 563–572.

[56] Petersen, P., Riemannian Geometry, New York, 1998, Springer.

[57] Polombo, A., Nombres characteristique d'une variete Riemannienne de dimension 4, *J. Diff. Geom.* 13 (1978), 145–162.

[58] Reed, M., Simon, B., Methods of Modern Mathematical Physics III, Moscow, 1982, Publishers Mir.

[59] Richards, I., On the classification of noncompact surfaces, *Trans. AMS* 106 (1963), 259–269.

[60] Roe, J., An index theorem on open manifolds I, *J. Diff. Geom.* 27 (1988), 87–113.

[61] Roe, J., Operator algebras and index theory on noncompact manifolds, *Contemp. Math.* 70 (1988), 229–240.

[62] Roe, J., Index theory, coarse geometry and topology of open manifolds, CBMS Reg. Conf. Series in Math. vol. 90, Providence, 1996.

[63] Roe, J., Coarse cohomology and index theory on complete Riemannian manifolds, *Memoirs of the AMS* 497, 1993.

[64] Rosenberg, S., Gauß–Bonnet theorems for noncompact surfaces, *Proc. AMS* 86 (1982), 184–185.

[65] Rosenberg, S., On the Gauß–Bonnet theorem for complete manifolds, *Trans. AMS* 287 (1985), 745–753.

[66] Salomonsen, G., Equivalence of Sobolev spaces, *Results Math.* 39 (2001), 115–130.

[67] Salomonsen, G., On the completeness of the spaces of metrics on an open manifold, *Results Math.* 29 (1996), 355–360.

[68] Schubert, H., Topologie, Teubner Publ., Stuttgart, 1964.

[69] Shi, W.X., Deforming the metric on complete Riemannian manifolds, *J. Diff. Geom.* 30 (1989), 223–301.

[70] Singer, I.M., Families of Dirac operators with applications to physics, Asterisque 1985, Numero Hors Serie, 324–340.

[71] Wladimirov, W.S., Equations of Mathematical Physics, Moscow, 1971, Nauka Publishers.

[72] Yafaev, D.R., Mathematical scattering theory, Transl. of Mathematic Monographs, vol. 105, AMS, Providence 1992.

[73] Yafaev, D.R., Birman, M.S., The spectral shift function, *St. Petersburg Math. J.*, 4 (1993), 833–870.

[74] Yu, G., K-theoretic indices of Dirac type operators on complete manifolds and the Roe algebra, *K-Theory* 11 (1997), 1–15.

List of notations

	page
$\tilde{b}^{q,2}(M)$	30
$\tilde{\beta}^{q,2}(M)$	31
(B_k)	65
$\mathcal{B}^{N,n}_{L,diff}(I, B_k)$	131
$\mathcal{B}^{N,n,p,r}_{L,diff}(I, B_k)$	131
$\mathcal{B}^{N,n,p,r}_{L,diff,F}(I, B_k)$	135
$\mathcal{B}^{N,n,p,r}_{L,diff,rel}(I, B_k)$	134
$\text{comp}^{p,r}(g)$	109
$^{b,m}\text{comp}(g)$	105
$^{b,2}\text{comp}^{p,2}(g)$	16
$\text{comp}^{p,r}(\varphi)$	95
$\text{CL}\mathcal{B}^{N,n}(I, B_k)$	139
$\text{CL}\mathcal{B}^{N,n,p,r}_{L,diff}$	140
$\text{CL}\mathcal{B}^{N,n,p,r}_{L,diff,F}(I, B_k)$	140
$\text{CL}\mathcal{B}^{N,n,p,r}_{L,diff,F,rel}(I, B_k)$	141
$C^{\infty,m}(M, N)$	121
$C^{r,p,r}(M_1, M_2)$	125
$^{b,1}C^{p,1}(B_0, f, p)$	7
d^Z_H	110
d_{GH}	111
d_L	114
$d_{L,h}$	118
$d_{L,h,rel}$	119
$d_{L,top}$	115
$\tilde{d}^{p,r}_{L,diff,rel}(M_1, M_2)$	126
$\tilde{d}^{p,r}_{L,diff}(E_1, E_2)$	130
$\tilde{d}^{p,r}_{L,diff,rel}(E_1, E_2)$	133
$\tilde{d}^{p,r}_{L,diff,F}(E_1, E_2)$	135
$d^{p,r}_{L,diff,F,rel}(E_1, E_2)$	136
$\det(D^2, \tilde{D}'^2)$	268
$\mathcal{D}^{p,r}(M, N)$	123
$\mathcal{D}^{p,r}(M)$	123
$\mathcal{D}^{p,r}_{vb}(E_1, E_2)$	124
$\mathcal{D}^{p,r}_{vb,rel}(E_1, E_2)$	124
$\eta(s, D, \tilde{D}')$	275
$\text{gen comp}^{p,r}_{L,diff,rel}(M, g)$	129
$\text{gen comp}^{p,r}_{L,diff,rel}(E)$	134
$\text{gen comp}^{p,r}_{L,diff,F,rel}(E)$	136
$\text{gen comp}^{p,r}_{L,diff}(E)$	140
$\text{gen comp}^{p,r}_{L,diff,F}(E)$	136
HX^*	143
$H^*_{GH}, H^*_{L,b}$	145
$H^{q,p}(M)$	10
$H^{q,p,\{d\}}(M)$	10
$H^r(E)$	66
$H^{2,r}(E)$	66
$H^{p,r}(E)$	64
$H^r_\xi(E)$	310
(I)	65
$\text{ind}(D, \tilde{D}')$	243
$^b_m\mathcal{M}, {}^{b,m}\mathcal{M}$	104
$\mathcal{M}(I), \mathcal{M}(B_k), \mathcal{M}(I, B_k)$	106
$\mathcal{M}^p_r, \mathcal{M}^{p,r}$	108
$^{b,2}\mathcal{M}^{p,2}(B_0, f, p)$	16
$\mathfrak{M}_{GH}$	111
$\mathfrak{M}_{L,top}$	116

$\mathfrak{M}_L(nc), \mathfrak{M}_{L,h}(nc)$ 116
$\mathfrak{M}_{L,top,rel}$ 120
$\mathfrak{M}_{L,h,rel}$ 119
$n^c(\lambda, D, \tilde{D}'), n^c(D, \tilde{D}')$ ...251
$\mathcal{N}(\varrho), \mathcal{N}_K(\varrho)$ 57
$\Omega_r^{q,p}(E), \Omega^{q,p,r}(E), \Omega^{p,r}(E)$ .73
$\mathring{\Omega}^{q,p,r}(E), \mathring{\Omega}^{p,r}(E)$ 73
$\overline{\Omega}^{q,p,r}(E), \overline{\Omega}^{p,r}(E)$ 73
$_m^b\Omega(E), {}^{b,m}\Omega(E)$ 74
$\Omega_\xi^{2,r}(E), \overline{\Omega}_\xi^{2,r}(E)$,
$\mathring{\Omega}_\xi^{2,r}(E), \Omega_{r,\xi}^2(E)$ 309
$\Omega^{p,\varrho}(E, \nabla)$ 73
$\Omega_{2s}^2(E, \Delta)$ 85
$\mathring{\Omega}^{2,2s}(E, \Delta)$ 85
$\overline{\Omega}^{2,2s}(E, \Delta)$ 85
$\Omega^{2,2s}(E, \Delta)$ 85
$\Omega_r^2(E, D), \mathring{\Omega}^{2,r}(E, D)$ 87
$\overline{\Omega}^{2,r}(E, D), \Omega^{2,r}(E, D)$ 87
$\Omega^{p,r}(C^\infty(E))$ 94
$^{b,m}\Omega(C^\infty(E))$ 97
$\Omega^{p,r}(M, N)$ 122
Ω_n^{nc} 150
$\Omega_n^{nc}(cs)$ 151
$\Omega_n^{nc}(I, B_k)$ 150
$\Omega_n^{nc}(ne)$ 162
$\mathcal{P}_m(g)$ 98
$\mathcal{P}_m^{p,r}(g)$ 98
$S(\lambda)$ 188
$\sigma(A)$ 184
$\sigma_p(A), \sigma_{c,R}(A)$ 184
$\sigma_e(A), \sigma_c(A), \sigma_{p,f}(A)$ 184
$\sigma_{pp}(A), \sigma_{ac}(A), \sigma_{sc}(A)$ 184
$\sigma_{pd}(A)$ 184
$\tau^a(M, M')$ 272
$\xi(\lambda)$ 188
$W_\pm(A, B)$ 185
$W^{p,r}(E)$ 64
$W(t, m, p)$ 178
$\zeta(s, D^2, \tilde{D}'^2)$ 266

Index

	page
analytic torsion	272
asymptotic expansion	246, 252
bordism group	150, 151
bounded geometry	65
characteristic classes	12, 13
characteristic numbers	13, 14
Clifford bundles	63
Clifford multiplication	63
coarse cohomology	143
coarse structure	143
coarse map	142
collared end	277
cusp manifold	55
Duhamel's principle	181
end	159
Gaffneys theorem	15
generalized component	129, 134, 136, 140
generalized Dirac operator	63
geodesic ray	159
GH–cohomology	145
GH–component	112
GH–distance	111
GH–uniform structure	112
heat kernel	178
hyperbolic space	284
index theorems	49, 50, 51, 54, 56
K–groups	61
Lipschitz uniform structure	115
– cohomology	145
– component	117, 118
– distance	114
– map	114
L_p–cohomology	10
mapping manifold	122
non–expanding end	160
non–parabolic	56
relative characteristic number	155
– determinant	268
– eta function	275
– index	243
– index theorem	247
– zeta function	266
scattering matrix	188
spectrum	184
– absolutely continuous	184
– components of	184
– continuous	184
– essential	184
– purely discrete	184
– resolvent continuous	184
– singular continuous	184
spectral shift function	188
Sobolev spaces	73
– embedding theorems	68, 75
– module structure theorems	67, 75
spaces of connections	5, 7

– metrics 104, 108
supersymmetric scattering
   system 243
uniform structure 90

warped product metric 277
wave operator 185
weighted Sobolev space
   .... 309, 310